M·A·T·H·S INVESTIGATIONS

TEACHERS' GUIDE

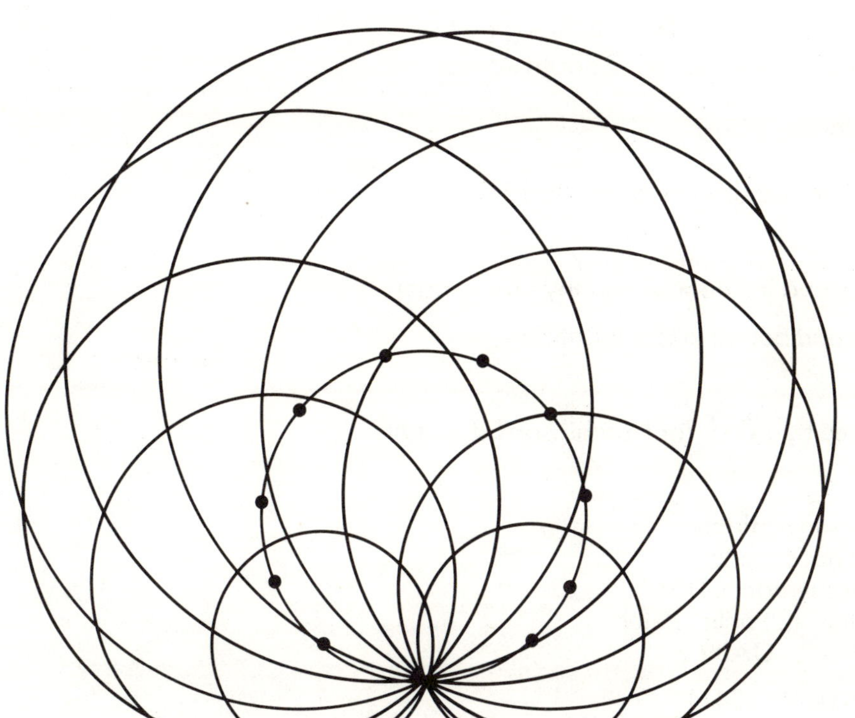

DAVID KIRKBY and PETER PATILLA

Hutchinson Education

Hutchinson Education

An imprint of Century Hutchinson Ltd
62–65 Chandos Place, London WC2N 4NW

Century Hutchinson Australia Pty Ltd
PO Box 496, 16–22 Church Street, Hawthorn,
Victoria 3122, Australia

Century Hutchinson New Zealand Ltd
PO Box 40-086, Glenfield, Auckland 10, New Zealand

Century Hutchinson South Africa (Pty) Ltd
PO Box 337, Bergvlei, 2012 South Africa

First published 1987

Typeset in 12pt Metrolight by
The Pen and Ink Book Company Ltd, London

Printed and bound in Great Britain

Produced by AMR for Hutchinson Education

Kirkby, Dave
Maths investigations.
Teacher's guide.
1. Mathematics–1961-
I. Title II. Patilla, Peter
510 QA39.2

ISBN 0 09 170411 1

CONTENTS

The Cards

Maths Investigations contains 80 cards based on ten themes:

1 Squares   **6 Circles**
2 Games   **7 Cubes**
3 Polygons   **8 Digits**
4 Grids   **9 Calculator**
5 Number I   **10 Number II**

There are eight cards for each theme.

The theme title provides an indication of a common identity for the eight cards in the set. Although, for example, a set of cards has the theme title 'Squares' investigations involving squares will arise from cards with other theme titles. There is no order to the eight cards within each theme.

Using the cards

Maths Investigations will compliment any school's mathematics scheme and is very flexible:

△ It fits naturally into any individualised scheme.
△ It can act as a resource bank to suit the needs of a small group of children at any particular time.
△ It can be used in investigations where:

all pupils are involved in the same investigation

pupils are involved in a variety of investigations based on the same theme

pupils are involved in investigations of their own choice.

In whatever way the cards are used it must be stressed that they merely provide a *starting point* for some mathematical study. Each activity can be developed in many different directions involving mathematics at various levels.

The cards are not a set that need to be 'worked through'. It is possible that several lessons work may arise from any particular card. It is hoped that pupils will become sufficiently interested in an investigation to continue in depth. It is better for a pupil to have completed one or two investigations in depth rather than to have tried several superficially.

It is intended that pupils will learn to develop their own lines of enquiry. the ability to do this does not come naturally to most pupils and therefore needs to be developed by the teacher through discussion and suggestions. The teachers' notes provide some possibilites for points of discussion, questions and extensions.

Althought the teachers' notes suggest some *possible* solutions, they are not *answers*. It is important that pupils are not manoeuvred towards particular solutions. The value of investigational work is to enable pupils to devlop thought processes and strategies for themselves.

Supplementing the cards

The cards provide a resource bank of starting points, which can be easily supplemented by the addition of teacher-made cards. Sources of ideas for these include:

The Investigator
SMILE CENTRE, Middle School Row, Kensal Raod, London

Investigation Bank Books
D. Kirkby: Eigen Publications, 39, Den Bank Crescent, Sheffield, S10 5PB

Peak Plus (Books 3 and 4)
A. Brighouse, D. Godber, P. Patilla: Thomas Nelson and Sons Ltd

Mathematical Activities and *More Mathematical Activities*
Brian Boult: Cambridge University Press

Sources of Mathematical Discovery and *Investigations in Mathematics*
Lorraine Mottershead: Basil Blackwell

Thinking Things Through
Leone Burton: Basil Blackwell

Mathematics Teaching and other publications
ATM: Kings Chambers, Queen Street, Derby

Mathematics in School
Mathematics Association: 259, London Road, Leicester

Pupils Work

An important aspect of investigation work is the opportunity provided for pupils to communicate their mathematics: by 'write-ups' describing how they went about the investigation and by short periods of class discussion where ideas can be pooled. It is important that pupils develop the ability to write and explain their thinking clearly.

It is a good idea for each pupil to have a loose leaf 'Investigations Log Book' which has a pocket for the various pieces of simple apparatus which accompany some of the investigations and which allow the inclusion of different forms of stationery as well as allowing the pupil to add to investigations previously attempted. These log books can be simply made by the pupils themselves:

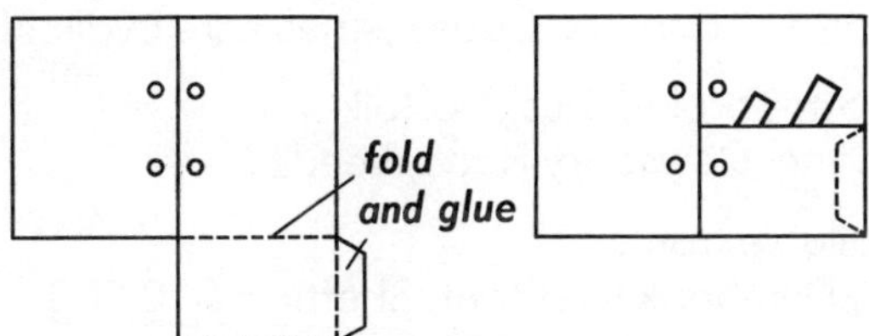

It is important that pupils have the opportunity of returning to an investigation.

Creative stimulus can arise from 'Investigation Corners' in which a starting point is given and the resulting work displayed.

Materials

The cards are activity based and require the pupils to use simple and inexpensive materials as an integral part of the investigation. The importance of allowing pupils the opportunity of handling the materials cannot be over emphasised.

Materials and stationery should be easily accessible to pupils.

Apparatus required

Dice	Squares (about 2 cm) card or plastic
Geoboards	Calculators
Counters	Cubes
Compasses	Matchsticks
Table-tennis balls	Number cards
Card	Scissors
Scrap paper	Plain paper
Tracing paper	Glue (glue-sticks are ideal)
Sticky tape	Special papers (see Copy masters)

Number cards: These are small pieces of card (approx 5 x 8 cm) on which numbers are written, usually in the 0 – 25 range.

Copy master

There are eleven A4 Copy master sheets to accompany the cards:

1 1 cm isometric	7 Hexagon dot
2 2 cm isometric	8 Circles (4 and 5 dot)
3 1 cm square	9 Circles (6 and 7 dot)
4 2 cm square	10 Circles (8 and 9 dot)
5 Table of primes	11 Circles (12 dot)
6 1 cm square dot	

These sheets are free from copyright and may be duplicated in whatever form the teacher wishes.

Suppliers

Dice; Squares; Geoboards; Counters; Cubes – these are available from:

Tarquin Publications, Stradbroke, Diss, Norfolk
E.J. Arnold, Parkside Lane, Dewsbury Road, Leeds

Number Cards are available from:
Eigen Publications, 39, Den Bank Crescent, Sheffield S10 5PB

SQUARES

	APPARATUS	ACTIVITY
1 Take-Away	Matchsticks	Arrangement of sticks
2 Cut-Outs	Squared paper (2 cm)	Shapes made by cutting squares
3 Lines	Squared paper (1 cm)	Drawing grids with straight lines
4 Perimeters	Squared paper (1 cm)	Fixed perimeters
5 Frames	Squares	Arranging squares inside a frame
6 Joining	Squares: squared paper (1 cm)	Arrangement with squares
7 Halving	Compasses: squared paper (1 cm or 2 cm)	Halving the area of a square
8 Chunks	(Tracing paper – optional)	Dissecting a square with two lines

These investigations tend to fall into two categories: dissecting squares or square arrangements. Most require a systematic approach in order to find all possible combinations.

Natural extensions to many of these investigations lead to consideration of other shapes eg. rectangles, triangles.

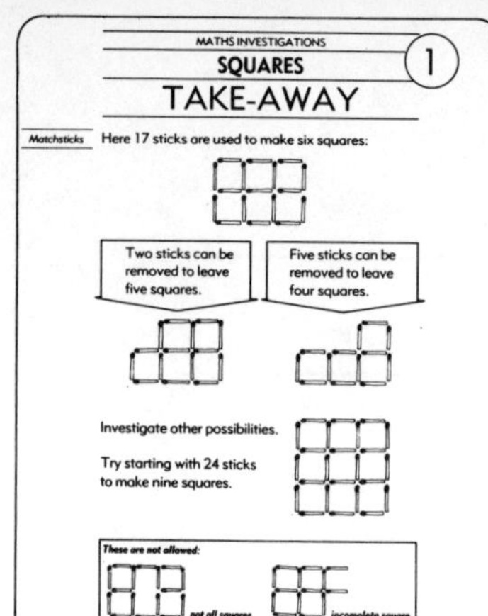

MATHS INVESTIGATIONS

SQUARES

TAKE-AWAY

1

Matchsticks

Here 17 sticks are used to make six squares:

Two sticks can be removed to leave five squares.

Five sticks can be removed to leave four squares.

Investigate other possibilities.

Try starting with 24 sticks to make nine squares.

These are not allowed:

not all squares

incomplete square

MATHS INVESTIGATIONS

SQUARES

TAKE-AWAY

1

Matchsticks

Squared or dotty paper

△ Squared or dotty paper can be used to record results.

△ Pupils can be systematic by:

removing one match, two matches, three matches ...
leaving five squares, four squares, three squares ...

△ Some solutions:

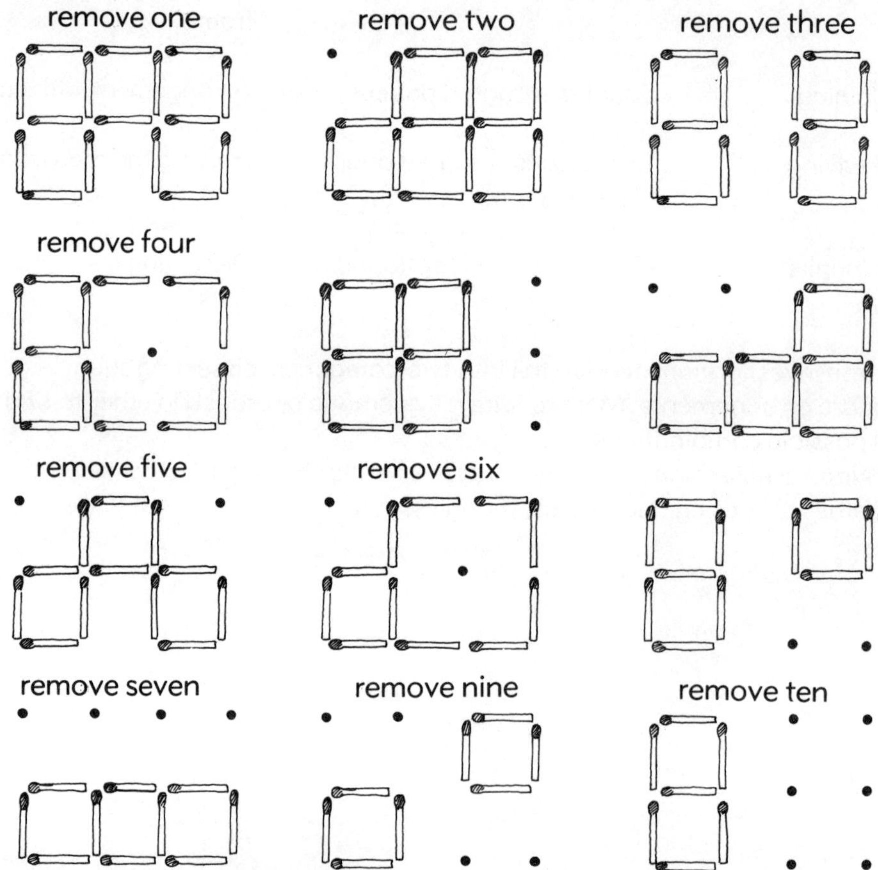

△ Try different starting arrangements.

△ Try to leave rectangles.

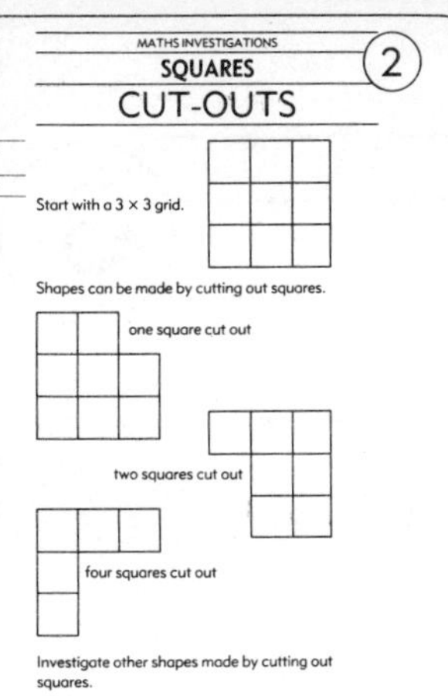
MATHS INVESTIGATIONS

SQUARES 2

CUT-OUTS

Start with a 3 × 3 grid.

Shapes can be made by cutting out squares.

one square cut out

two squares cut out

four squares cut out

Investigate other shapes made by cutting out squares.

MATHS INVESTIGATIONS

SQUARES 2

CUT-OUTS

2 cm Squared paper

Scissors

△ Pupils can start with a number of 3 × 3 grids and experiment by cutting out squares.

△ One approach is to find all shapes possible when one square is removed, then two squares, ...

One square cut-out

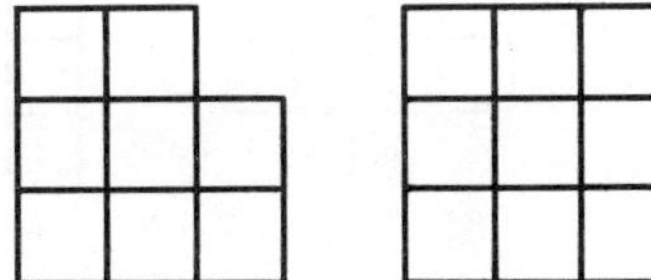

Can the middle square be cut out?

Two square cut-outs

Two corners

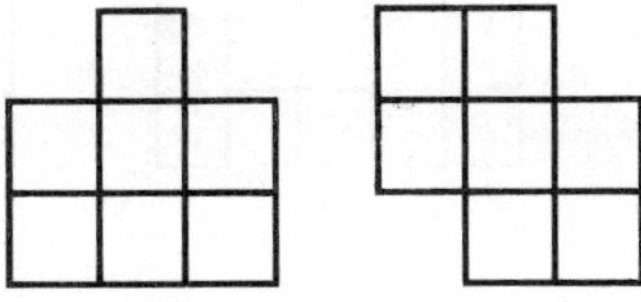

Two centre edges

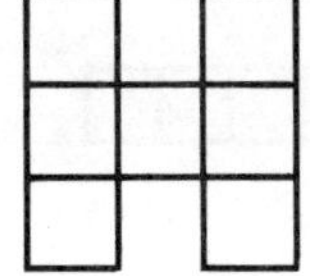

One corner, One centre edge

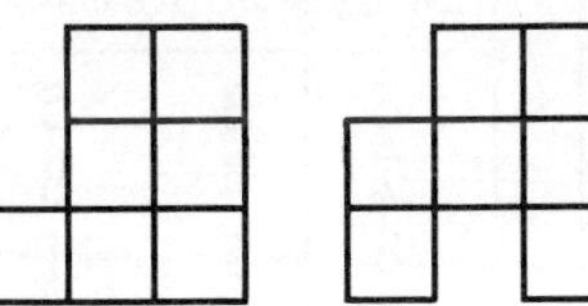

One centre edge, One middle

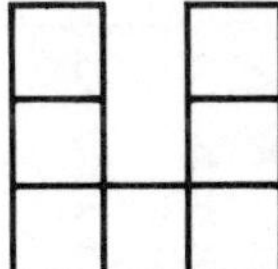

△ Start with different grids.

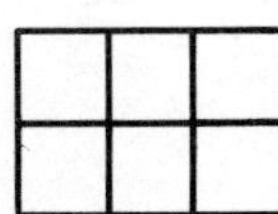

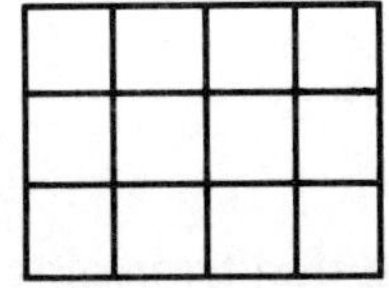

△ Cut out rectangles instead of squares.

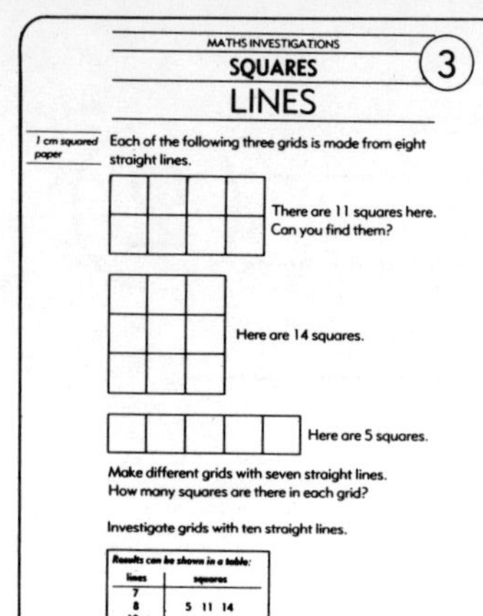

MATHS INVESTIGATIONS

SQUARES

LINES

3

1 cm squared paper

Each of the following three grids is made from eight straight lines.

There are 11 squares here. Can you find them?

Here are 14 squares.

Here are 5 squares.

Make different grids with seven straight lines. How many squares are there in each grid?

Investigate grids with ten straight lines.

Results can be shown in a table:

lines	squares
7	
8	5 11 14
10	

MATHS INVESTIGATIONS

SQUARES

LINES

3

1cm Squared paper

△ Some solutions include:

Six lines

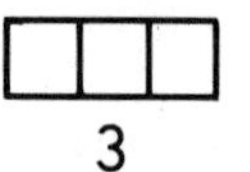
3

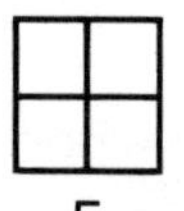
5

Seven lines

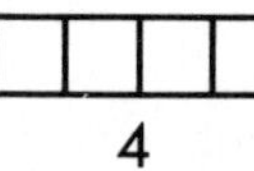
4

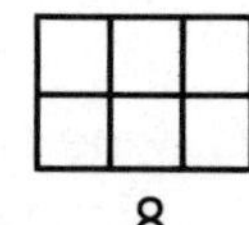
8

Eight lines

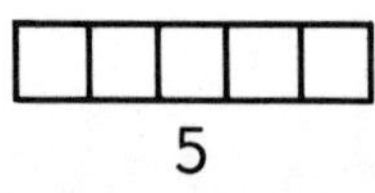
5

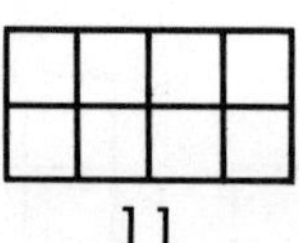
11

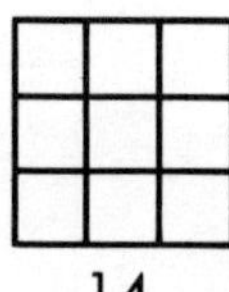
14

Nine lines

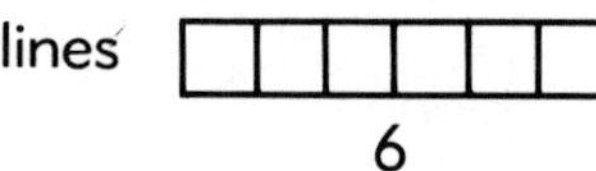
6

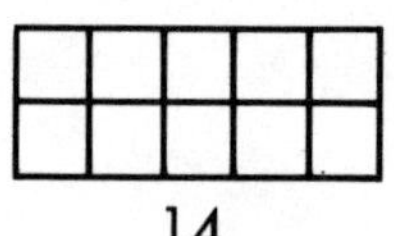
14

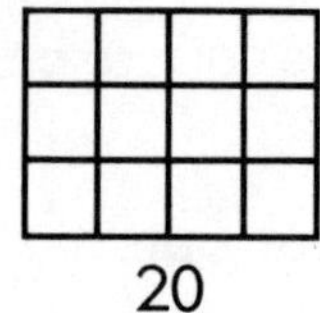
20

Ten lines

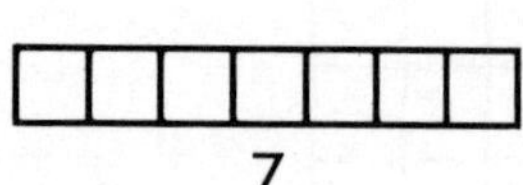
7

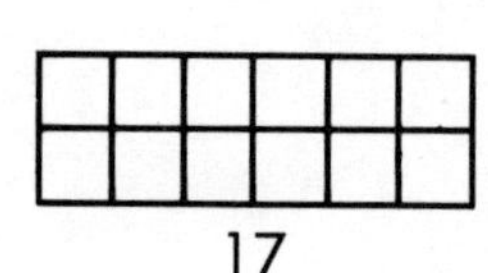
17

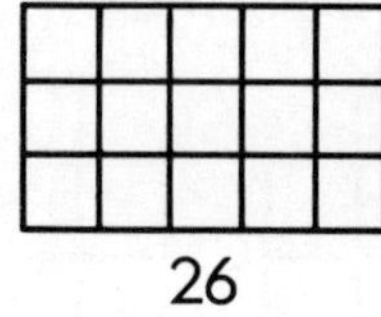
26

30

Lines	Available rectangles			
6	1 × 3	2 × 2		
7	1 × 4	2 × 3		
8	1 × 5	2 × 4	3 × 3	
9	1 × 6	2 × 5	3 × 4	
10	1 × 7	2 × 6	3 × 5	4 × 4
11	1 × 8	2 × 7	3 × 6	4 × 5

Lines	Squares			
6	3	5		
7	4	8		
8	5	11	14	
9	6	14	20	
10	7	17	26	30
11	8	20	32	40

△ Investigate triangles.
Six straight lines can make five triangles.

MATHS INVESTIGATIONS
SQUARES 4
PERIMETERS

These shapes have **perimeters** of ten units.

Can you find some more shapes with perimeters of ten units?
Investigate shapes which have a perimeter of 12 units.
Investigate shapes with other perimeters.

MATHS INVESTIGATIONS

SQUARES

PERIMETERS

4

1 cm Squared paper

△ Plastic or card squares can be used to try different arrangements.

△ One approach is to move one square to different positions ensuring that an equal perimeter is maintained.

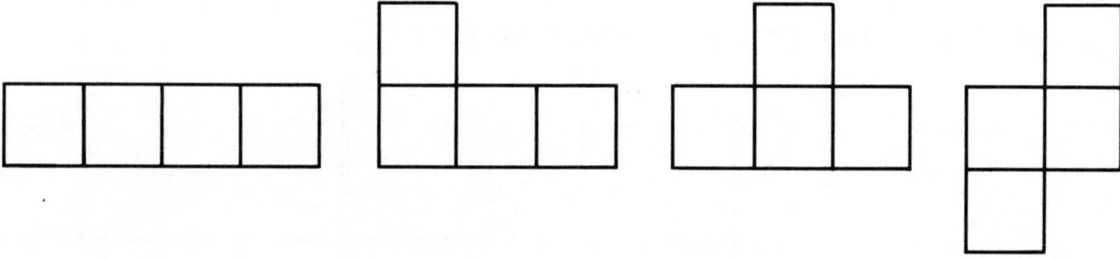

△ Some other solutions include:

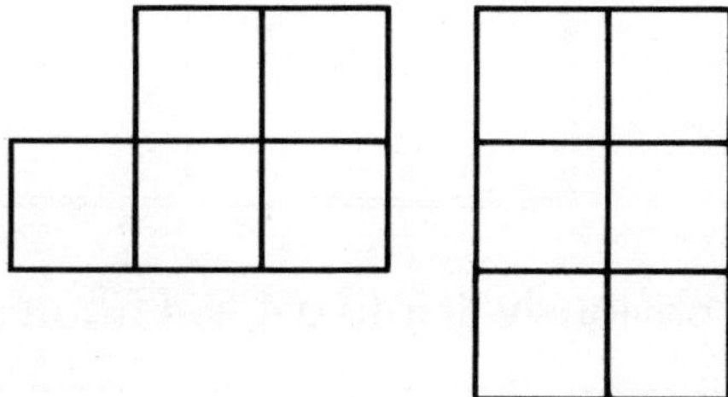

△ Which of the shapes has maximum/minimum area?

△ Explore perimeters on isometric paper.

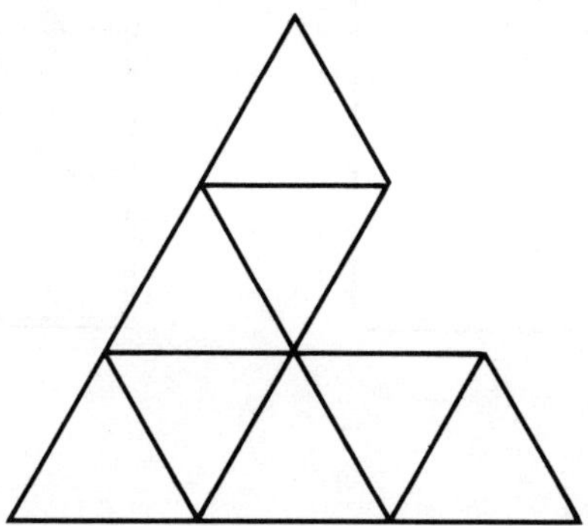

MATHS INVESTIGATIONS
SQUARES 5
FRAMES

Squares

Five squares have to be arranged into a **square frame**.

The squares must fit tightly with no overlapping.

How many different sized square frames can you find?

Try finding square frames for ten squares. Investigate for different numbers of squares.

Ten squares in a square frame.

MATHS INVESTIGATIONS

SQUARES

FRAMES

5

Squares

1 cm. Squared paper

△ The sizes of the **square frames** can be compared by drawing, tracing or calculation.

It may be helpful to work on 1 cm squared paper.

△ With five squares the smallest **square frame** is:

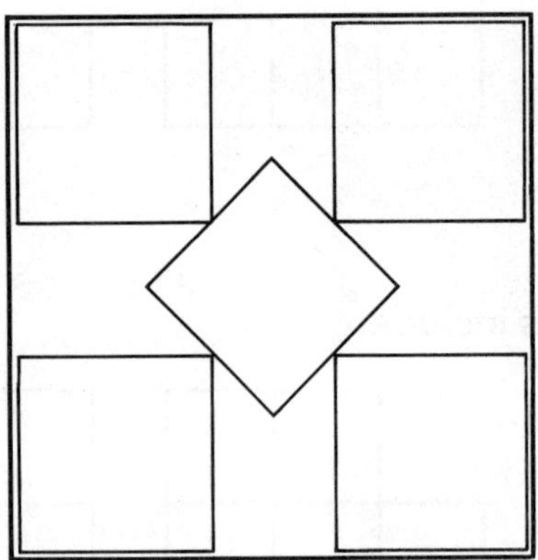

△ Ten squares will obviously fit into a 4 × 4 frame.

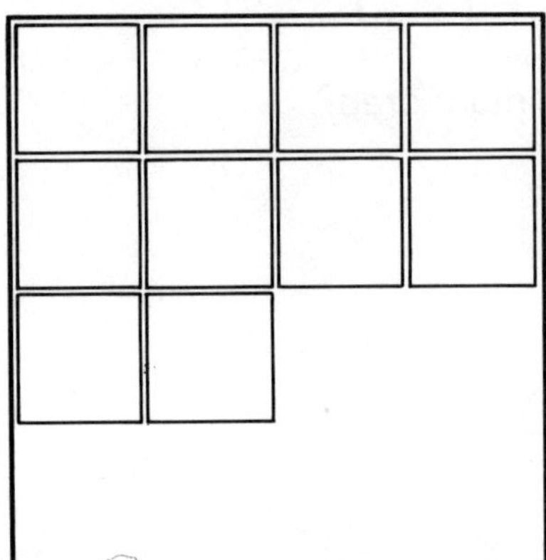

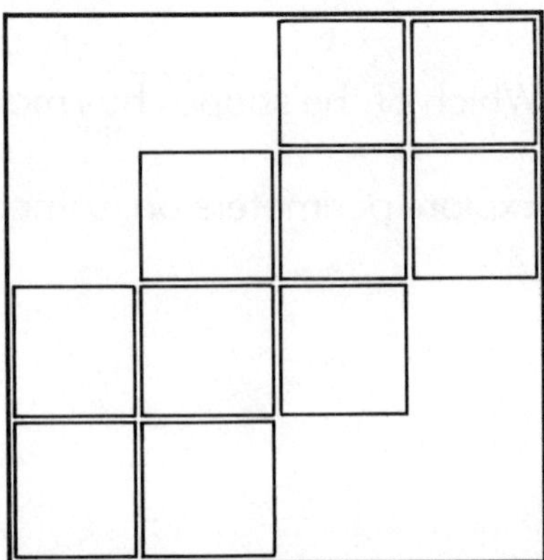

Find smaller frames.

△ Try rectangle frames.

MATHS INVESTIGATIONS
SQUARES
JOINING
6

Investigate how many different shapes can be made by joining squares edge to edge.

Here is a four-square shape:

Here is a five-square shape:

These are not different.
They are the same shape but in a different position.

MATHS INVESTIGATIONS 6
SQUARES
JOINING

Squares
1 cm Squared paper

Two-square shape

Three-square shapes

Four-square shapes

Five-square shapes

The five-square shapes can be obtained by considering each of the four-square shapes in turn and adding one square in different positions.

There are 34 different six-square shapes.

Number of squares	2	3	4	5	6
Number of shapes	1	2	5	12	34

△ Try joining squares corner to corner.

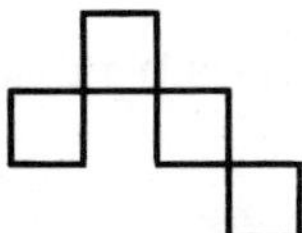

△ Try joining triangles  or hexagons.

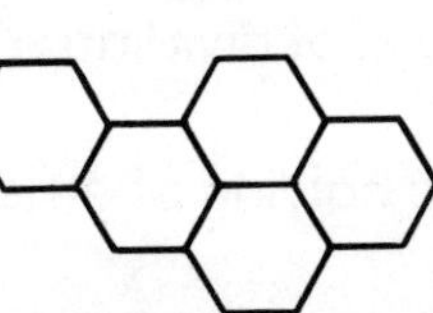

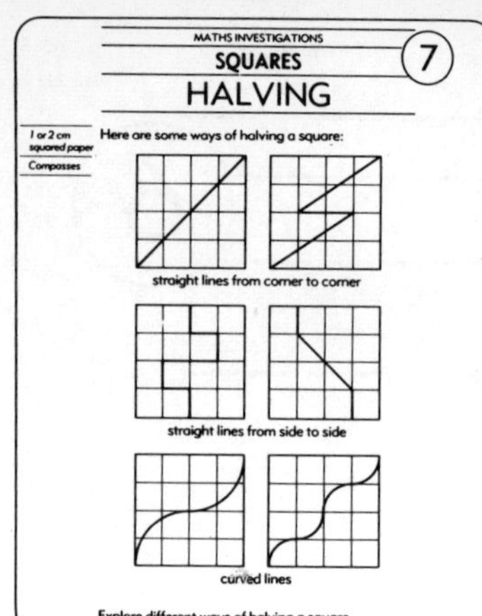
MATHS INVESTIGATIONS
SQUARES (7)
HALVING

1 or 2 cm squared paper
Compasses

Here are some ways of halving a square:

Explore different ways of halving a square.

MATHS INVESTIGATIONS

SQUARES

HALVING

7

1 or 2 cm
Squared paper
Compasses

△ Corner to corner
One approach is to measure equal points on opposite sides of the square:

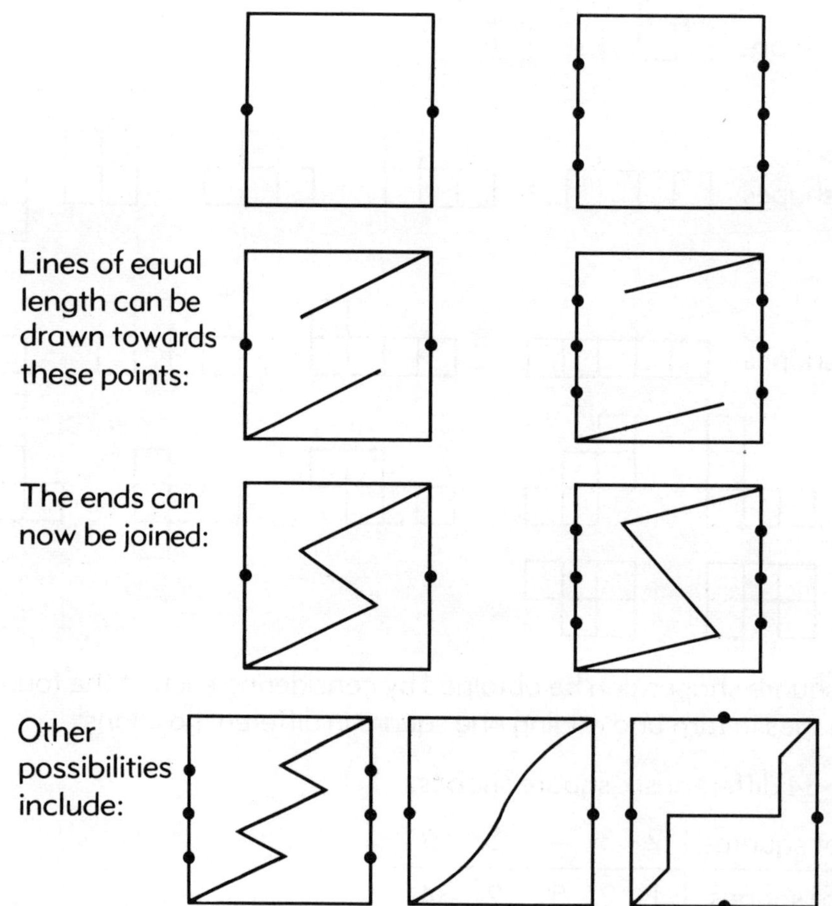

Lines of equal length can be drawn towards these points:
The ends can now be joined:
Other possibilities include:

△ Designs can be created using tracing paper and rotating a half-turn about the centre of the square.

△ Investigate ways of dividing a square into three or four identical pieces.

△ Investigate the halving of other shapes.

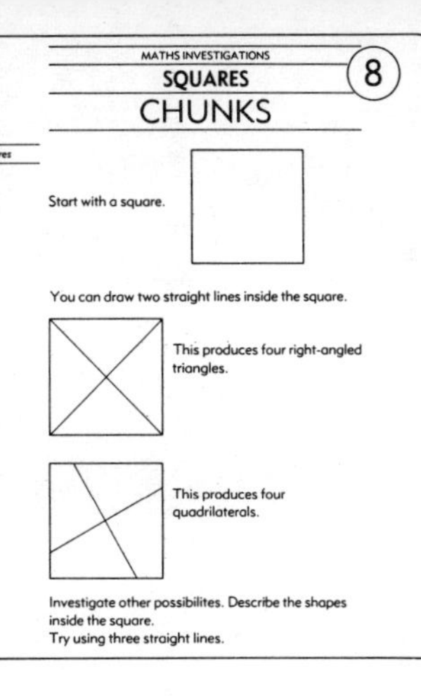
MATHS INVESTIGATIONS

SQUARES 8

CHUNKS

Squares

Start with a square.

You can draw two straight lines inside the square.

This produces four right-angled triangles.

This produces four quadrilaterals.

Investigate other possibilites. Describe the shapes inside the square.
Try using three straight lines.

MATHS INVESTIGATIONS

SQUARES

CHUNKS

8

Tracing paper

△ A possible approach is for pupils to use two pieces of tracing paper, each containing a straight line, to be placed on top of the square.

△ Some possibilities include:

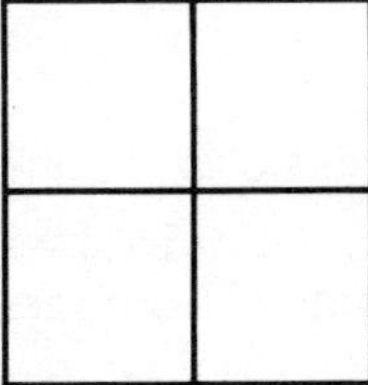

Four squares

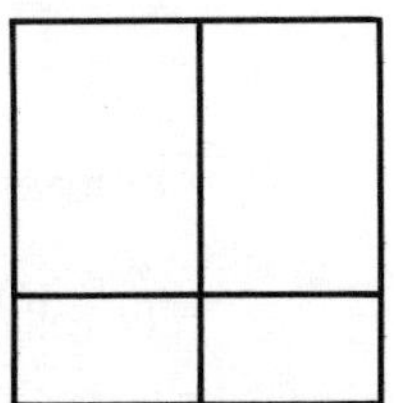

Four rectangles

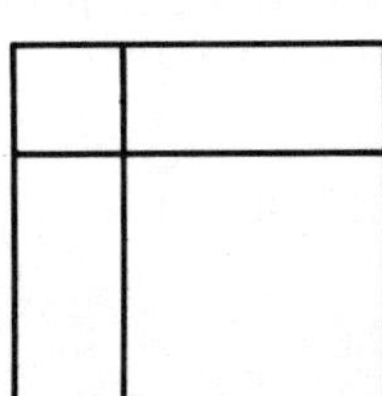

Two squares
Two rectangles

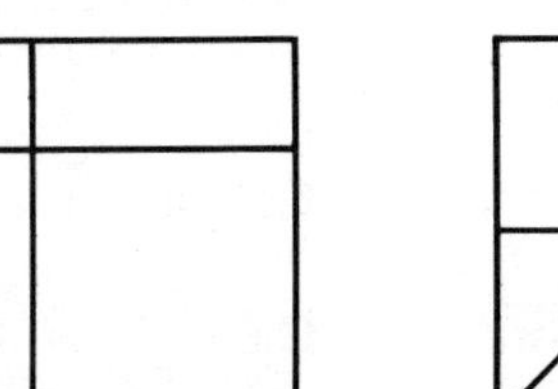

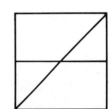

Two right-angled triangles
Two right-angled trapezia

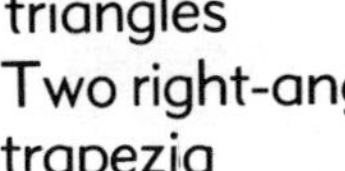

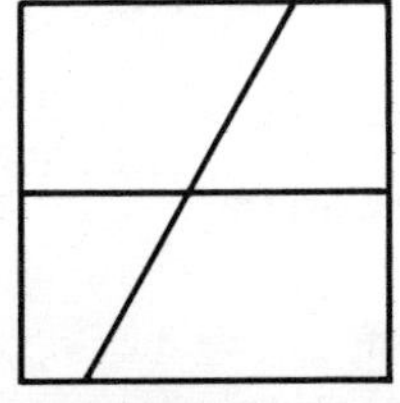

Four right-angled trapezia

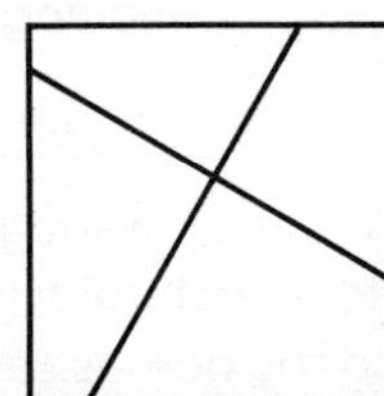

Four quadrilaterals

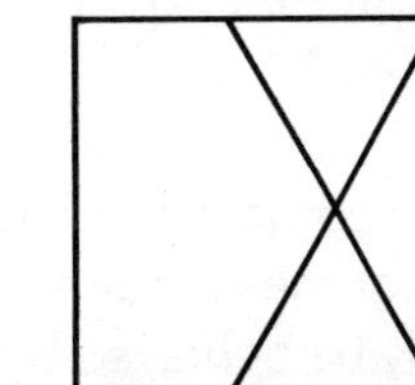

Three triangles
One pentagon

△ Pupils should be encouraged to describe, as accurately as possible, the different shapes created.

△ Try increasing the number of straight lines.

△ Start with a different shape.

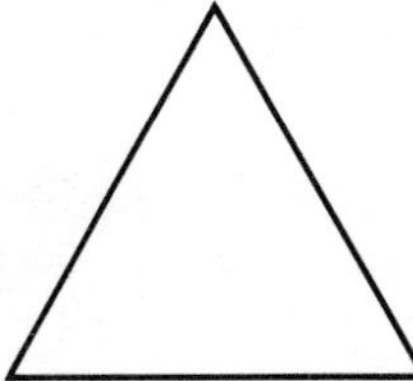

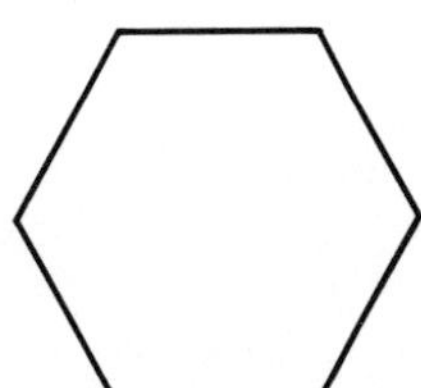

GAMES

	APPARATUS	ACTIVITY
9 Track	Dice (6-spot)	Totalling dice scores
10 Fifteens	Number cards (1 – 9)	Totalling fifteen
11 Sticks	Matchsticks	NIM type game
12 Two Card Swap	Number cards (1 – 25)	Totalling to 25
13 Hangman	Calculator	Numerical expressions
14 Low Score	Dice	Totalling and special numbers
15 Attraction	Counters: squared paper	Board game involving direction
16 Griddle	Counters: squared paper (1 cm)	Placing counters on a grid

It is assumed that each game will involve two players although some can be played by more. Some games require players to search for winning strategies; this can only be achieved by playing the games several times.

In some instances this game is merely a starting point for a mathematical investigation.

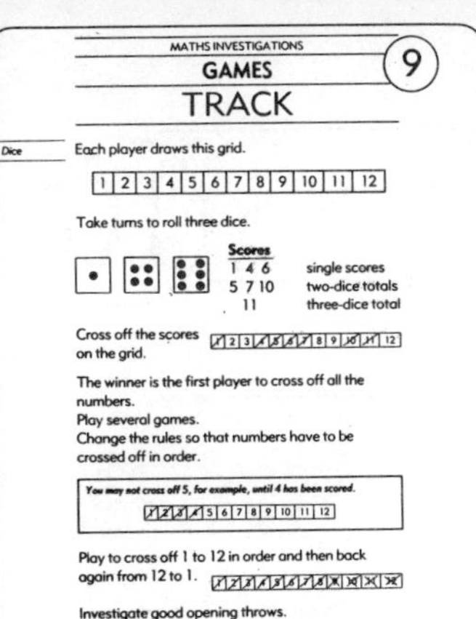
MATHS INVESTIGATIONS

GAMES 9

TRACK

Dice

Each player draws this grid.

1	2	3	4	5	6	7	8	9	10	11	12

Take turns to roll three dice.

Scores	
1 4 6	single scores
5 7 10	two-dice totals
11	three-dice total

Cross off the scores on the grid.

The winner is the first player to cross off all the numbers.
Play several games.
Change the rules so that numbers have to be crossed off in order.

You may not cross off 5, for example, until 4 has been scored.

Play to cross off 1 to 12 in order and then back again from 12 to 1.

Investigate good opening throws.
What is the fewest number of throws possible to win the game?

MATHS INVESTIGATIONS
GAMES
TRACK 9

Dice

△ When numbers have to be crossed off in order, the best opening throw is:

 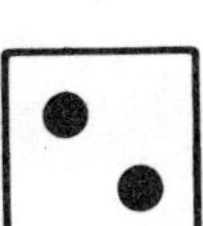 

This allows scores of 1, 2, 3, 4, 5, 6 and 7.

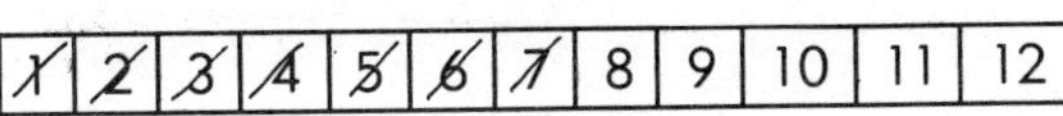

These all score 8 and 9.

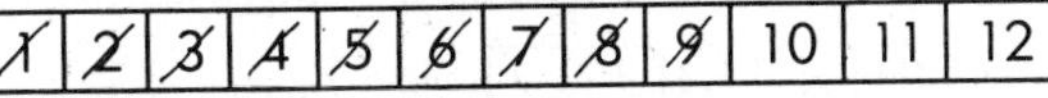

This scores 11 and 12.

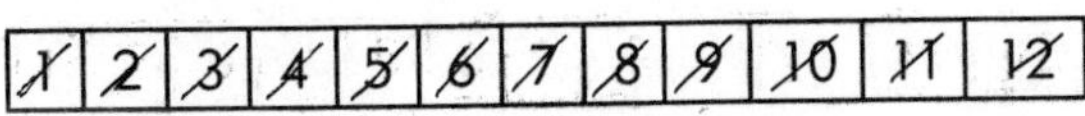

△ Suppose the numbers can be crossed off in any order and pupils have to select one of the possible dice combinations. For example:

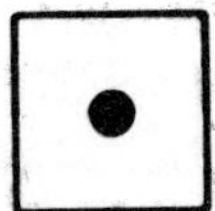  

Cross off: 1, 2 and 4
or 3 and 4
or 5 and 2
or 1 and 6
or 7

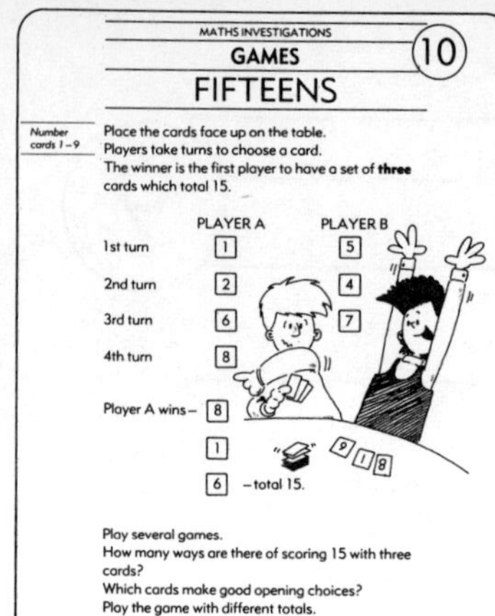

MATHS INVESTIGATIONS

GAMES 10

FIFTEENS

Number cards 1–9

Place the cards face up on the table.
Players take turns to choose a card.
The winner is the first player to have a set of **three** cards which total 15.

	PLAYER A	PLAYER B
1st turn	1	5
2nd turn	2	4
3rd turn	6	7
4th turn	8	

Player A wins – 8
1
6 – total 15.

Play several games.
How many ways are there of scoring 15 with three cards?
Which cards make good opening choices?
Play the game with different totals.

MATHS INVESTIGATIONS

GAMES

FIFTEENS

10

Number cards 1–9

△ There are eight ways of obtaining a total of 15 with three cards:

(8,1,6) (3,5,7) (4,9,2) (8,3,4)
(1,5,9) (6,7,2) (4,5,6) (2,5,8)

These numbers could be arranged to form a magic square:

8	1	6
3	5	7
4	9	2

Now it is possible to use a 'noughts and crosses' strategy to play the game:

i.e. [8] [6] [5] [4] [2] are good opening choices.

Played strategically each game ends in a draw.

△ Try the game using cards numbered 1 to 16 with sets of four totalling 34.

MATHS INVESTIGATIONS

GAMES

STICKS

11

Matchsticks

Start with 20 matchsticks.
Players take turns to remove one, two, three or four sticks.
The player who takes the last stick, or sticks, wins.

Play several games.
Work out a plan for winning.

Try starting with a different number of sticks.
What happens if players are allowed to take as many as six sticks?

MATHS INVESTIGATIONS

GAMES

STICKS

11

Matchsticks

△ Counters may be used instead of matchsticks.

△ Pupils quickly realise that if they eventually leave five sticks they will win. Working backwards from this, it will be seen that a multiple of five must be left at any stage of the game to be sure of winning.
Starting with 20 sticks allows player B to win.
Starting with a non-multiple of five allows player A to win.

△ When up to six sticks may be removed, seven or a multiple of seven must be left to win.

△ Change the rules so that the player who takes the last stick loses.
Investigate winning strategies.

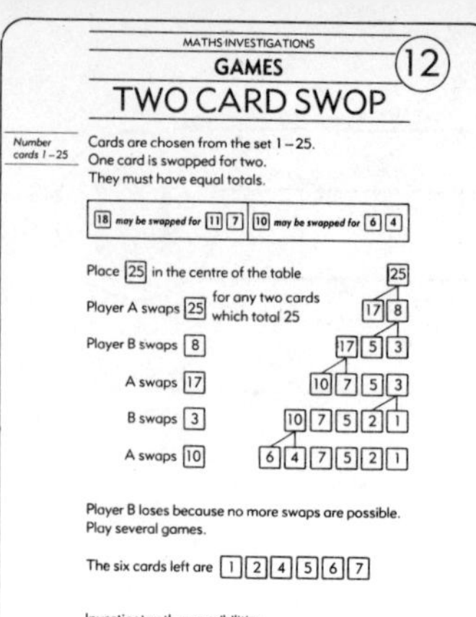
MATHS INVESTIGATIONS

GAMES 12

TWO CARD SWOP

Number cards 1 – 25

Cards are chosen from the set 1 – 25.
One card is swapped for two.
They must have equal totals.

18 *may be swapped for* 11 7 | 10 *may be swapped for* 6 4

Place 25 in the centre of the table — 25
Player A swaps 25 for any two cards which total 25 — 17 8
Player B swaps 8 — 17 5 3
A swaps 17 — 10 7 5 3
B swaps 3 — 10 7 5 2 1
A swaps 10 — 6 4 7 5 2 1

Player B loses because no more swaps are possible.
Play several games.

The six cards left are 1 2 4 5 6 7

Investigate other possibilities.
Is it possible to have five cards left?

MATHS INVESTIGATIONS

GAMES

12

TWO CARD SWOP

Number cards 1 – 25

△ Cards must be selected from the set 1 – 25. Once a card has been selected it cannot be used again.

△ Most games result in six cards left at the end. These must sum to 25.
It is possible for the game to end with five cards left:

e.g.

25	25	25
17 8	9 16	8 17
17 5 3	5 4 16	8 6 11
10 7 5 3	5 4 10 6	8 2 4 11
9 1 7 5 3	2 3 4 10 6	1 7 2 4 11

Again, the five cards sum to 25.
Is it possible for the game to end with seven cards?

△ Try changing the rules.
Suppose you swap one card for two when the sum **or** difference is the same?

Then it is possible to use all the cards.

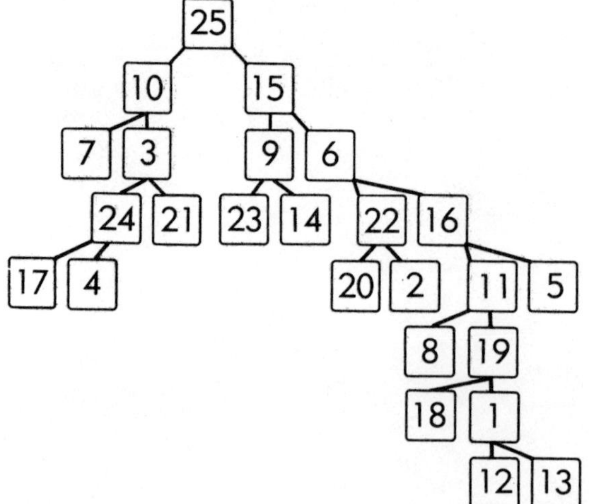

MATHS INVESTIGATIONS

GAMES (13)

HANGMAN

Calculator

Start with a scaffold.
Player A invents an equation to fit these **seven** spaces.

_ _ _ _ = _ _ _

Player B guesses digits (0 to 9) or signs (+, −, ×, ÷).

GUESS

1 YES _ _ 1 _ = 1 _ _

5 NO

6 NO

4 YES _ _ 1 _ = 1 4 4

+ NO

8 YES 8 _ 1 8 = 1 4 4

× YES 8 × 1 8 = 1 4 4 Player B wins.

Six wrong guesses lead to: Then Player A wins.

Investigate different possible solutions to _ _ _ _ = 1 4 4

Play with **eight** spaces. _ _ _ _ _ = _ _ _

MATHS INVESTIGATIONS

13

GAMES

HANGMAN

Calculator

△ Calculators may be helpful.

△ Other solutions to _ _ _ _ = 144 include:

2 × 72 = 144
3 × 48 = 144
4 × 36 = 144
6 × 24 = 144

Many solutions to _ _ _ _ _ = 144 are possible:

involving multiplication 12 × 12 = 144
addition 143 + 1 = 144
division 432 ÷ 3 = 144
subtractions 153 − 9 = 144

How many of each type are possible?

△ Invent equations to fit _ _ _ _ _ = _ _ _
Do not allow digits to be repeated.

△ Try with nine spaces.

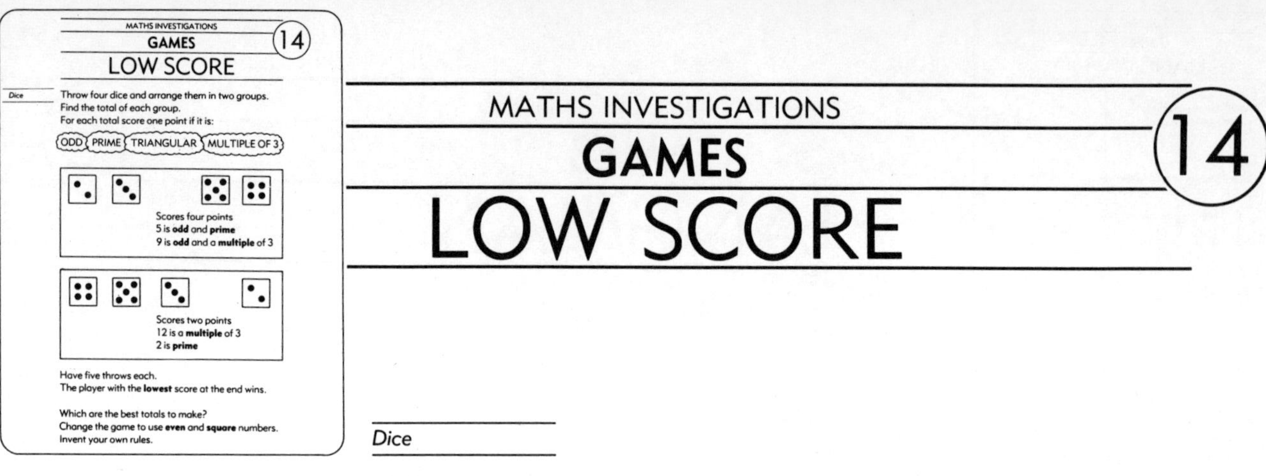

MATHS INVESTIGATIONS

GAMES

LOW SCORE

14

Dice

Throw four dice and arrange them in two groups.
Find the total of each group.
For each total score one point if it is:

ODD PRIME TRIANGULAR MULTIPLE OF 3

Scores four points
5 is **odd** and **prime**
9 is **odd** and a **multiple** of 3

Scores two points
12 is a **multiple** of 3
2 is **prime**

Have five throws each.
The player with the **lowest** score at the end wins.

Which are the best totals to make?
Change the game to use **even** and **square** numbers.
Invent your own rules.

MATHS INVESTIGATIONS

GAMES

LOW SCORE

14

Dice

△ Pupils must make two groups with the dice and use both totals to calculate their points score.

Totals of 1 to 18 can be made. The scores can be tabulated:

	1	2	3	4	5	6	7	8	9	10	11	12	13	14	15	16	17	18
Odd	✓		✓		✓		✓		✓		✓		✓		✓		✓	
Prime		✓	✓		✓		✓				✓		✓				✓	
Triangular	✓		✓			✓				✓					✓			
Multiple of 3			✓			✓			✓			✓			✓			✓

Some totals score a maximum of four points, others zero points.

△ Variations include:

- Highest score wins.
- Use a different set of number properties e.g. even, square, multiples of 3, multiples of 2 (How do you obtain 0 points here?).
- Try with a different number of dice.
- Vary the points awarded (ODD: 1 Point, PRIME: 2 points).

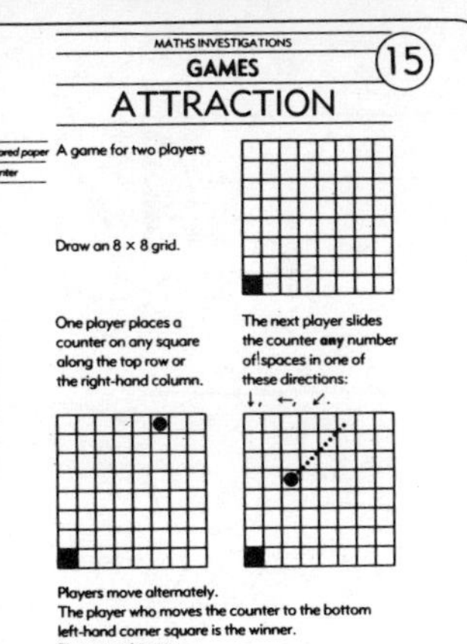

MATHS INVESTIGATIONS

GAMES

ATTRACTION

15

Squared paper

Counter

△ Ensure pupils realise that moves are south, west, or south-west **any** number of squares. The slide must be in a straight line.

△ The player who moves first will win by moving the counter to the 'winning squares'.

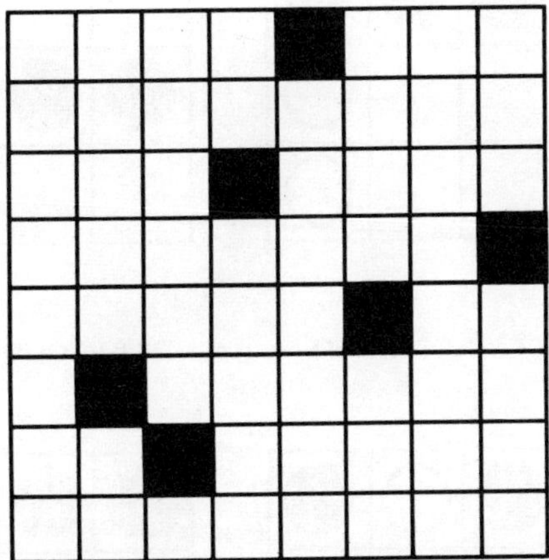

There are two 'winning squares' for the opening move. Irrespective of an opponent's move, the starting player will always be able to move to another 'winning square'.

△ Investigate playing on different sized grids: e.g. 12 × 12, 10 × 6

△ What happens if the counter can move differently? e.g. knight's moves.

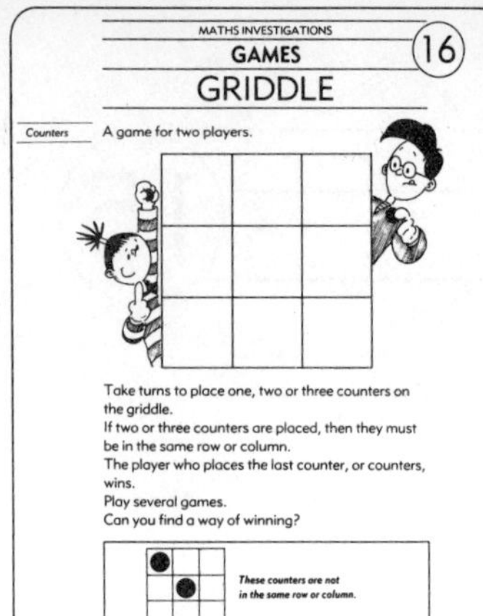
MATHS INVESTIGATIONS

GAMES 16

GRIDDLE

Counters

A game for two players.

Take turns to place one, two or three counters on the griddle.
If two or three counters are placed, then they must be in the same row or column.
The player who places the last counter, or counters, wins.
Play several games.
Can you find a way of winning?

These counters are not in the same row or column.

MATHS INVESTIGATIONS

GAMES 16

GRIDDLE

Counters

△ Encourage pupils to look for winning counter-moves to any opening.

Black opening with three counters – a winning counter-move is to place two counters like this:

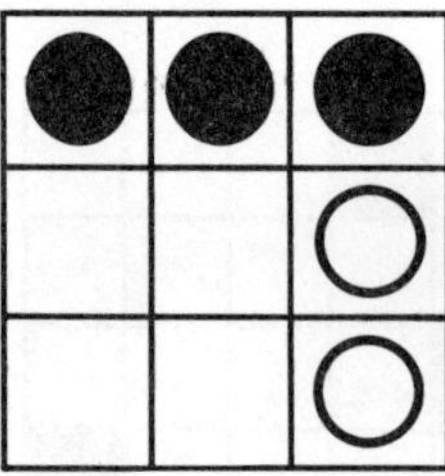
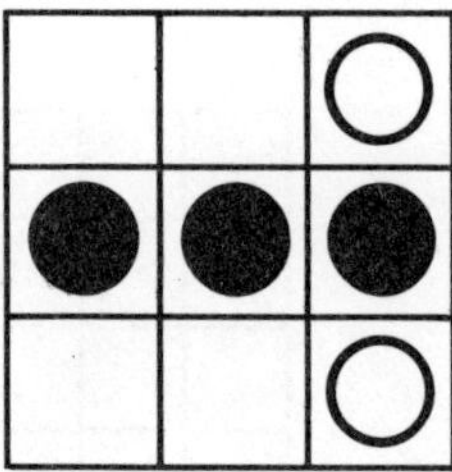

Black opening with two counters – a winning counter-move is to place three counters like this:

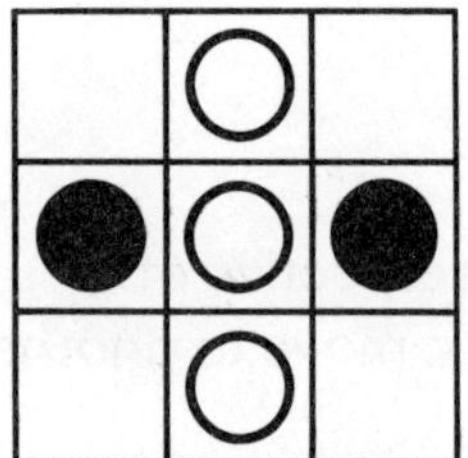
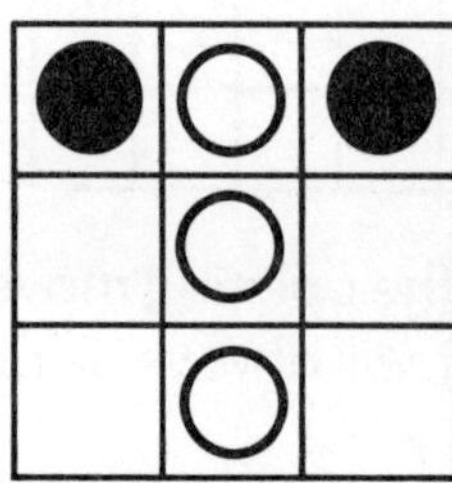
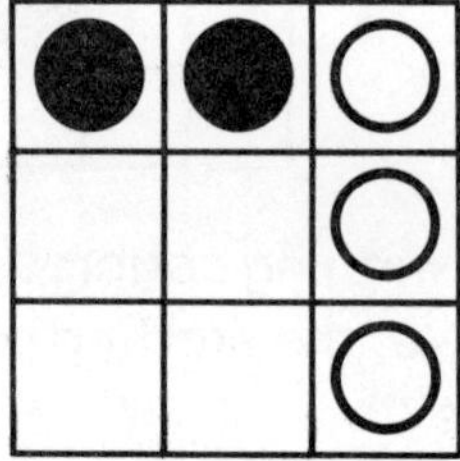
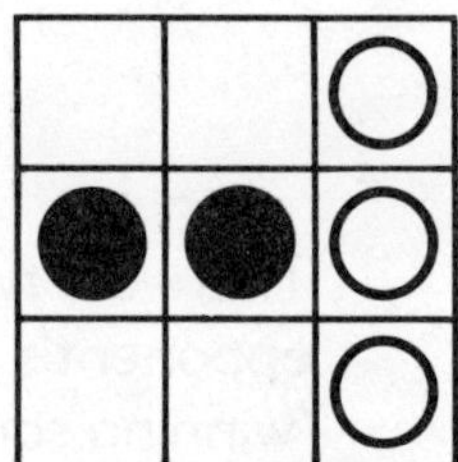

What can happen if black opens with one counter?

△ Try different sized griddles.

POLYGONS

	APPARATUS	ACTIVITY
17 Dots	Dotty paper	Different triangles and quadrilaterals on square grids
18 Paper Cuts	Scissors: scrap paper	Cutting polygons out of folded paper
19 Holes	Squares: (regular polygons)	Arrangement of squares to leave regular gaps
20 Hexa-dots	Hexagon dotty paper	Different polygons on a hexagon grid
21 Square Cuts	Scissors: Card: plain paper	Two and three piece dissections of a square
22 Rotations	Dotty paper	Polygons with rotational symmetry on a square grid
23 Sides	Dotty paper: (geoboards)	Different sided polygons on a square grid
24 Areas	Dotty paper: (geoboards)	Areas of polygons on a square grid

Several cards provide the opportunity for pupils to consolidate ideas associated with angle and area.

Emphasis on making clear descriptions of the polygons will help the pupils to become familiar with the language and classification of shape.

Congruent shapes which are rotations and reflections can be checked by the use of tracing paper.

Activities involving the use of dotty paper will be enhanced by the use of geoboards.

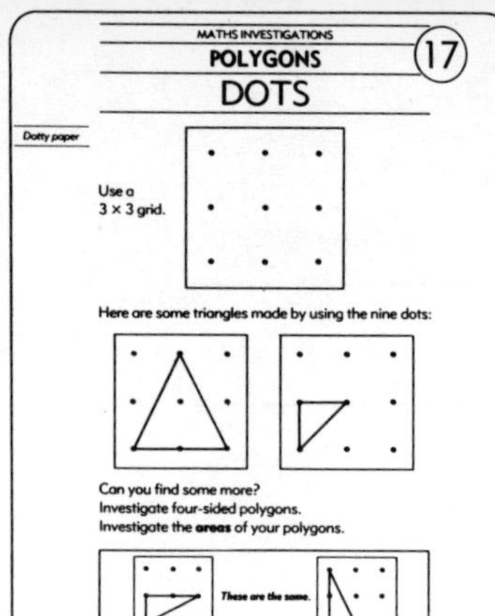
MATHS INVESTIGATIONS
POLYGONS (17)
DOTS

Dotty paper

Use a 3 × 3 grid.

Here are some triangles made by using the nine dots:

Can you find some more?
Investigate four-sided polygons.
Investigate the **areas** of your polygons.

These are the same.

MATHS INVESTIGATIONS

POLYGONS

DOTS

(17)

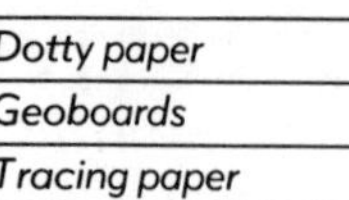
Dotty paper
Geoboards
Tracing paper

△ Geoboards are helpful.

△ Eight different triangles are available.

One strategy for finding these is to fix one side:

Then find all the triangles which can be made containing this side.

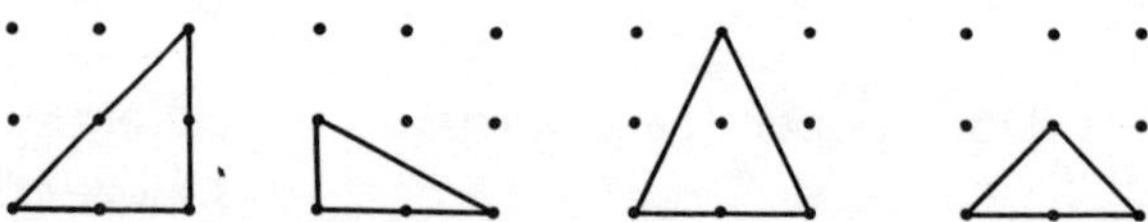

Tracing paper can be used to check if two shapes are identical. Sixteen four-sided polygons can be found. Here are four of them, each of which has an area of two units:

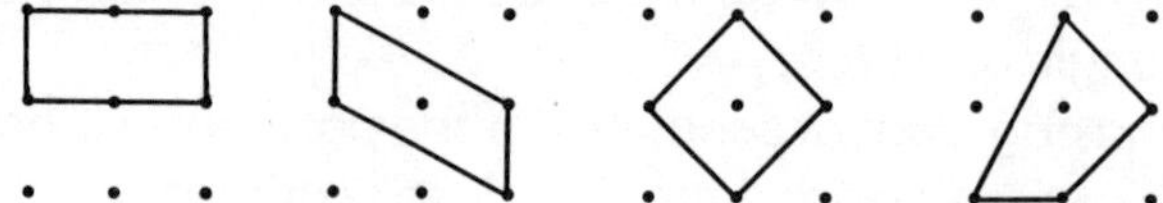

△ The areas of the polygons are ½, 1, 1½, 2, 2½, 3, 3½ or 4 square units. Which triangle has the maximum area? Which has the minimum area? Which shapes have equal areas?

△ Investigate five-sided polygons, six-sided polygons, and so on. It is possible to find 23 pentagons, 22 hexagons, and 5 heptagons.

△ Investigate polygons on rectangular grids e.g. 3 × 4.

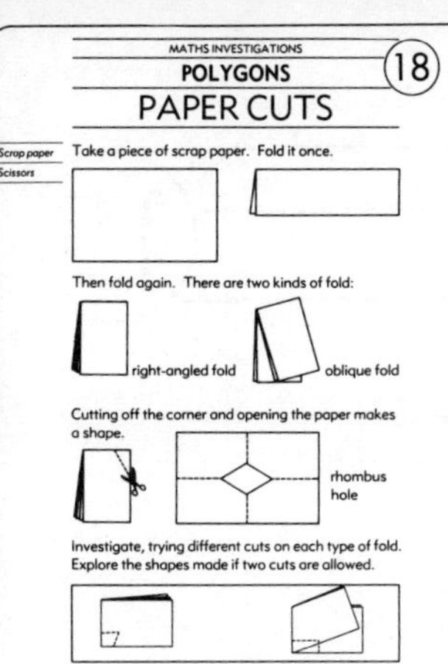

MATHS INVESTIGATIONS

POLYGONS 18

PAPER CUTS

Scrap paper
Scissors

Take a piece of scrap paper. Fold it once.

Then fold again. There are two kinds of fold:

right-angled fold   oblique fold

Cutting off the corner and opening the paper makes a shape.

rhombus hole

Investigate, trying different cuts on each type of fold. Explore the shapes made if two cuts are allowed.

MATHS INVESTIGATIONS

POLYGONS

PAPER CUTS

18

Scrap paper

Scissors

△ With a right-angled fold most straight cuts produce a rhombus.

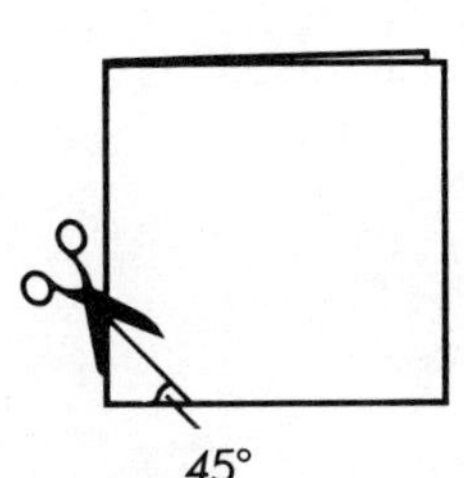

This cut produces a square:

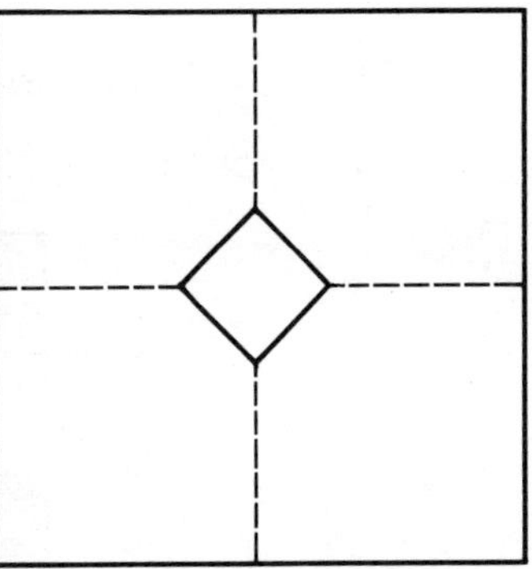

△ With oblique folds most straight cuts produce a kite.

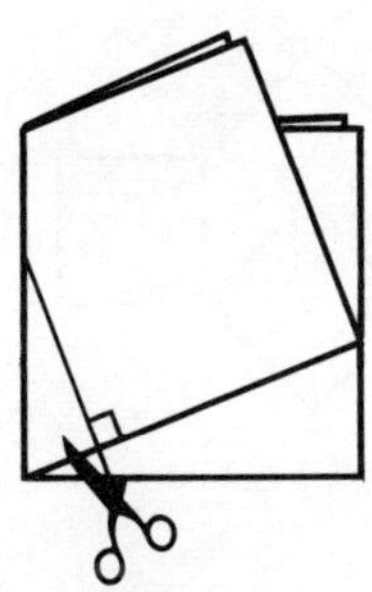

This cut produces an isosceles triangle:

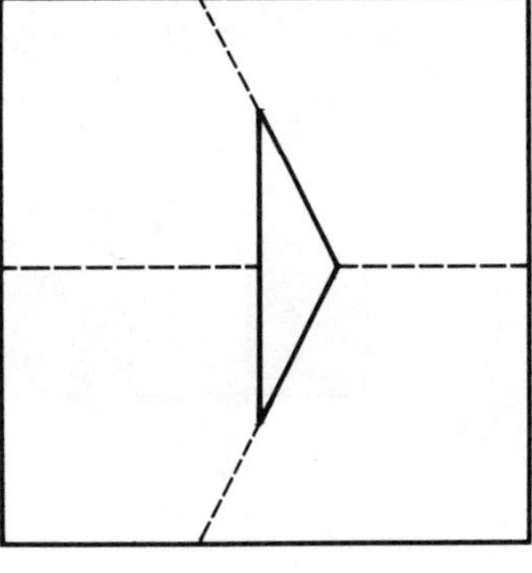

△ Which cut and fold produces an arrowhead?

△ Experiment with two cuts on each fold.

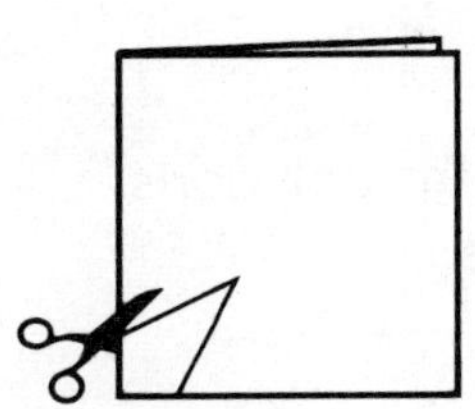

This cut produces:

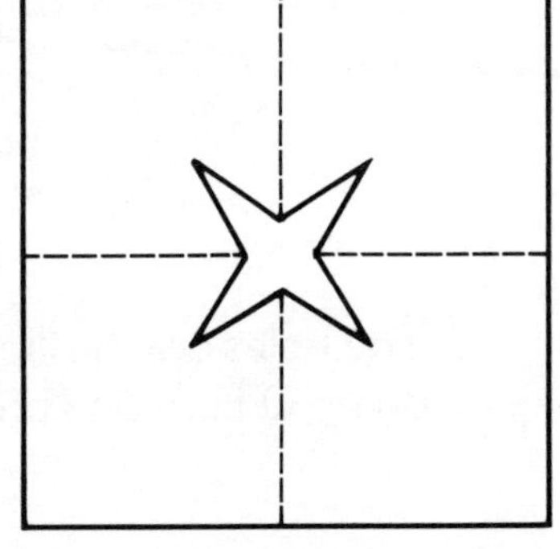

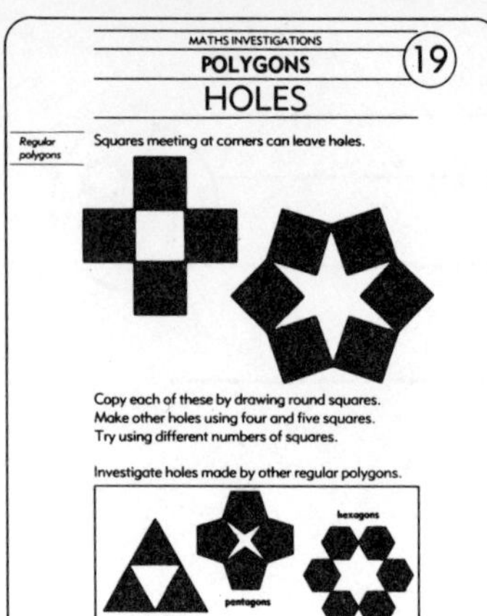
MATHS INVESTIGATIONS

POLYGONS (19)

HOLES

Regular polygons

Squares meeting at corners can leave holes.

Copy each of these by drawing round squares.
Make other holes using four and five squares.
Try using different numbers of squares.

Investigate holes made by other regular polygons.

MATHS INVESTIGATIONS

POLYGONS (19)

HOLES

Regular polygons

△ Holes with four squares – two types are possible:

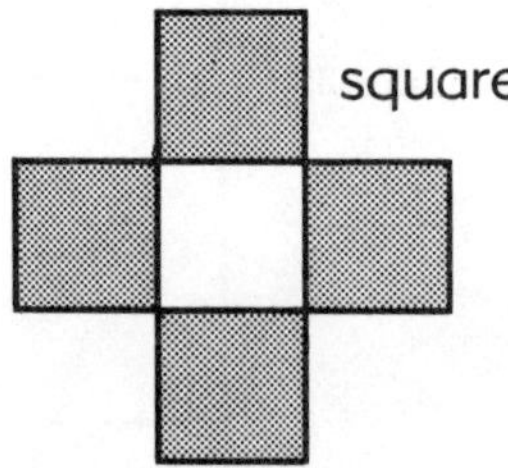

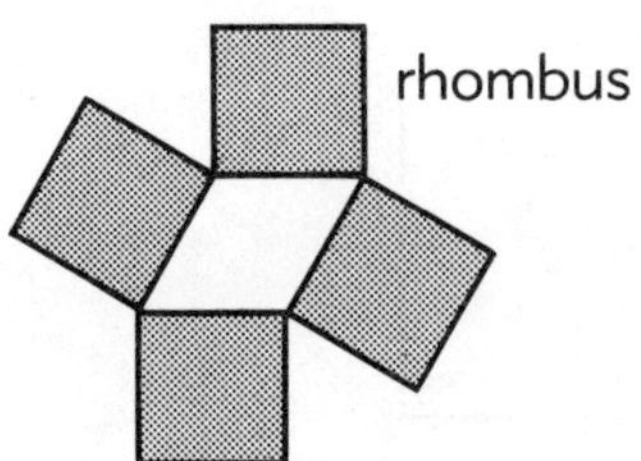

△ Holes with five squares – several types are possible:

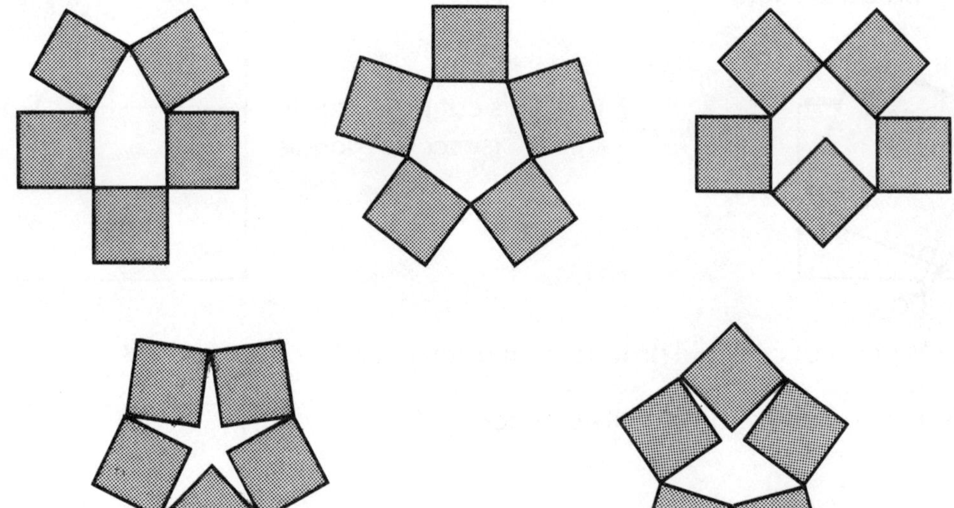

△ The holes can be illustrated in an interesting way by placing the polygons on an O.H.P. and projecting the patterns onto a wall or screen.

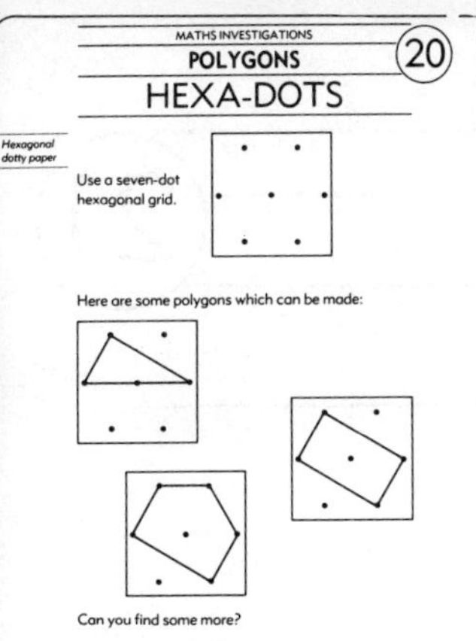
MATHS INVESTIGATIONS
POLYGONS
HEXA-DOTS
20

Hexagonal dotty paper

Use a seven-dot hexagonal grid.

Here are some polygons which can be made:

Can you find some more?

Investigate the **angles** of the polygons.

MATHS INVESTIGATIONS

POLYGONS

HEXA-DOTS

20

Hexagonal dotty paper

△ Nineteen shapes are possible:

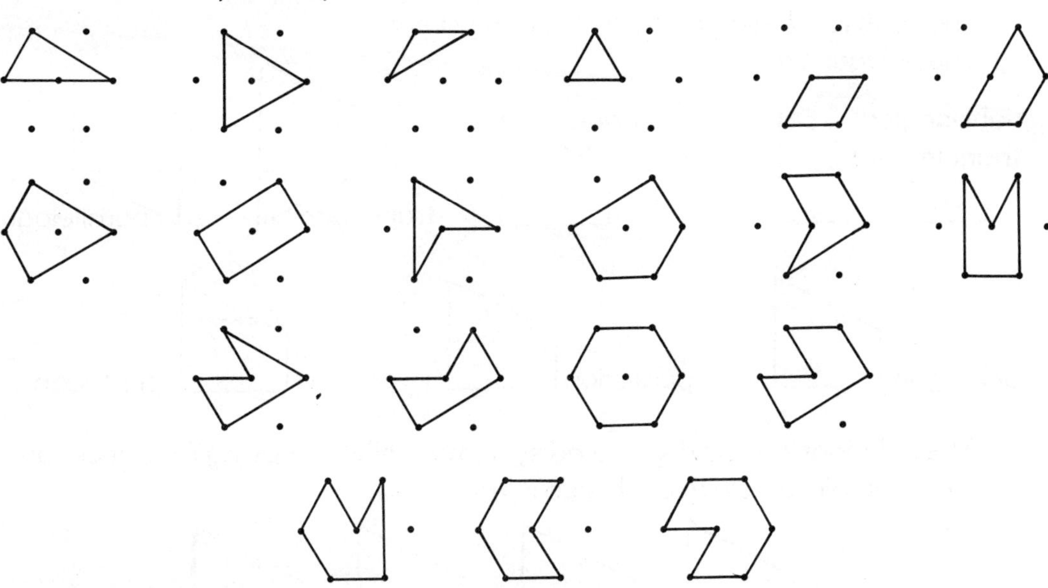

The interior angles of the polygons are all multiples of 30°.

- Which shape has one right-angle, more than one right-angle?
- Which shapes have equal angles?
- Which shape has six equal angles?
- Explore the sums of the interior angles of the polygons.

△ The polygons can be sorted in several ways:

- How many three-sided shapes, four-sided shapes, ...?
- How many shapes have two equal sides, three equal sides, ...?
- Which shapes have line symmetry, rotational symmetry?
- Which shapes have equal areas?

△ Try a six-dot pentagonal arrangement.

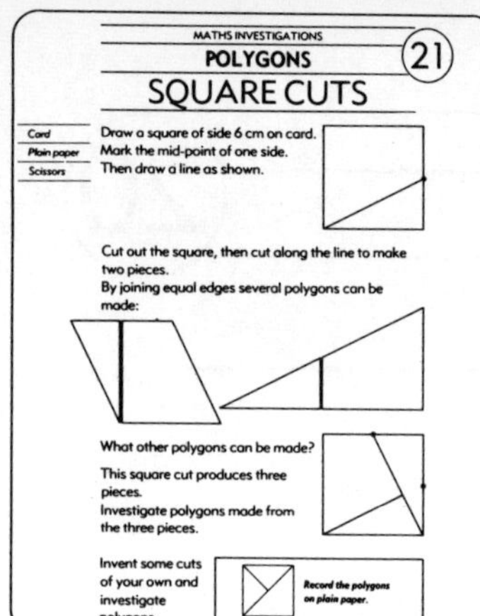
MATHS INVESTIGATIONS
POLYGONS 21
SQUARE CUTS

Card
Plain paper
Scissors

Draw a square of side 6 cm on card.
Mark the mid-point of one side.
Then draw a line as shown.

Cut out the square, then cut along the line to make two pieces.
By joining equal edges several polygons can be made:

What other polygons can be made?

This square cut produces three pieces.
Investigate polygons made from the three pieces.

Invent some cuts of your own and investigate polygons.

Record the polygons on plain paper.

MATHS INVESTIGATIONS

POLYGONS

SQUARE CUTS

21

Card
Plain paper
Scissors

△ Pupils should be encouraged to name the shapes.
They can label them on their recording sheets.
The polygons are:

right-angled triangle

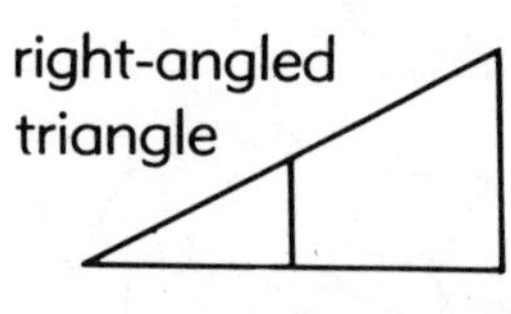

pentagon

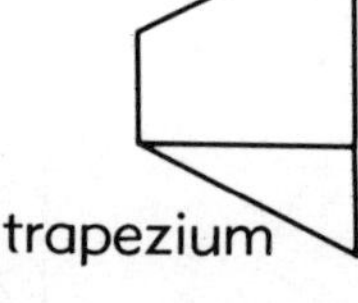

trapezium

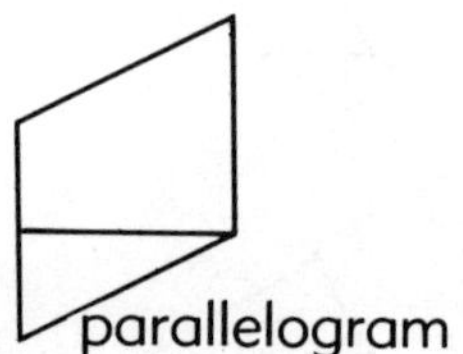

parallelogram

pentagon

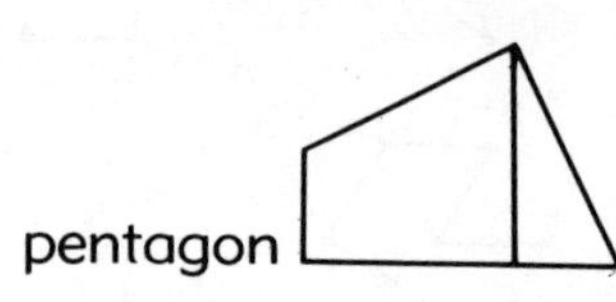

pentagon

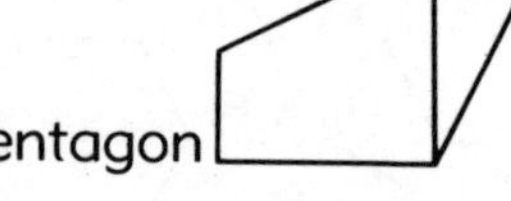

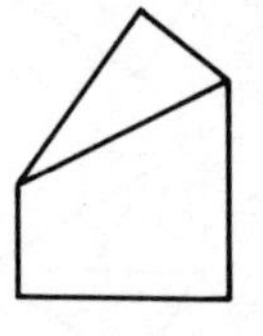

pentagon

△ The polygons can be discovered systematically by moving one piece round the other piece, taking each side in turn.
i.e.

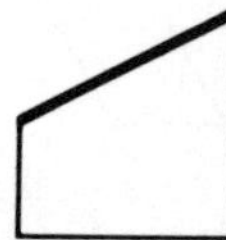
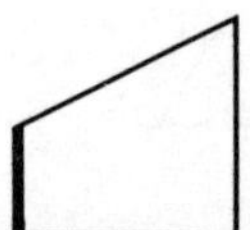
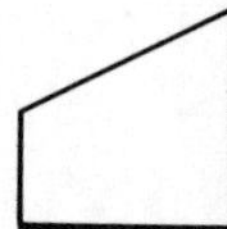
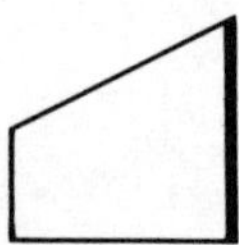

△ Many more polygons are available when the square is cut into three pieces.

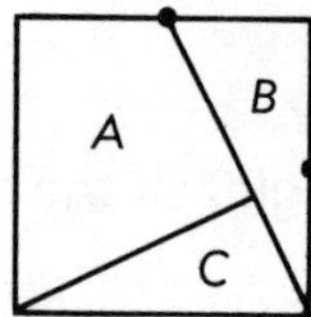

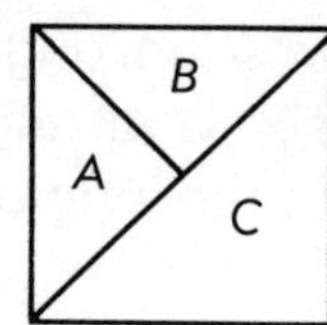

Polygons can be found using all three pieces A, B and C.
Other polygons are possible using any two pieces e.g. A and B, B and C, A and C.

△ Investigate cuts on shapes other than squares.
e.g.

rectangles

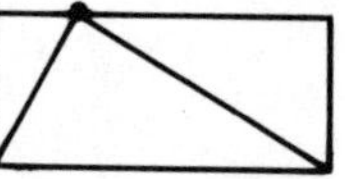

triangles

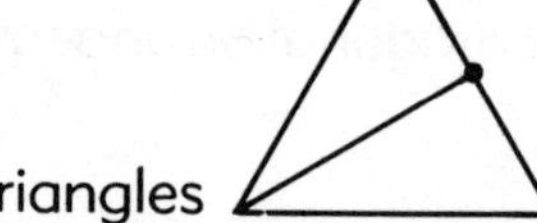

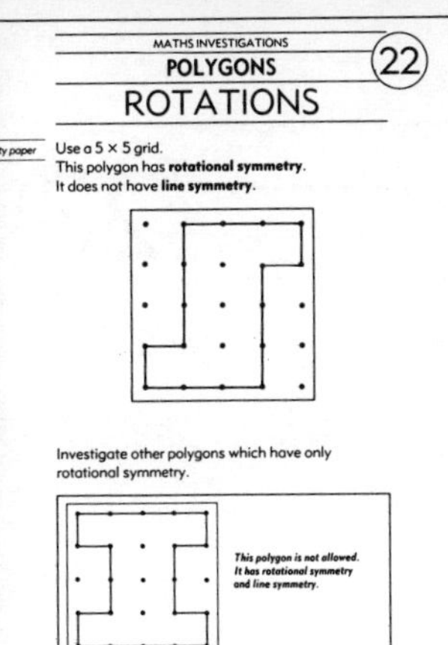
MATHS INVESTIGATIONS
POLYGONS
22
ROTATIONS
Use a 5 × 5 grid.
This polygon has rotational symmetry.
It does not have line symmetry.
Investigate other polygons which have only rotational symmetry.
This polygon is not allowed. It has rotational symmetry and line symmetry.

MATHS INVESTIGATIONS

POLYGONS

ROTATIONS

22

Dotty paper

△ Pupils may need some discussion about the meaning of rotational symmetry.

△ Some examples include:

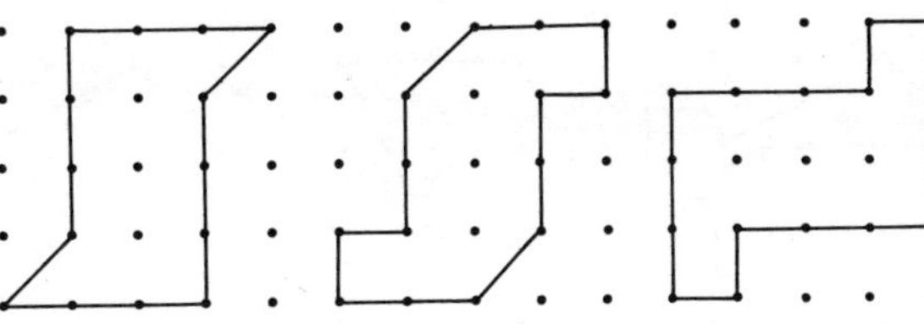

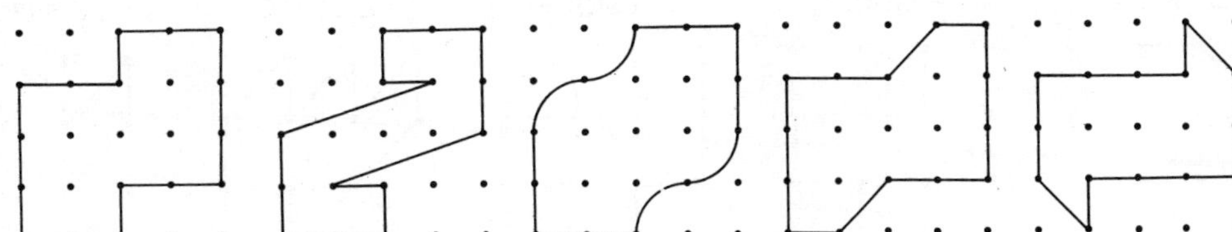

A possible strategy is to build up rotational patterns about the centre dot:

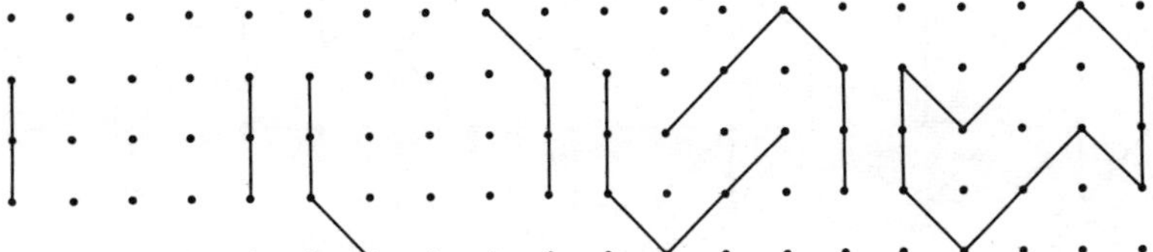

This shape has rotational symmetry of order 2.

A similar method can produce a shape with rotational symmetry of order 4.

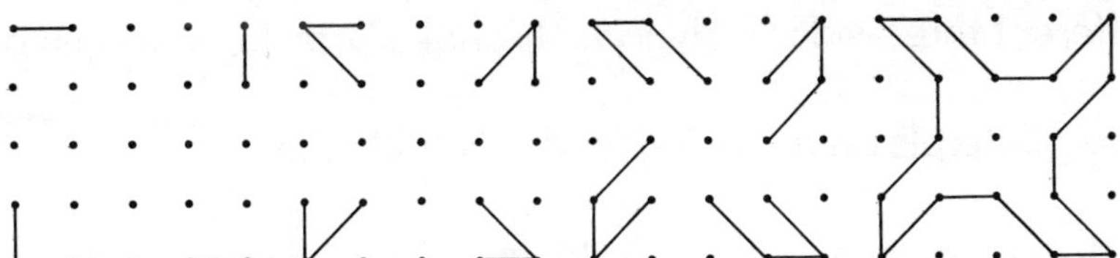

△ To test if a shape has rotational symmetry it can be traced and then rotated about the centre.

△ Find shapes which have: line symmetry **and** rotational symmetry
line symmetry **and no** rotational symmetry

△ Change the size of the grid.

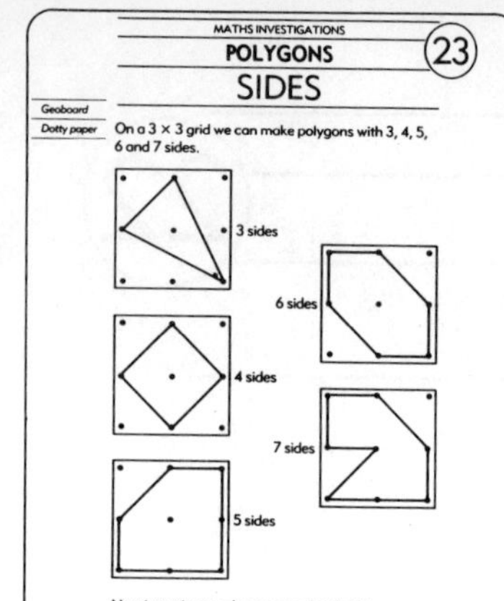
MATHS INVESTIGATIONS

POLYGONS 23

SIDES

Geoboard

Dotty paper

On a 3 × 3 grid we can make polygons with 3, 4, 5, 6 and 7 sides.

3 sides

6 sides

4 sides

7 sides

5 sides

Now investigate polygons on a 4 × 4 grid.
Try using grids of different sizes and shape.

MATHS INVESTIGATIONS

POLYGONS

SIDES (23)

Geoboard

Dotty paper

△ 4 × 4 grids
Some examples include:

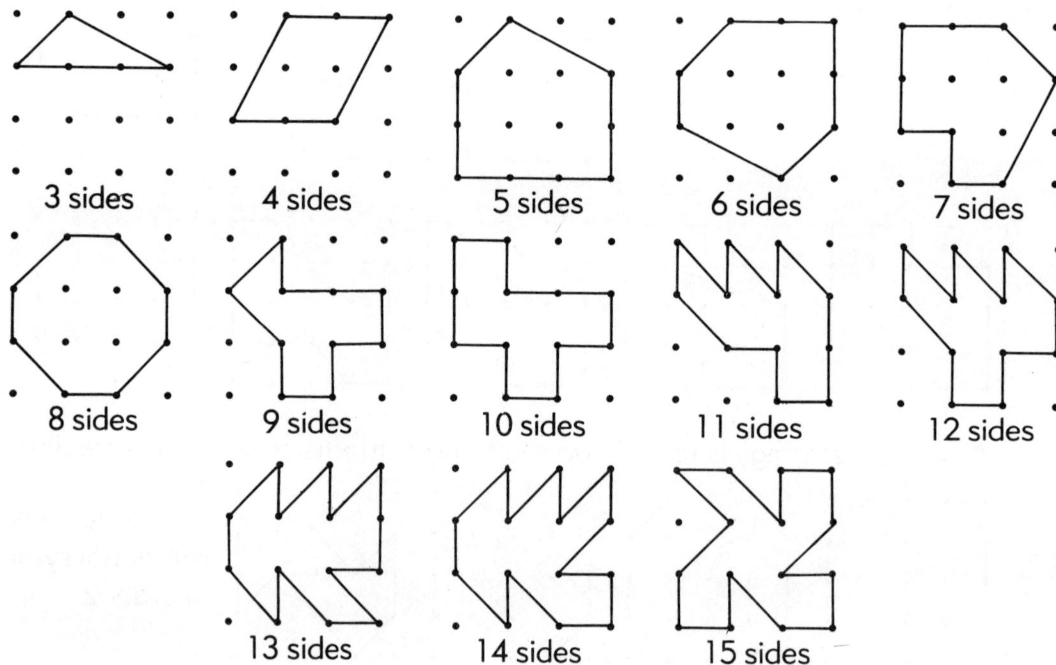

△ Is 15 sides the maximum for a polygon on a 4 × 4 grid?
What is the maximum number of sides for a 5 × 5, 6 × 6 grid?
How many different three-sided polygons (triangles) can be found on a 4 × 4 grid?
How many quadrilaterals on a 3 × 3 grid, 4 × 4 grid, ...?

△ Investigate rectangular grids, e.g. 3 × 5

△ Investigate triangular grids.

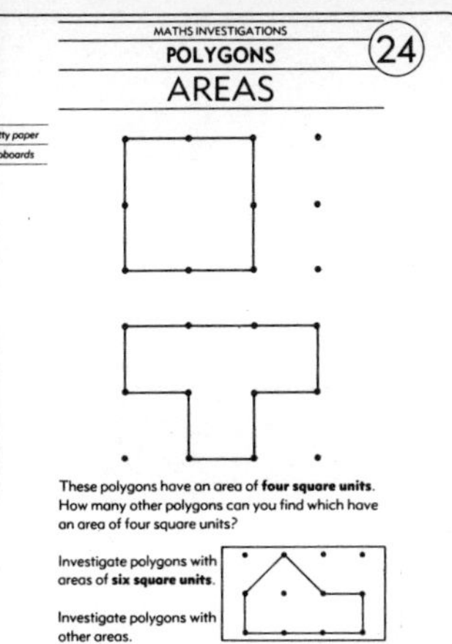
MATHS INVESTIGATIONS
POLYGONS
AREAS
24

Dotty paper
Geoboards

These polygons have an area of **four square units**. How many other polygons can you find which have an area of four square units?

Investigate polygons with areas of **six square units**.

Investigate polygons with other areas.

MATHS INVESTIGATIONS

POLYGONS

AREAS

24

Dotty paper

Geoboards

△ With an area of four square units, here are some examples using horizontal and vertical lines only:

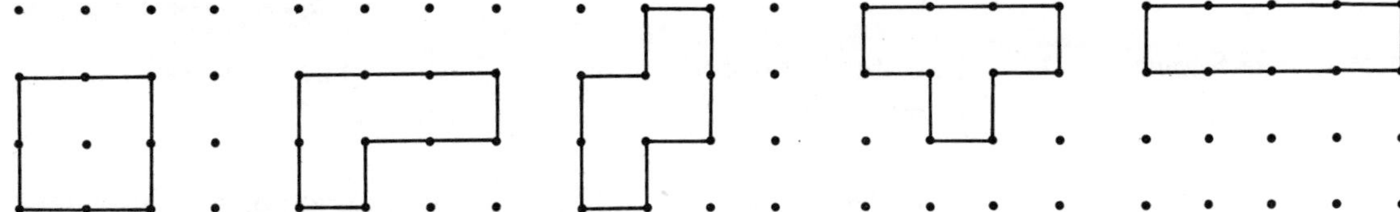

Some examples where diagonal lines are permitted:

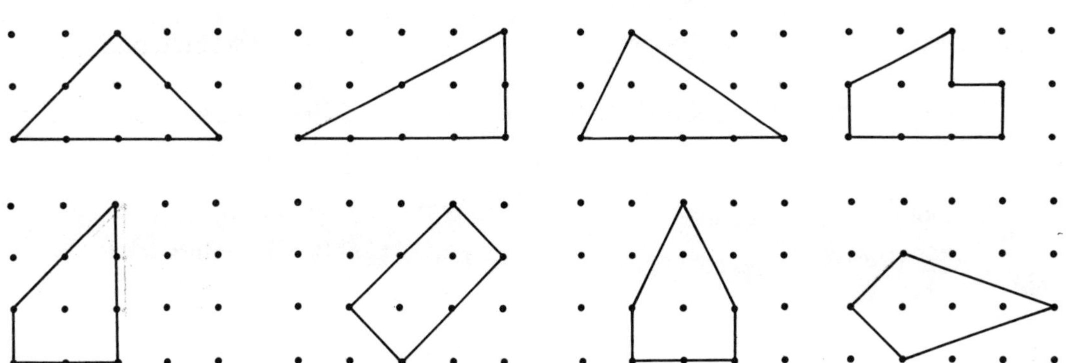

△ A game for two players on a 4 × 4 grid:
Player A draws any shape.
Player B draws a different shape with the same area.
Continue until either player repeats a shape. This player loses.

GRIDS

	APPARATUS	ACTIVITY
25 Route 22	Squared paper (1 cm)	Adding numbers on a grid to produce a known total
26 Lines of Three	Counters: squared paper (1 cm)	Arranging counters on a grid
27 Snails	Counters: squared paper (1 cm or 2 cm)	Moving counters on a square grid
28 Even Lines	Counters: squared paper (1 cm)	Arranging counters in even lines on a grid
29 Knight shift	Counters: squared paper (1 cm)	Moving a counter to different positions with a knight's move
30 Links	Dotty paper	Joining dots on a grid
31 Word Squares	Geoboard: dotty paper	Different sized squares on a grid
32 Colour Exchange	Counters	Exchanging positions of two sets of counters

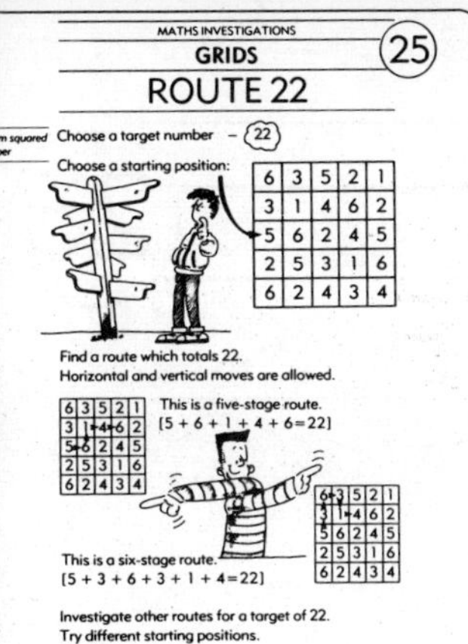

MATHS INVESTIGATIONS

GRIDS

ROUTE 22

25

Choose a target number – 22

Choose a starting position:

6	3	5	2	1
3	1	4	6	2
5	6	2	4	5
2	5	3	1	6
6	2	4	3	4

Find a route which totals 22.
Horizontal and vertical moves are allowed.

This is a five-stage route.
(5 + 6 + 1 + 4 + 6 = 22)

This is a six-stage route.
(5 + 3 + 6 + 3 + 1 + 4 = 22)

Investigate other routes for a target of 22.
Try different starting positions.
Try different targets.

MATHS INVESTIGATIONS

GRIDS

ROUTE 22

25

1 cm Squared paper

△ Starting with 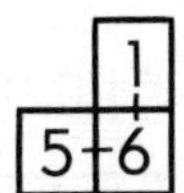, the possibilities include:

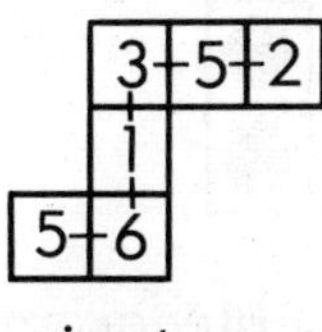

six-stage

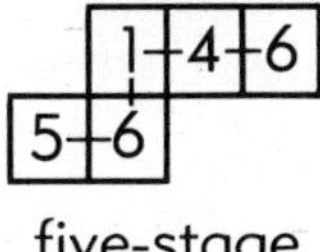

five-stage

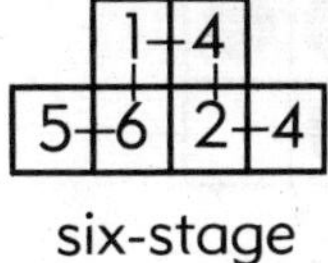

six-stage

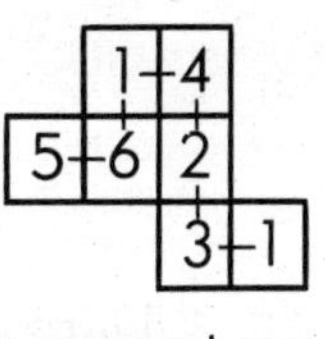

seven-stage

△ Starting with 5+6+2, other possibilities are:

5+6+2+4+5

five-stage

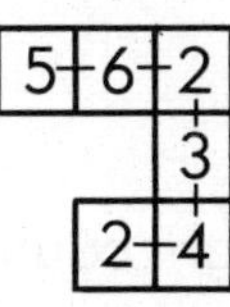

six-stage

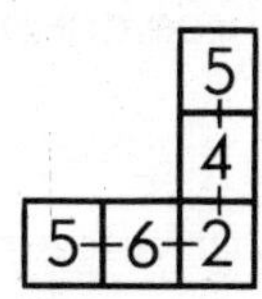

five-stage

△ How many different routes are there starting with the five?
△ What is the most common length of route (i.e. how many stages?)
△ Which is the longest/shortest route?
△ Are there more available routes if we start in the centre?
△ What happens if we start at a corner square?
△ Try different grids e.g. rectangular, triangular.
△ Try different numbers.

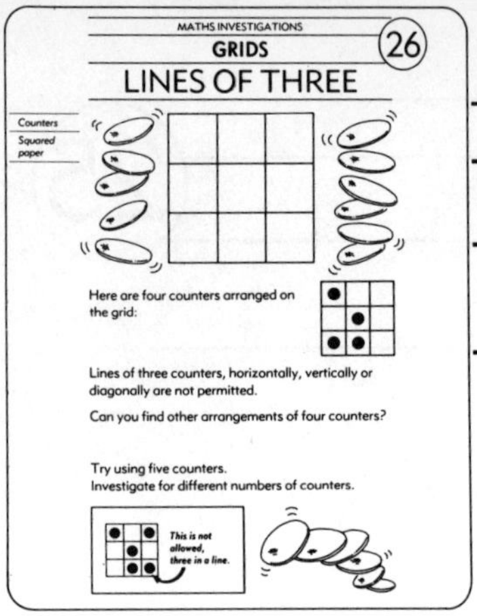
MATHS INVESTIGATIONS
GRIDS
26
LINES OF THREE
Counters
Squared paper
Here are four counters arranged on the grid:
Lines of three counters, horizontally, vertically or diagonally are not permitted.
Can you find other arrangements of four counters?
Try using five counters.
Investigate for different numbers of counters.
This is not allowed, three in a line.

GRIDS

LINES OF THREE

26

Counters
Squared paper

△ It is not difficult to find arrangements which involve up to five counters.

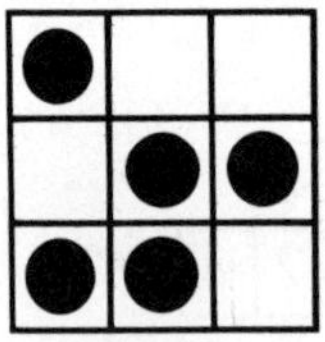

Here is a five-counter example

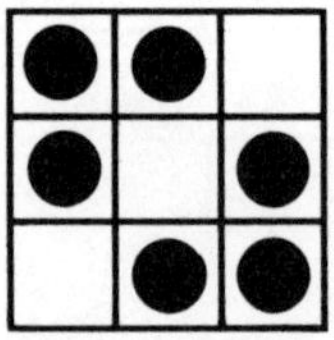

Here is a six-counter example

It is impossible to place seven counters without completing a line of three.

△ How many counters can be placed on different sized grids so that no line contains three counters?

e.g. 4 × 4

Here is an arrangement of eight counters.

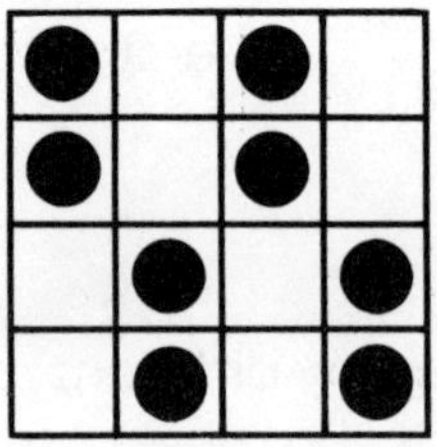

△ How many counters can be placed on a 4 × 4 grid so that no line contains four counters?

Here is an arrangement which places 12 counters on the grid.

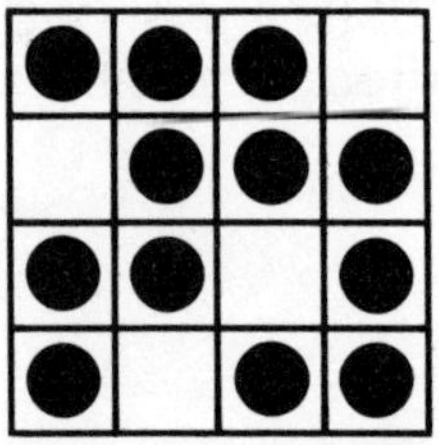

△ Play a two-person game. Players decide on a grid and alternately place counters in available squares. The first player to complete a line of three loses.

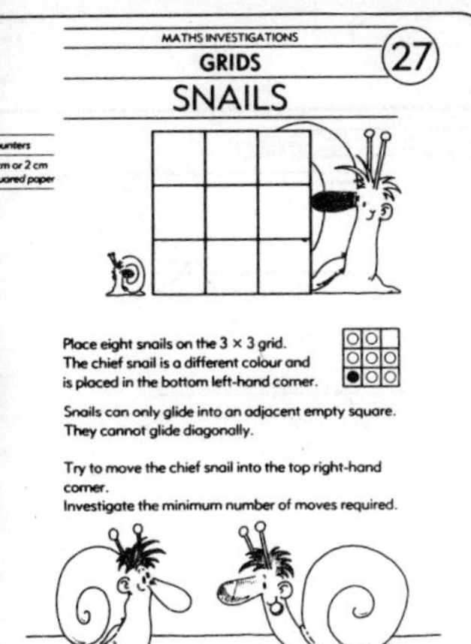
MATHS INVESTIGATIONS
GRIDS
27
SNAILS
Place eight snails on the 3 × 3 grid.
The chief snail is a different colour and is placed in the bottom left-hand corner.
Snails can only glide into an adjacent empty square.
They cannot glide diagonally.
Try to move the chief snail into the top right-hand corner.
Investigate the minimum number of moves required.

MATHS INVESTIGATIONS

GRIDS

SNAILS

27

Counters

1 cm or 2 cm Squared paper

△ 13 moves are required.

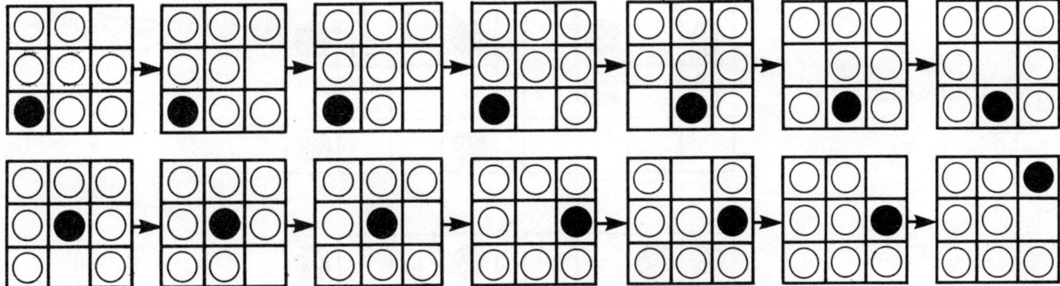

△ How many moves are required to reach other points on the grid?

Solution

Note the symmetry

9	10	13
4	7	10
	4	9

△ Solutions for a 4 × 4 and 5 × 5 grid are:

16	17	18	21
11	12	15	18
6	9	12	17
	6	11	16

23	24	25	26	29
18	19	20	23	26
13	14	17	20	25
8	11	14	19	24
	8	13	18	23

△ Number of moves required to finish in the top right-hand corner

size	moves
2 × 2	5
3 × 3	13
4 × 4	21
5 × 5	29
⋮	
$n \times n$	$8n - 11$

△ Try rectangular grids.

△ Start with the gap in a different position.

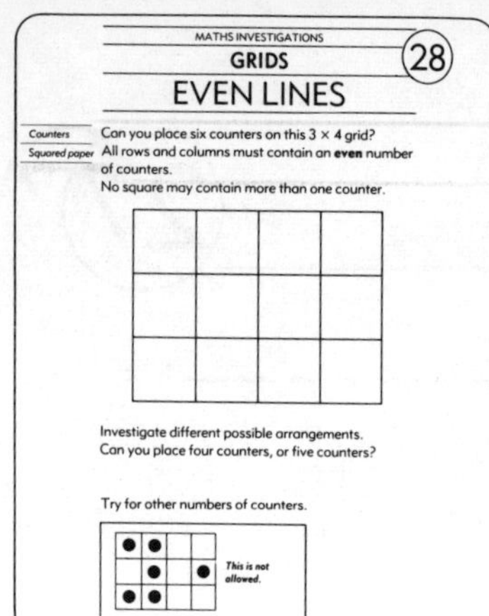
MATHS INVESTIGATIONS

GRIDS

28

EVEN LINES

Counters

Squared paper

Can you place six counters on this 3 × 4 grid?
All rows and columns must contain an **even** number of counters.
No square may contain more than one counter.

Investigate different possible arrangements.
Can you place four counters, or five counters?

Try for other numbers of counters.

This is not allowed.

MATHS INVESTIGATIONS

GRIDS

EVEN LINES

28

Counters

Squared paper

△ 3 × 4 grid – some possible solutions:

four counters

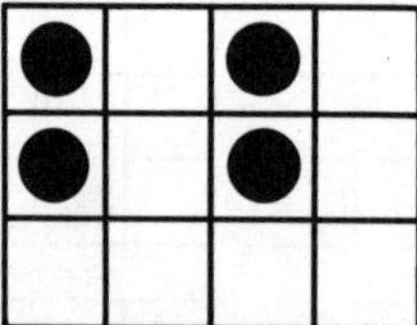
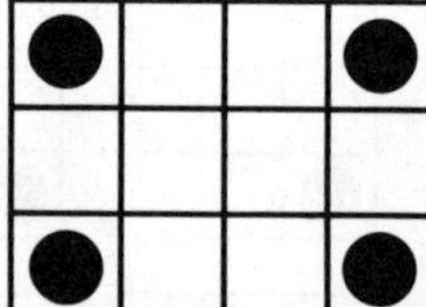

six counters

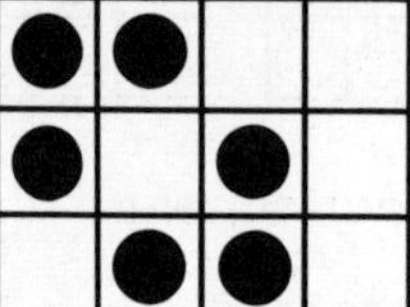
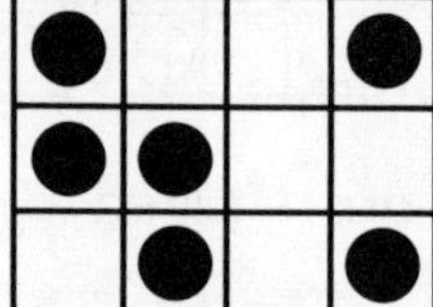

eight counters

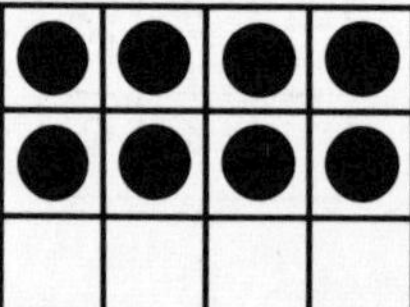
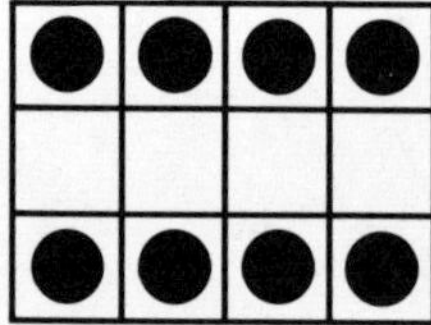

Four, six and eight are the only possibilites.

△ Investigate for different sized grids.

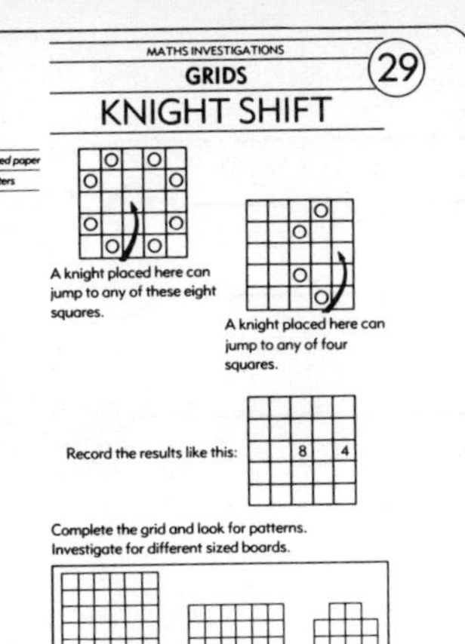

MATHS INVESTIGATIONS

GRIDS

KNIGHT SHIFT

29

Squared paper

Counters

△ Solution

2	3	4	3	2
3	4	6	4	3
4	6	8	6	4
3	4	6	4	3
2	3	4	3	2

Lines of symmetry 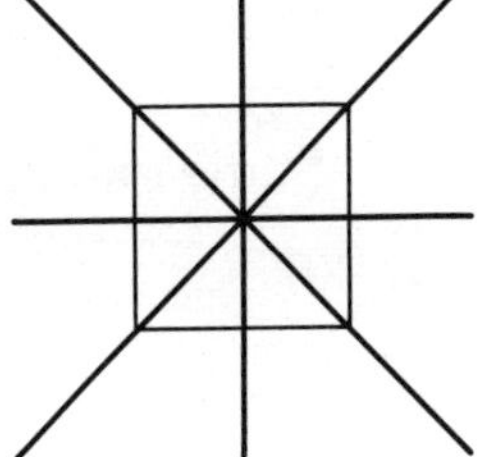

6 × 6 board

2	3	4	4	3	2
3	4	6	6	4	3
4	6	8	8	6	4
4	6	8	8	6	4
3	4	6	6	4	3
2	3	4	4	3	2

△ Is it possible for a square to contain a number greater than 8?

△ 3 × 6 board

2	3	4	4	3	2
2	2	4	4	2	2
2	3	4	4	3	2

△ Explore **bishop** shifts.

4	4	4	4	4
4	6	6	6	4
4	6	8	6	4
4	6	6	6	4
4	4	4	4	4

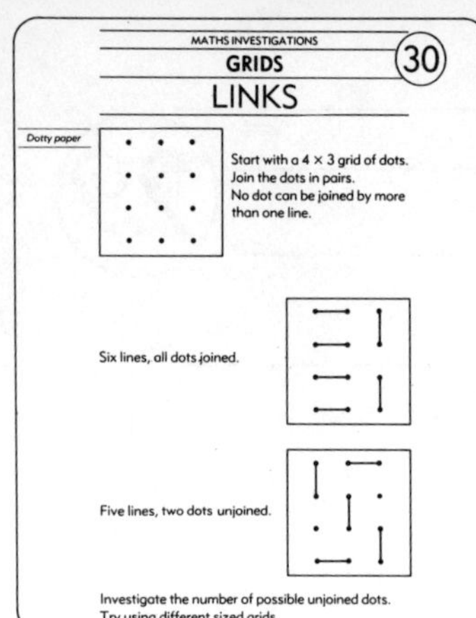
MATHS INVESTIGATIONS

GRIDS 30

LINKS

Dotty paper

Start with a 4 × 3 grid of dots.
Join the dots in pairs.
No dot can be joined by more than one line.

Six lines, all dots joined.

Five lines, two dots unjoined.

Investigate the number of possible unjoined dots.
Try using different sized grids.

MATHS INVESTIGATIONS

GRIDS

LINKS

30

Dotty paper

△ With a 4 × 3 grid, – suppose (a, b) represents a grid with a lines b unjoined dots.
Some examples:

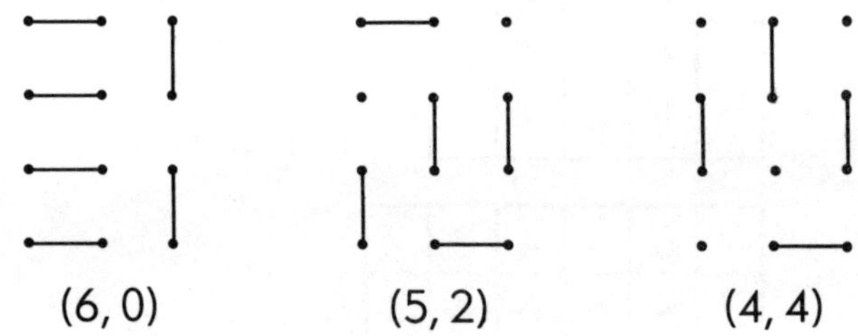

(6, 0) (5, 2) (4, 4)

△ 5 × 5 grid

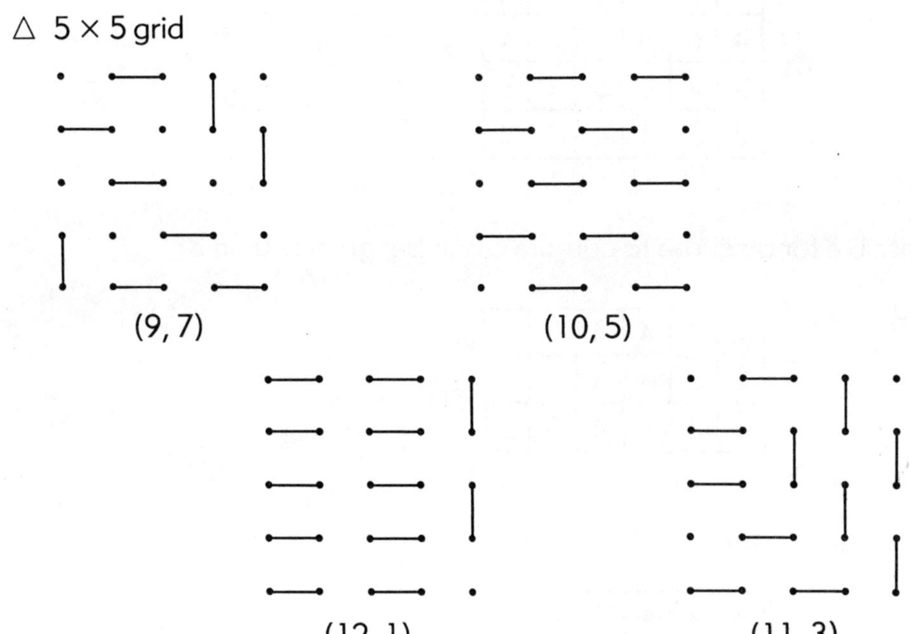

(9, 7) (10, 5)

(12, 1) (11, 3)

MATHS INVESTIGATIONS

GRIDS

GRID SQUARES

31

Geoboard

Dotty paper

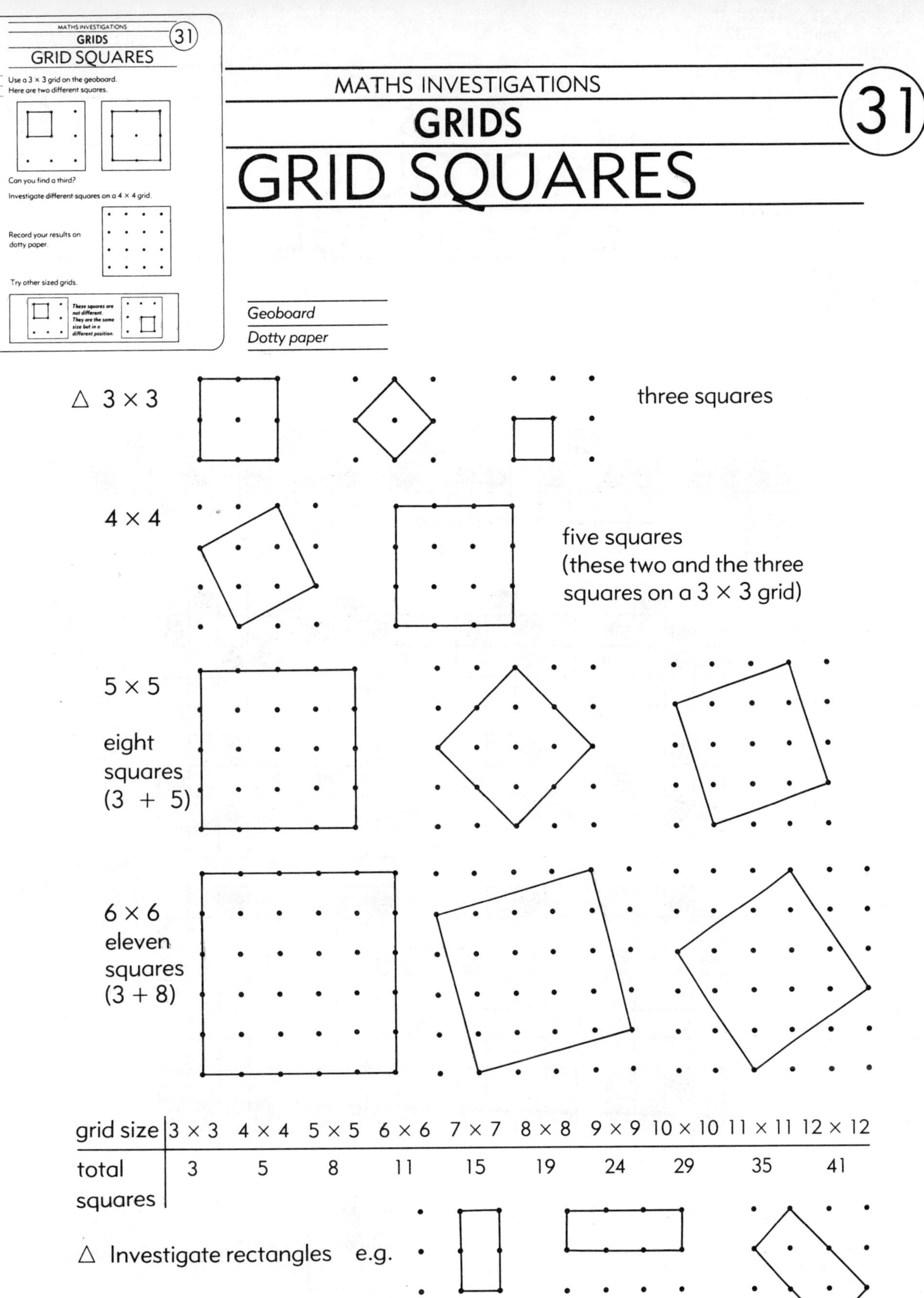

MATHS INVESTIGATIONS
GRIDS
31
GRID SQUARES
oboard
tty paper
Use a 3 × 3 grid on the geoboard.
Here are two different squares.
Can you find a third?
Investigate different squares on a 4 × 4 grid.
Record your results on dotty paper.
Try other sized grids.
These squares are not different. They are the same size but in a different position.

△ 3 × 3 three squares

4 × 4 five squares (these two and the three squares on a 3 × 3 grid)

5 × 5 eight squares (3 + 5)

6 × 6 eleven squares (3 + 8)

grid size	3 × 3	4 × 4	5 × 5	6 × 6	7 × 7	8 × 8	9 × 9	10 × 10	11 × 11	12 × 12
total squares	3	5	8	11	15	19	24	29	35	41

△ Investigate rectangles e.g.

△ Investigate triangles.

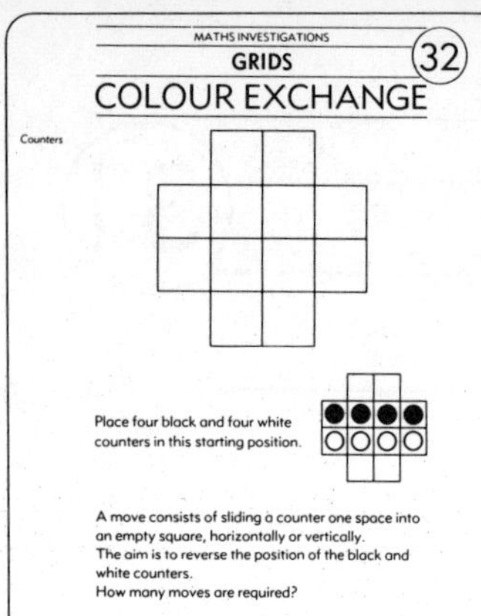

MATHS INVESTIGATIONS
GRIDS
COLOUR EXCHANGE
32

Counters

Place four black and four white counters in this starting position.

A move consists of sliding a counter one space into an empty square, horizontally or vertically.
The aim is to reverse the position of the black and white counters.
How many moves are required?

Investigate for different starting positions of the counters.

MATHS INVESTIGATIONS

GRIDS

COLOUR EXCHANGE

32

Counters

△ Twenty moves are required:

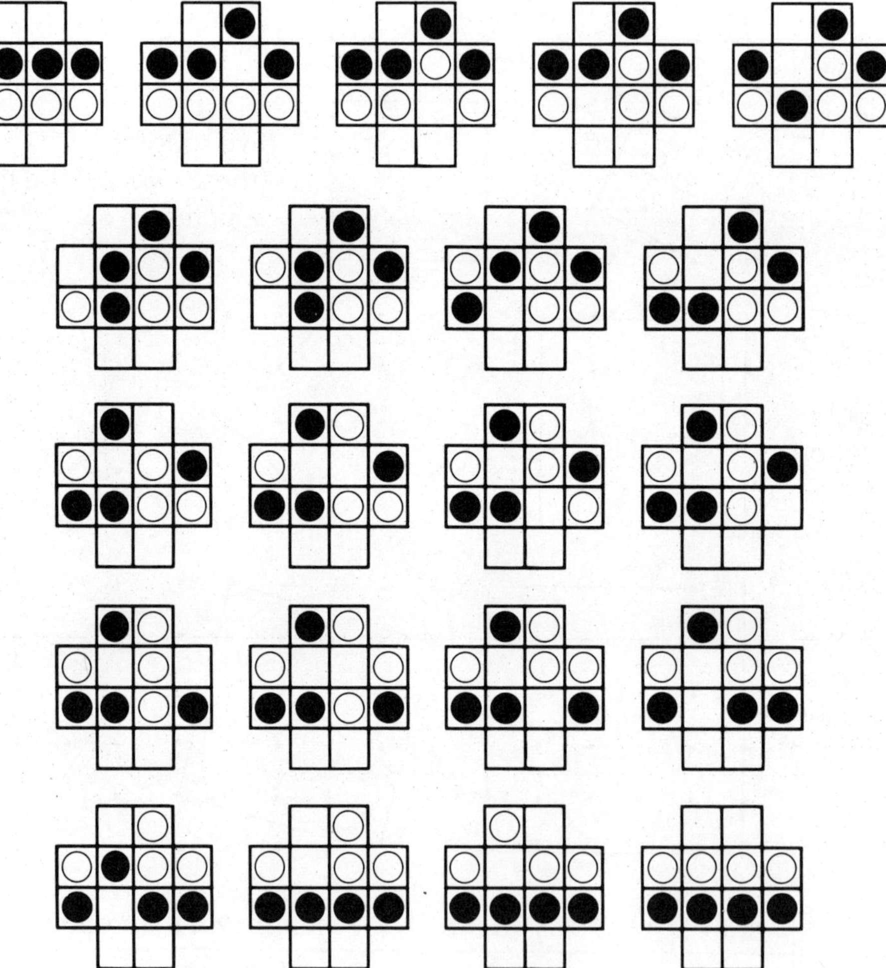

△ Consider a new starting position:

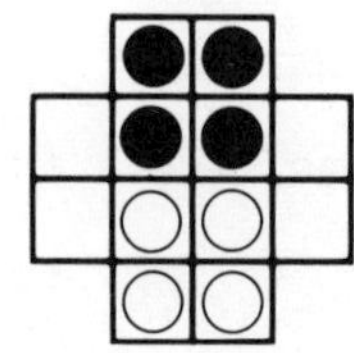

This requires 24 moves.

△ What happens if jumps are allowed?

NUMBER I

	APPARATUS	ACTIVITY
33 Very Odd		Summing odd numbers
34 Final Digits		Pattern of final digits of multiples
35 Ripe Pairs	Number cards (1 – 10)	Pairing numbers to produce square numbers
36 Positions	Squared paper (1 cm)	Patterns of numbers on rectangular grids
37 Square Sums		Summing pairs of square numbers
38 Island Spirals	Squared paper (1 cm)	Pattern of square numbers on number spirals
39 Prime Sums	Table of primes	Summing pairs of prime numbers
40 Express	Table of primes	Difference between consecutive primes

- Calculators could be used during these activities

These activities relate to particular sets of numbers; **square**, **prime**, **odd**, **multiples**, etc.

A table of prime numbers is available (copy master 5). Square numbers and multiples can be obtained by the use of a calculator.

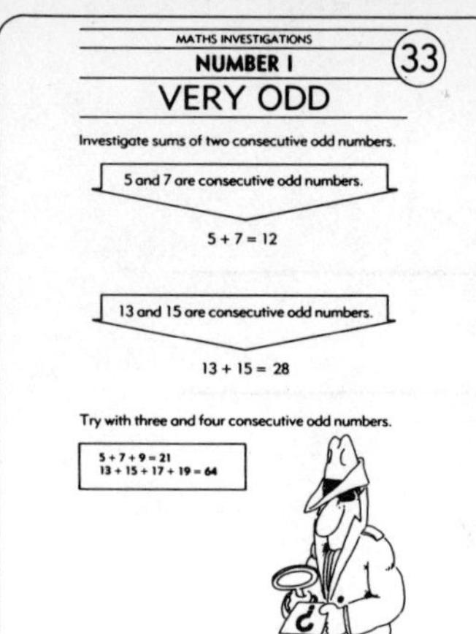

MATHS INVESTIGATIONS

NUMBER I

VERY ODD

(33)

Investigate sums of two consecutive odd numbers.

5 and 7 are consecutive odd numbers.

5 + 7 = 12

13 and 15 are consecutive odd numbers.

13 + 15 = 28

Try with three and four consecutive odd numbers.

5 + 7 + 9 = 21
13 + 15 + 17 + 19 = 64

MATHS INVESTIGATIONS

NUMBER 1

VERY ODD

△ A list of consecutive odd numbers may help.

△ Adding two consecutive odd numbers:

multiples of 4 are obtained.

△ Adding three consecutive odd numbers:

1 + 3 + 5 | 5 + 7 + 9 | 11 + 13 + 15

certain multiples of 3 are obtained. What are they?

△ Adding four consecutive odd numbers:

1 + 3 + 5 + 7 | 5 + 7 + 9 + 11

multiples of 8 are obtained.

△ What happens with five and six consecutive odd numbers?

△ What happens if consecutive even numbers are summed?

MATHS INVESTIGATIONS
NUMBER 1 (34)
FINAL DIGITS

These are multiples of 4.

4 8 12 16 20 24 28 32 36...

The final digits are

4 8 2 6 0 4 8 2 6...

What do you notice about the final digits?

Investigate the final digits of multiples of 6.
Investigate patterns for other multiples.

NUMBER 1

FINAL DIGITS

△ Repeating cycles of final digits occur:

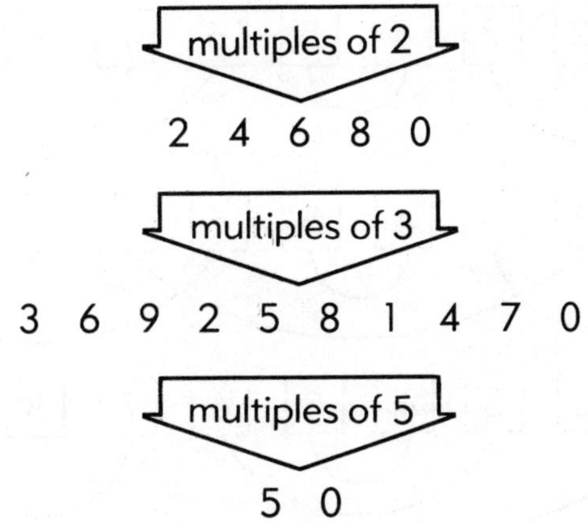

△ Explore final digits of: square numbers
prime numbers
odd numbers
triangular numbers

△ Investigate repeating cycles formed by divisions:

e.g. $3 \div 7 = 0.428571$

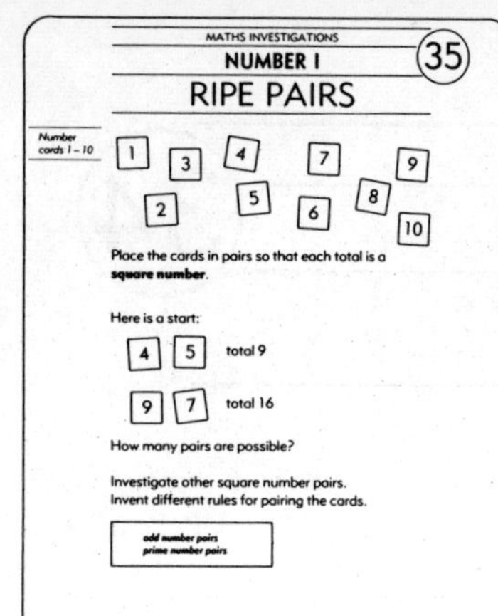
MATHS INVESTIGATIONS

NUMBER 1

35

RIPE PAIRS

Number cards 1 – 10

1 3 4 7 9 2 5 6 8 10

Place the cards in pairs so that each total is a **square number**.

Here is a start:

4 5 total 9

9 7 total 16

How many pairs are possible?

Investigate other square number pairs.
Invent different rules for pairing the cards.

odd number pairs
prime number pairs

MATHS INVESTIGATIONS

NUMBER 1

RIPE PAIRS

35

Number cards 1 – 10

△ Some possible sets of **square number** pairs are:

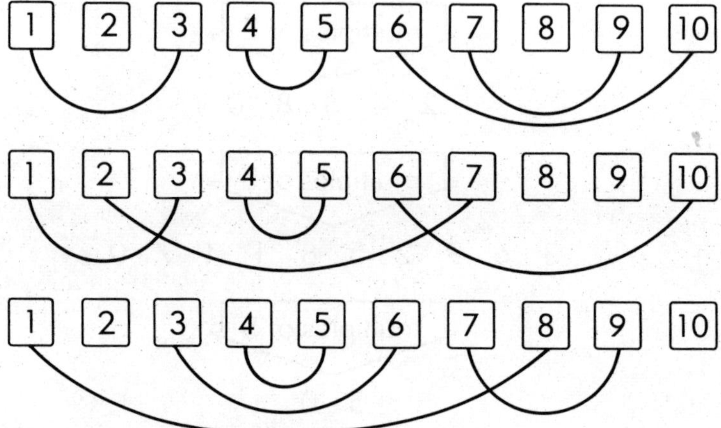

△ Are four pairs the maximum?

△ Can you avoid using 4, 5, 6 or 7 in a set of four pairs?

△ Try this investigation when triples are allowed as well as pairs.

△ Alternatives for pairing cards include: odd numbers
even numbers
prime numbers
triangular numbers
multiples

MATHS INVESTIGATIONS
NUMBER 1
POSITIONS
36

This is a four-column grid.

1	2	3	4
5	6	7	8
9	10	11	12
13	14	15	16
17	18	19	20
21			

Copy and continue the sequence of numbers.

Investigate the position of some special numbers on the grid.

Try: odd numbers
even numbers
multiples of five
prime numbers.

This is a six-column grid.

1	2	3	4	5	6
7	8	9	10	11	12
13	14				

Continue this sequence.

What are the positions of the special numbers now?

Investigate other grids.

MATHS INVESTIGATIONS

NUMBER 1

POSITIONS

36

1 cm Squared paper

△ The number grids could be pre-duplicated if preferred.

△ Pupils should be encouraged to describe the positions. of the numbers under investigation.

- Where are the multiples of four?
- What do you notice about the multiples of five?
- Where are the square numbers?
- In which column will 236 appear?
- Which multiples produce diagonal patterns?

△ Note the position of prime numbers in the six-column grid.

1	2	3	4	5	6
7	8	9	10	11	12
13	14	15	16	17	18
19	20	21	22	23	24
25	26	27	28	29	30
31	32	33	34	35	36

This illustrates that all prime numbers, except two and three, are adjacent to multiples of six.

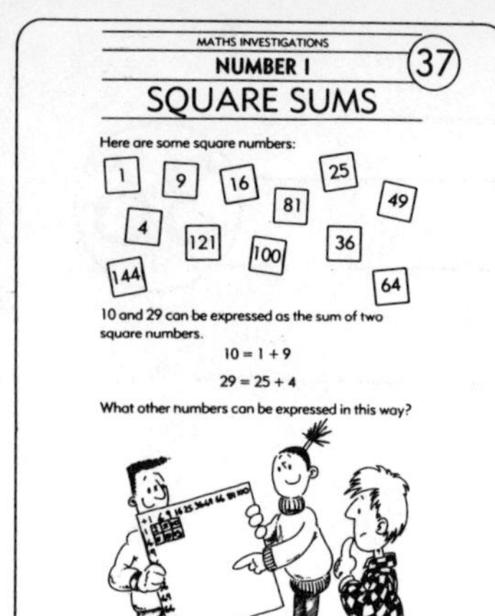
MATHS INVESTIGATIONS
NUMBER 1
37
SQUARE SUMS

Here are some square numbers:

1 9 16 25 81 49 4 121 100 36 144 64

10 and 29 can be expressed as the sum of two square numbers.

$10 = 1 + 9$

$29 = 25 + 4$

What other numbers can be expressed in this way?

MATHS INVESTIGATIONS

NUMBER 1 (37)

SQUARE SUMS

△ Pupils could construct an addition table for the square numbers:

+	1	4	9	16	25 →
1	2	5	10		
4	5	8	13		
9					
16					
25 ↓					

△ Which numbers can be expressed in more than one way?

△ Which numbers cannot be expressed as the sum of two squares? Can they be expressed as the sum of three squares?

△ Do some numbers require the sum of more than three squares?

MATHS INVESTIGATIONS

NUMBER I

ISLAND-SPIRALS

38

1 cm squared paper

Copy this number spiral and continue it to 100.

	10	11	12	13
	9	2	3	14
	8	1	4	15
↑	7	6	5	16
21	20	19	18	17

Colour the **square numbers**.
Describe the position of the square numbers.

Copy these spirals and continue them to 100.

14	15	16	17	18	19
13	2	3	4	5	20
12	1			6	↓
11	10	9	8	7	

13	14	15	16	17
12	1	2	3	18
11			4	↓
10			5	
9	8	7	6	

Describe the positions of the square numbers.
Invent some spirals of your own. Investigate the position of square numbers.

MATHS INVESTIGATIONS

NUMBER 1

ISLAND SPIRALS

38

1 cm Squared paper

△ It may be necessary to continue some spirals beyond 100 before a pattern emerges.

△ There is no need to write every number in the spiral, only the square numbers, e.g.

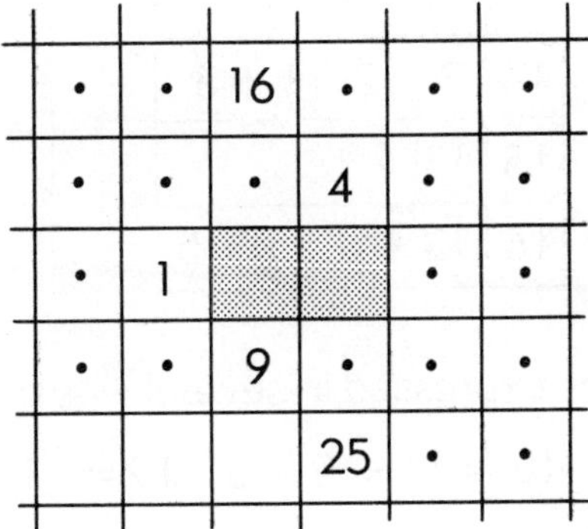

△ Eventually diagonal patterns of odd and even square numbers appear.

△ Given an island and a starting point, is it possible to predict the position of the diagonal pattern?

△ Explore spirals which start with numbers other than one.

△ Suppose the island is not rectangular?

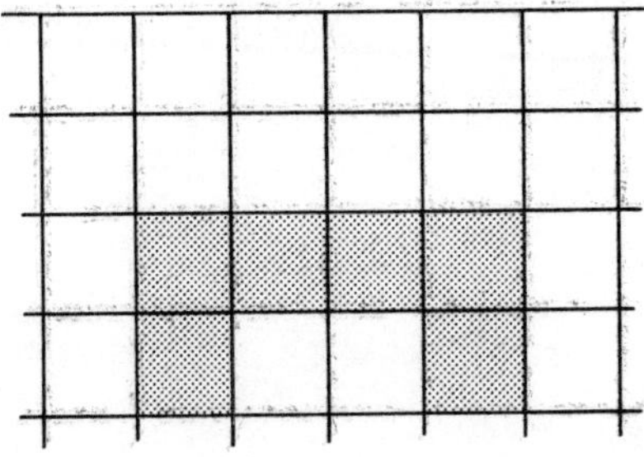

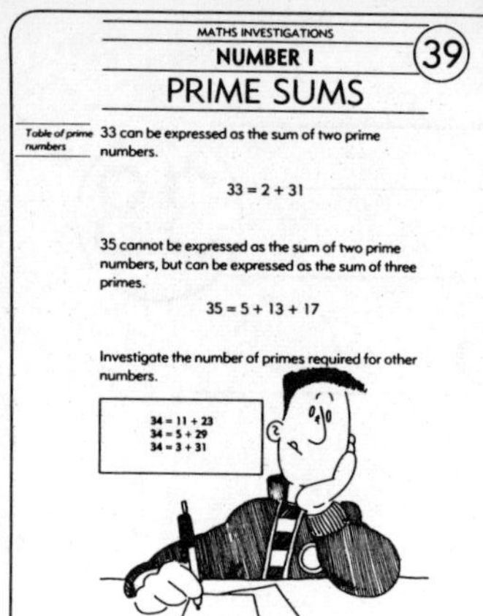
MATHS INVESTIGATIONS
NUMBER 1 (39)
PRIME SUMS

Table of prime numbers

33 can be expressed as the sum of two prime numbers.

33 = 2 + 31

35 cannot be expressed as the sum of two prime numbers, but can be expressed as the sum of three primes.

35 = 5 + 13 + 17

Investigate the number of primes required for other numbers.

34 = 11 + 23
34 = 5 + 29
34 = 3 + 31

MATHS INVESTIGATIONS

NUMBER 1

39

PRIME SUMS

Table of prime numbers

△ Results can be tabulated:

5	2 + 3
6	3 + 3
7	5 + 2
8	5 + 3
9	7 + 2
10	7 + 3; 5 + 5

11	7 + 2 + 2
12	7 + 5
13	11 + 2
14	7 + 7; 11 + 3
15	13 + 2
16	13 + 3; 11 + 5

17	13 + 2 + 2; 11 + 3 + 3; 7 + 5 + 5
	etc

△ Some numbers can be expressed in several ways:

19 = 19   19 = 17 + 2   19 = 11 + 5 + 3   19 = 11 + 3 + 3 + 2

△ Which numbers cannot be expressed as the sum of two primes?

△ Can you find a number which needs more than three primes?

△ Consider the set of triangular numbers:

1 3 6 10 15 21 ...

investigate triangular sums:

16 = 6 + 10
19 = 1 + 3 + 15

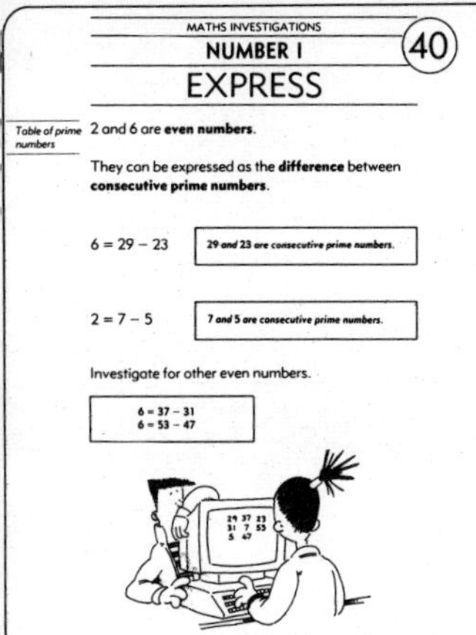

MATHS INVESTIGATIONS

NUMBER 1

EXPRESS

40

Table of prime numbers

△ Results could be tabulated in several ways for consecutive primes up to 100.

primes	2 3 5 7 11 13 ...
differences	1 2 2 4 2 ...

difference	consecutive primes
2	(5,3) (7,5) (13,11) (19,17) (31,29) (43,41) (61,59) (73,71)
4	(11,7) (17,13) (23,19) (41,37) (47,43) (71,67) (83,79)
6	(29,23) (37,31) (53,47) (59,53) (67,61) (79,73) (89,83)
8	(97,89)

△ What is the largest difference you can find?

△ Suppose the primes need not be consecutive,

e.g. $5 = 7 - 2$

$36 = 41 - 5$

CIRCLES

	APPARATUS	ACTIVITY
41 Loops	Compasses	Constructions with compass and ruler
42 Designs	Circle paper	Designing patterns inside circles
43 Ins and Outs	Compasses: tracing paper	Arrangements with three cirlces
44 Polygons	Circle paper	Polygons drawn inside cirlces
45 Patterns	Circle paper	Joining fixed points on the circumference
46 Chords	Compasses	Dividing a circle into regions with chords
47 No breaks	Circle paper	Network problem
48 Arc Forms	Compasses: card	Closed shapes made by arcs

Some of the cards require the use of circle paper (copy masters 8, 9, 10 and 11) although pupils can draw their own circles if preferred. It is assumed that pupils will be familiar with the terms circumference, radius, diameter, arcs and chords. Accurate compass work will also be expected.

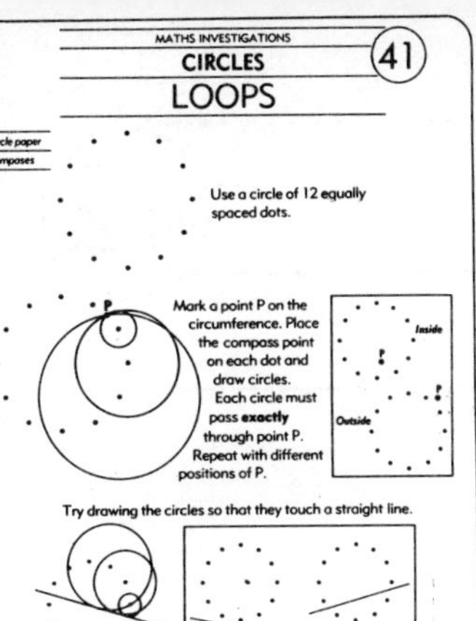
MATHS INVESTIGATIONS
CIRCLES
LOOPS
41

Circle paper
Compasses

Use a circle of 12 equally spaced dots.

Mark a point P on the circumference. Place the compass point on each dot and draw circles. Each circle must pass **exactly** through point P. Repeat with different positions of P.

Try drawing the circles so that they touch a straight line.

MATHS INVESTIGATIONS

CIRCLES

LOOPS

41

Circle paper

Compasses

△ Pupils can experiment with many different positions for the point P:

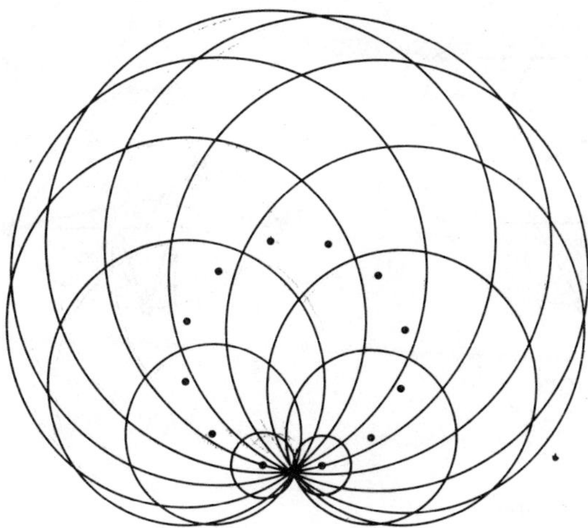

P on the circumference

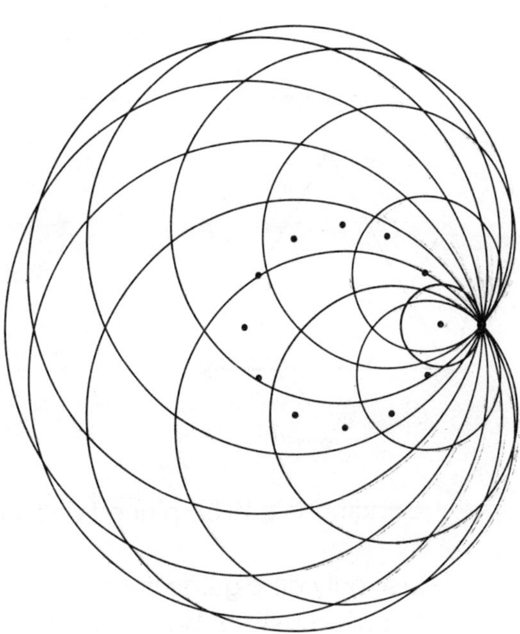

P outside the circle

△ What happens when P is at the centre of the circle?

△ Also, the straight line can assume several different positions:

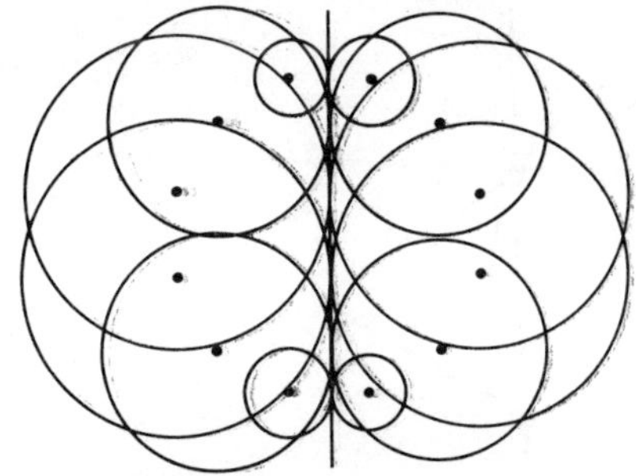

line as a diameter

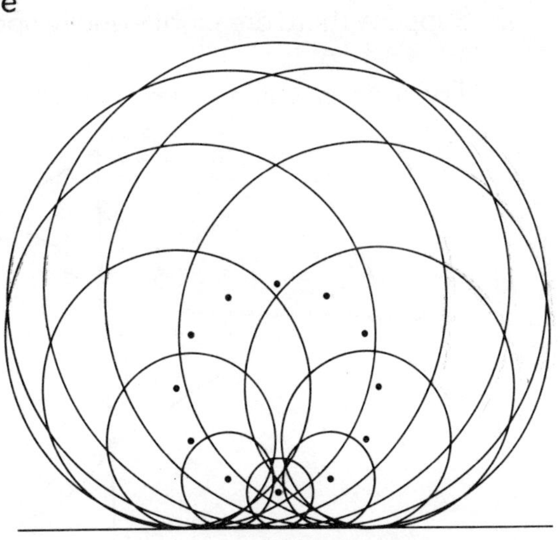

Line outside the circle

△ Suppose the line is curved.

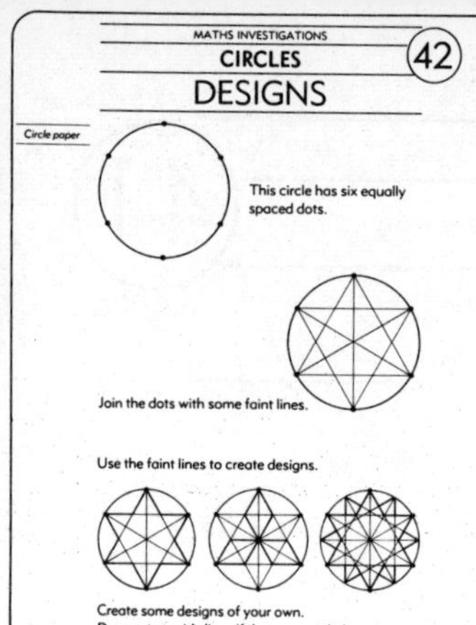

MATHS INVESTIGATIONS

CIRCLES

DESIGNS

42

Circle paper

△ Some other designs include:

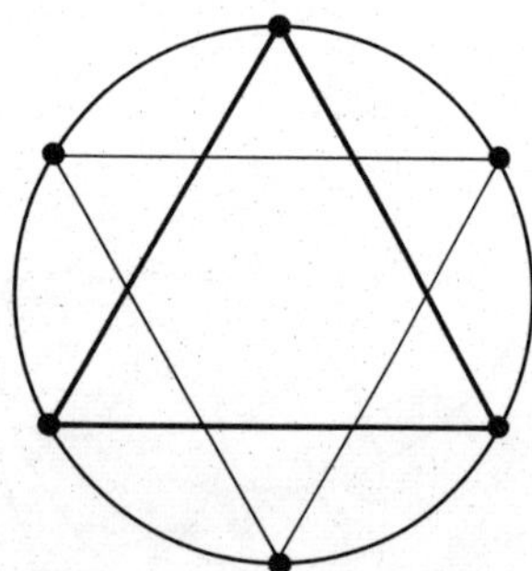

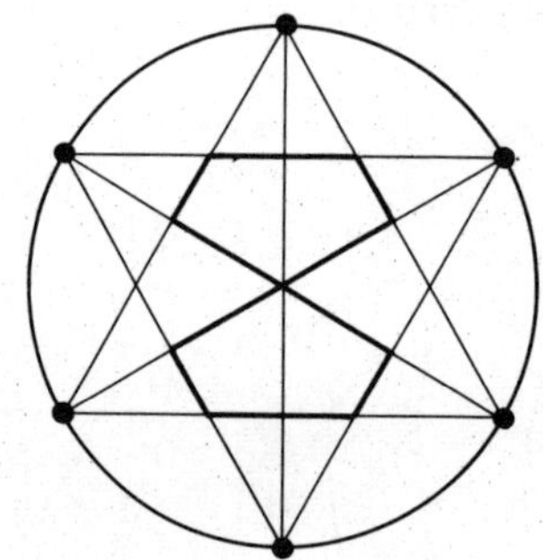

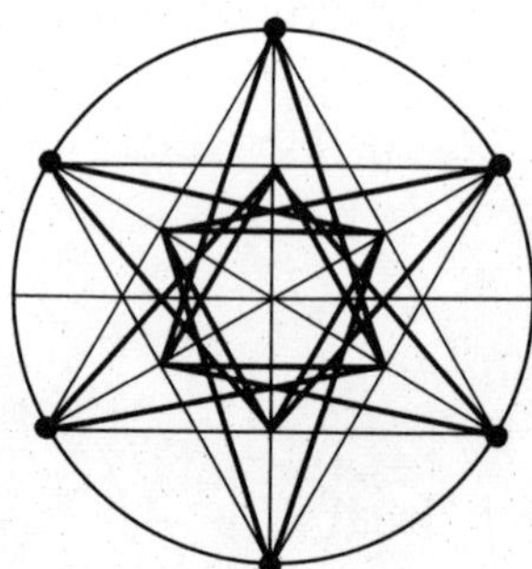

Conditions can be placed on the design.

Use only nine guide lines.
It must/must not have rotational symmetry.

△ Suppose curved lines can be used.

△ Suppose there are eight equally spaced dots.

△ Try designs which extend outside the circle.

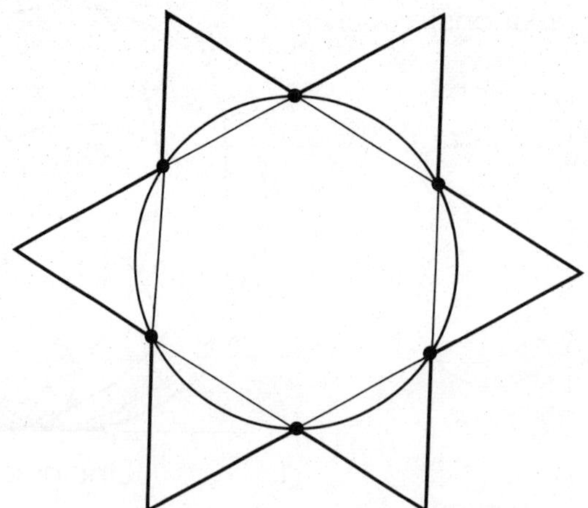

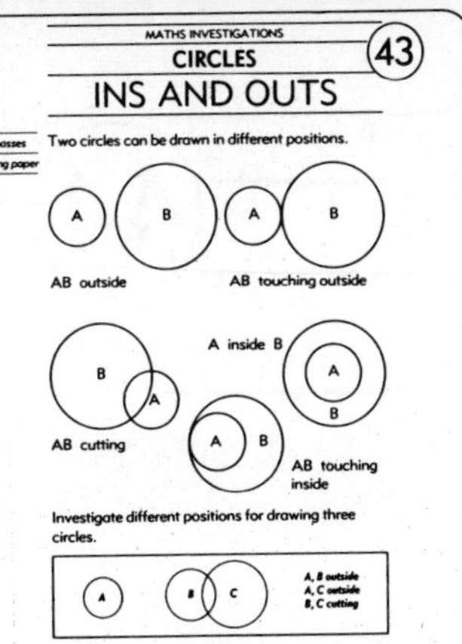
MATHS INVESTIGATIONS
CIRCLES
43
INS AND OUTS

Compasses
Tracing paper

Two circles can be drawn in different positions.

AB outside

AB touching outside

A inside B

AB cutting

AB touching inside

Investigate different positions for drawing three circles.

A, B outside
A, C outside
B, C cutting

Label the circles and describe their positions.

MATHS INVESTIGATIONS

CIRCLES

INS AND OUTS

43

Compasses

Tracing paper

△ Circles drawn on separate pieces of tracing paper allows for experimentation.

△ The circles can be equal in size, or of different sizes. With three different sized circles many variations are possible:

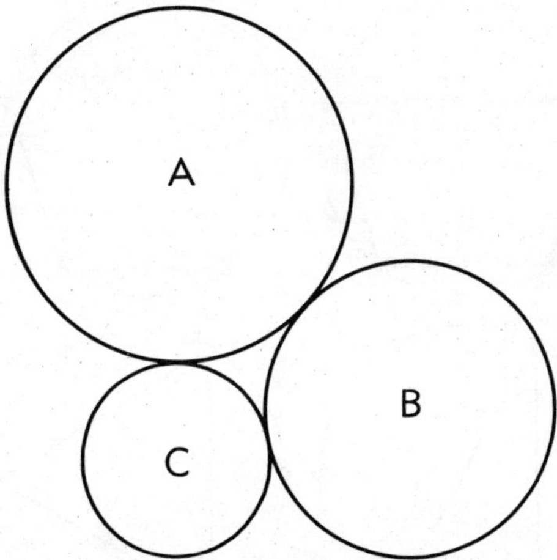

A
B
C

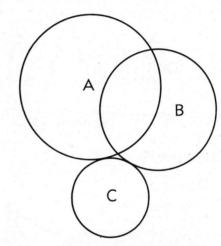

A, B touching outside A, C touching outside B, C touching outside	A, B cutting A, C touching outside B, C touching outside

△ Investigate different ways of drawing two circles and a straight line.

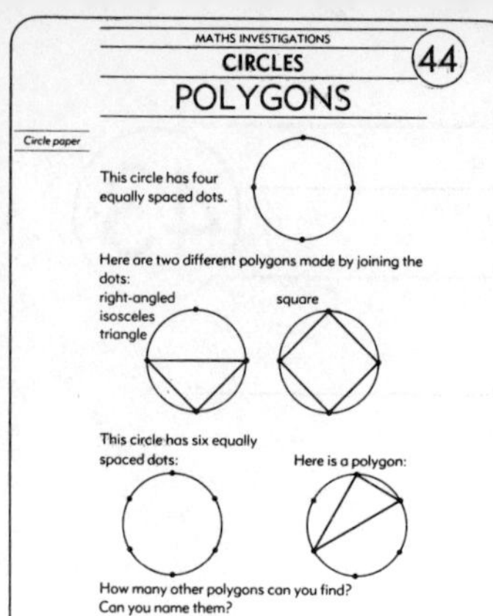
MATHS INVESTIGATIONS
CIRCLES 44
POLYGONS

Circle paper

This circle has four equally spaced dots.

Here are two different polygons made by joining the dots:
right-angled isosceles triangle
square

This circle has six equally spaced dots:
Here is a polygon:

How many other polygons can you find?
Can you name them?
Investigate polygons on circles with different numbers of equally spaced dots.

MATHS INVESTIGATIONS

CIRCLES

POLYGONS

44

Circle paper

△ One strategy is to start by exploring triangles, then quadrilaterals, then pentagons, and so on.
Another strategy is to fix one or two sides and vary the others.

△ With five dots the following are possible:

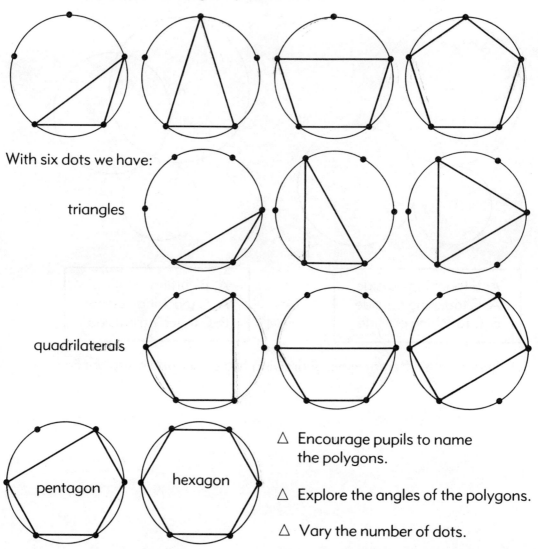

With six dots we have:

triangles

quadrilaterals

pentagon

hexagon

△ Encourage pupils to name the polygons.

△ Explore the angles of the polygons.

△ Vary the number of dots.

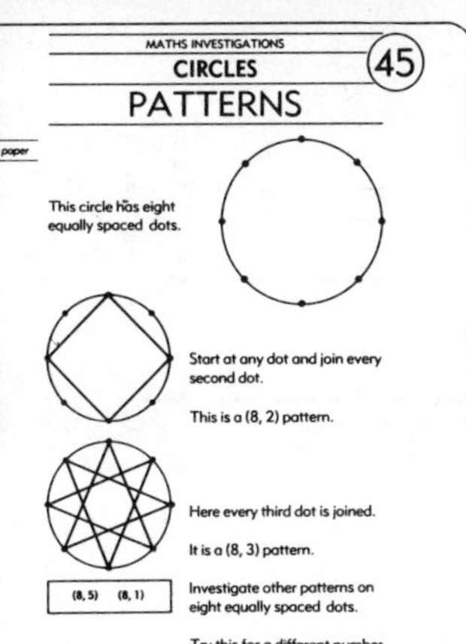

MATHS INVESTIGATIONS

CIRCLES

PATTERNS

45

Circle paper

△ With eight equally spaced dots, four different patterns can be drawn:

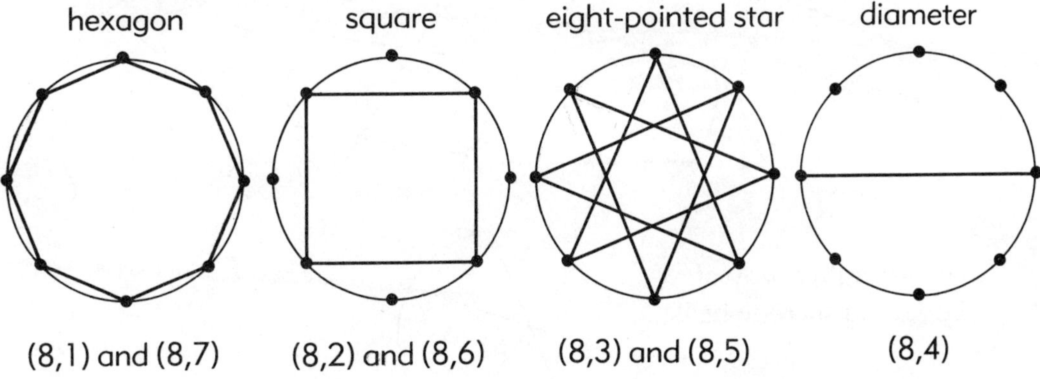

△ With seven equally spaced dots, there are three different patterns:

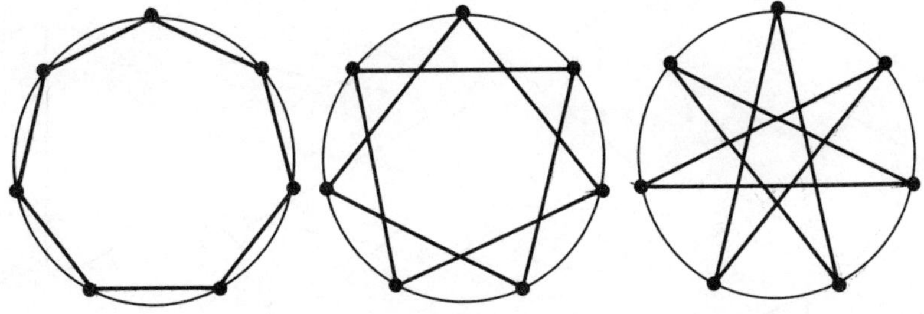

△ How many patterns are there for nine dots?

△ Is it possible to predict the pattern for a given ordered pair? e.g. (6,2) or (10,4).

△ Explore the angles at the vertices of the shapes.

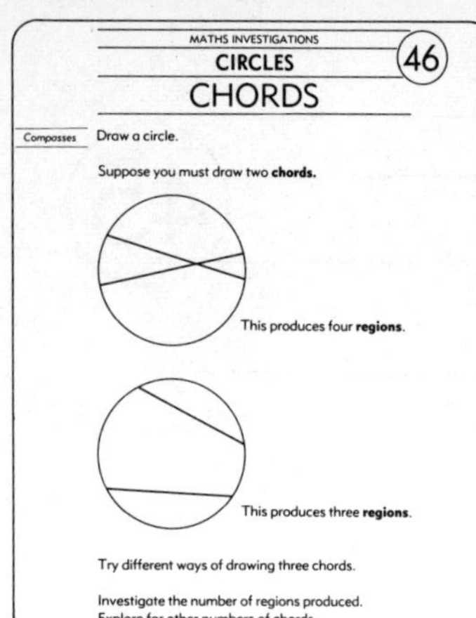
MATHS INVESTIGATIONS
CIRCLES (46)
CHORDS

Compasses

Draw a circle.

Suppose you must draw two **chords**.

This produces four **regions**.

This produces three **regions**.

Try different ways of drawing three chords.

Investigate the number of regions produced.
Explore for other numbers of chords.

MATHS INVESTIGATIONS

CIRCLES

CHORDS

46

Compasses

△ Drawing three chords can produce 4, 5, 6 and 7 regions.

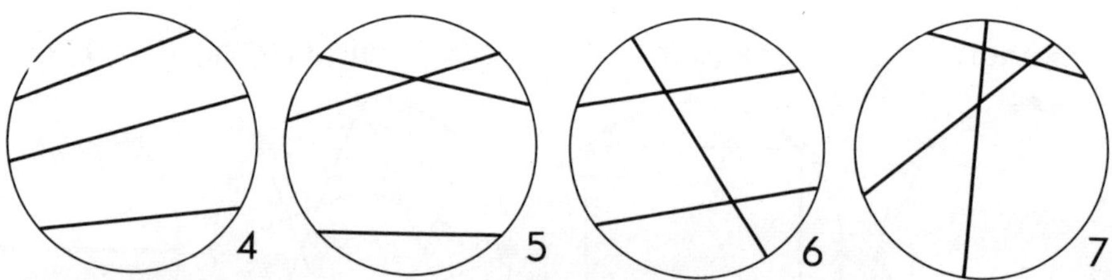

An alternative way of producing six regions is:

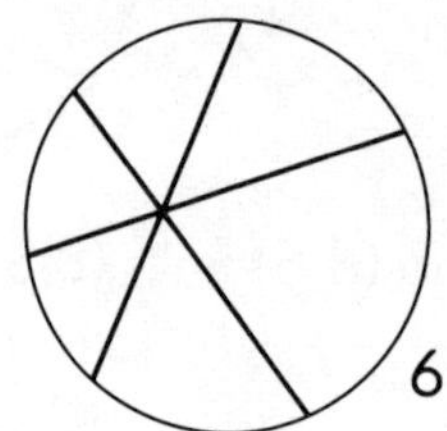

△ Four chord drawings include:

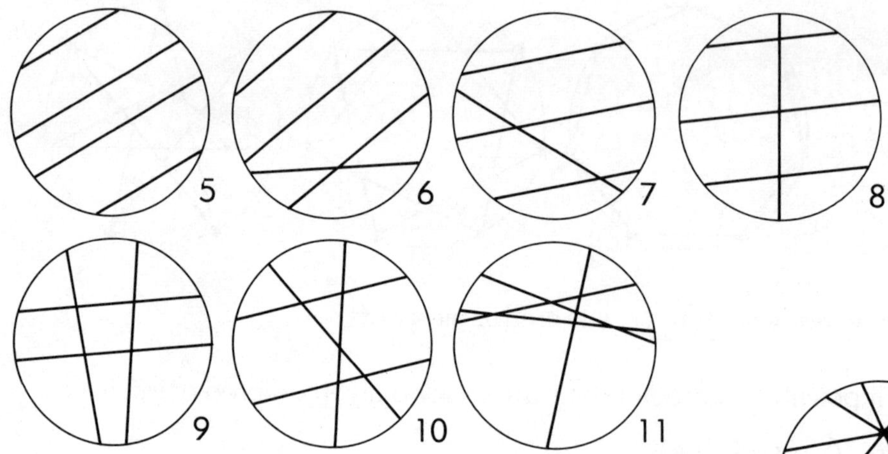

Other possibilities include situations in which several chords intersect at the same point:

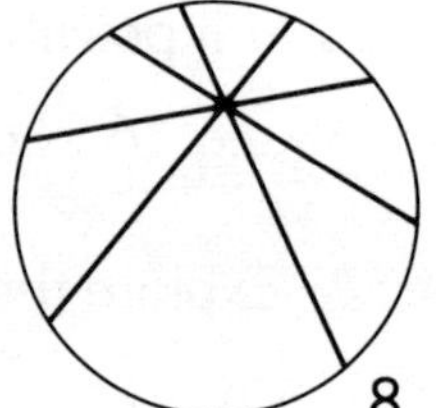

△ What is the maximum number of regions with five chords?
With n chords a maximum of $\frac{1}{2}(n^2 + n + 2)$ regions are possible.

MATHS INVESTIGATIONS

CIRCLES 47

NO BREAKS

ircle paper

Here are four equally spaced points on a circle.

The points can be joined by straight lines without lifting the pencil off the paper, and passing through each point **once only**.

Two different paths are possible:

Try joining five equally spaced points.

Investigate different paths.

Investigate for other equally spaced points.

These are the same path.

MATHS INVESTIGATIONS

CIRCLES

NO BREAKS

47

Circle paper

△ Instead of equally spaced points around a circle, the vertices of regular polygons can be used to mark the points.

△ Reflected and rotated paths are considered identical.

△ With five points, four different paths are possible:

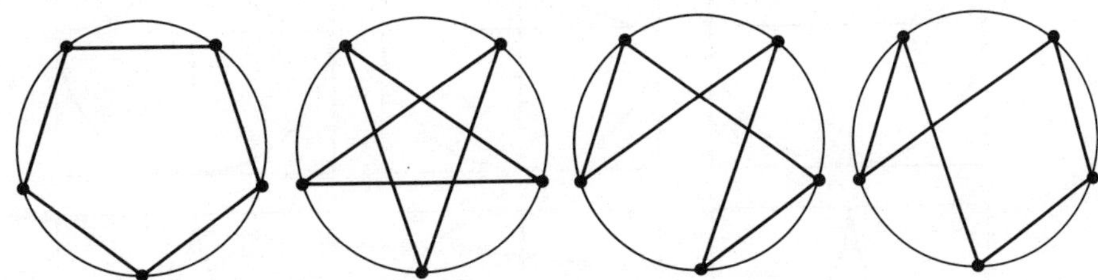

△ With six points:

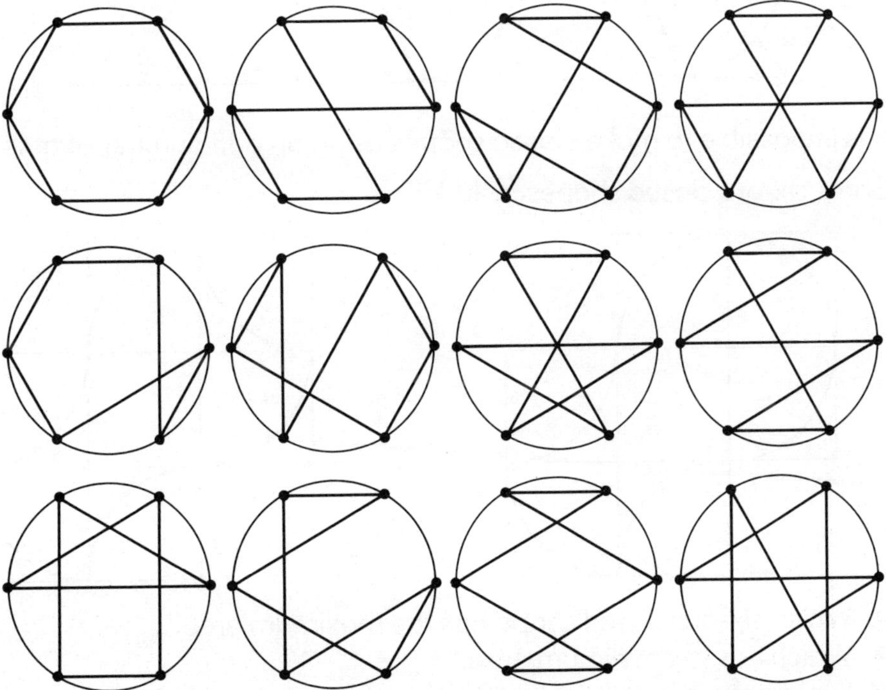

△ Which of the shapes have axes of symmetry?

△ Which have rotational symmetry?

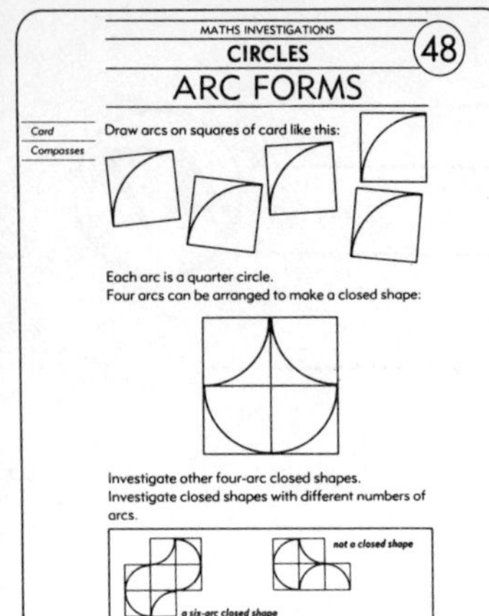
MATHS INVESTIGATIONS

CIRCLES 48

ARC FORMS

Card

Compasses

Draw arcs on squares of card like this:

Each arc is a quarter circle.
Four arcs can be arranged to make a closed shape:

Investigate other four-arc closed shapes.
Investigate closed shapes with different numbers of arcs.

MATHS INVESTIGATIONS

CIRCLES

ARC FORMS

(48)

Card

Compasses

△ The different possible four-arc closed shapes are:

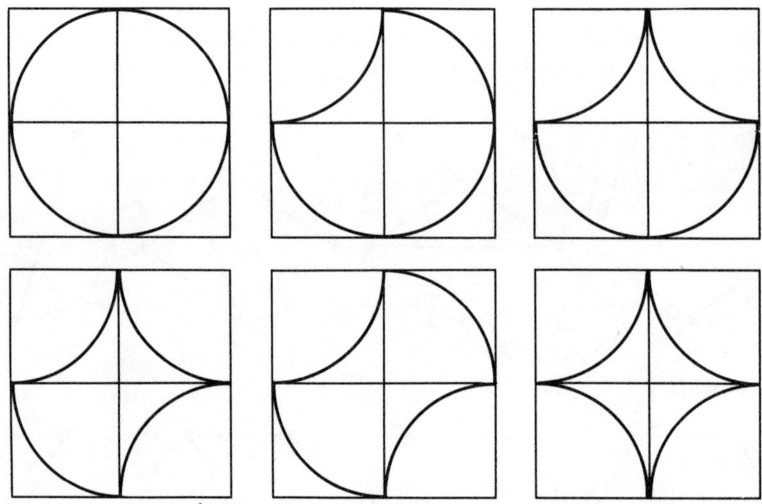

△ It is impossible to make closed shapes using an odd number of arcs.

△ Some six-arc closed shapes include:

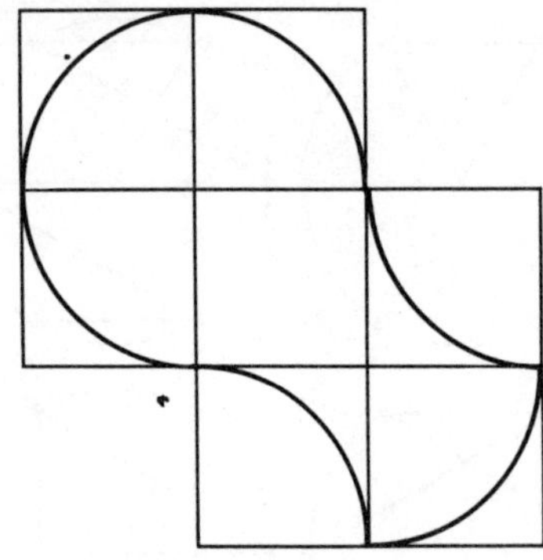

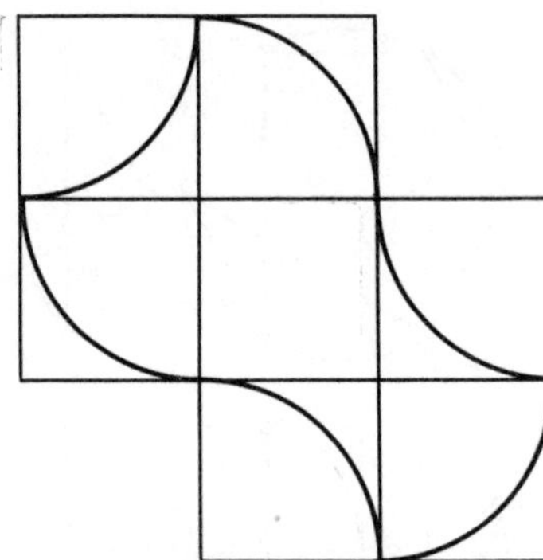

- Which six-arc closed shape has the maximum area?
- Which has the minimum area?
- Explore the symmetrical properties of the shapes.
- Which have line symmetry?
- Which have rotational symmetry?

CUBES

	APPARATUS	ACTIVITY
49 Dice Roll	Dice (6-spot)	Positioning dice on a grid
50 Blushing Cubes	(Cubes: isometric paper – optional)	Arrangement with cubes
51 Tidy Boxes	Table-tennis balls: card: sticky tape	Construction of nets
52 Blocks	Cubes	Arrangement of cubes
53 Two-Tone	Squares: squared paper (1 cm or 2 cm): sticky tape	Arrangement of squares to make a net
54 Net Designs	Scissors: card: glue	Designing nets
55 Corner Cuts	Scissors: squared paper (1 cm)	Volumes of open boxes
56 Faces	Dice (blank)	Arrangement of numbers on dice

Although the overall theme is **cubes** some activities relate to other three dimensional shapes such as cuboids and prisms.

Several of these investigations will involve the pupil in design and construction with card.

Since the investigations involve three dimensional objects there are inherent difficulties with recording. Overcoming this difficulty will create a challenge for the pupil.

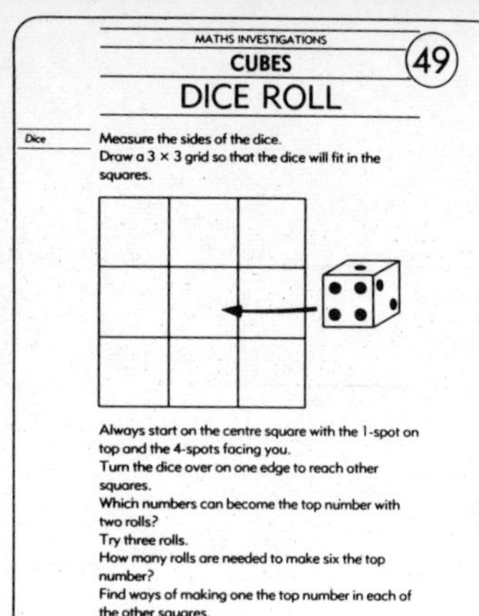
MATHS INVESTIGATIONS

CUBES

DICE ROLL

49

Dice

Measure the sides of the dice.
Draw a 3 × 3 grid so that the dice will fit in the squares.

Always start on the centre square with the 1-spot on top and the 4-spots facing you.
Turn the dice over on one edge to reach other squares.
Which numbers can become the top number with two rolls?
Try three rolls.
How many rolls are needed to make six the top number?
Find ways of making one the top number in each of the other squares.
Try this for other numbers.

MATHS INVESTIGATIONS

CUBES

DICE ROLL

49

Dice

△ Ensure that pupils know that rotating the dice is not allowed. It can only be moved by rolling on an edge.

△ Moves can be recorded by labelling the grid:

A	B	C
D	E	F
G	H	I

△ With two rolls the numbers 1 to 5 can become top numbers:

1	2	3	4	5
FE	HG	FI	FC	HI

Similarly with three rolls.

The 6-spot can become the top number in five rolls:

F I H E B

The 1-spot can become the top number on B, D, F and H in three rolls; it can be rolled to A, C, G and I in eight rolls:

H I F E H G D A

△ Is eight the fewest number of rolls?

△ Explore ways of making the 4-spot the top number in each of the squares.

DA	B	FC
FIH ED	BADE BCFE	DGH EF
DAB EHG	BCF IH	FCB EHI

△ Now try for the 3-spot.

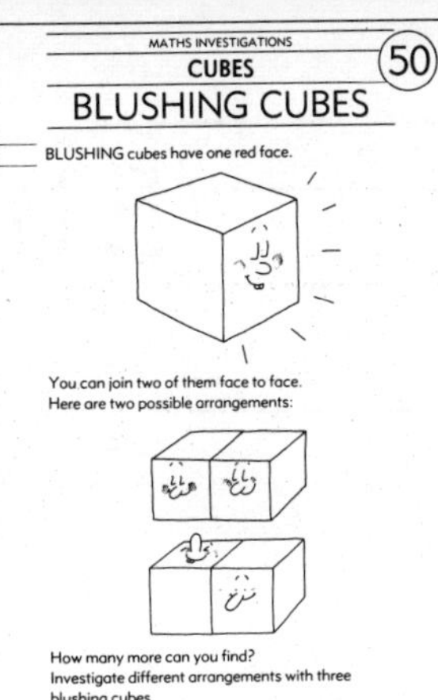
MATHS INVESTIGATIONS

CUBES 50

BLUSHING CUBES

BLUSHING cubes have one red face.

You can join two of them face to face.
Here are two possible arrangements:

How many more can you find?
Investigate different arrangements with three blushing cubes.

MATHS INVESTIGATIONS

CUBES

BLUSHING CUBES

50

Cubes

Isometric paper

△ Blushing cubes can be made by marking one face of each cube.

△ Pupils should be encouraged to create a method of recording solutions: physical representation using wooden or card cubes drawings, perhaps on isometric paper.

△

	Left-hand cube	Right-hand cube
Red on end		
Red on front		
Red on inside		

There are eight different solutions.

△ Extend the investigation to three blushing cubes.

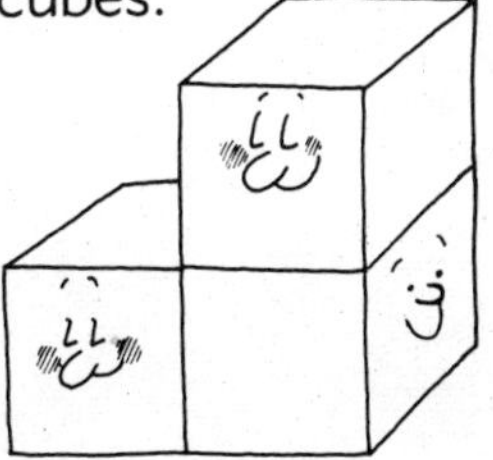

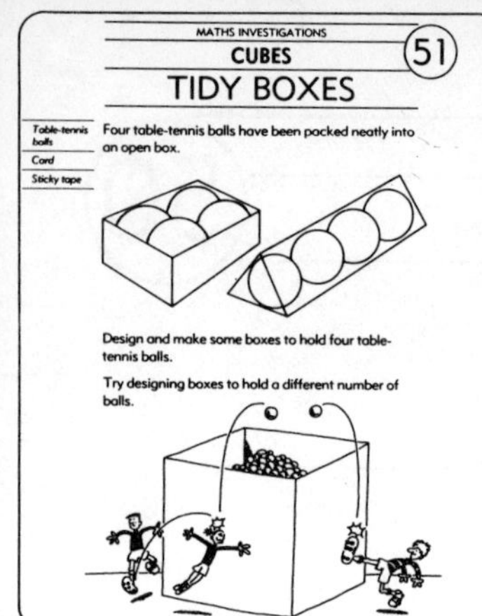
MATHS INVESTIGATIONS

CUBES (51)

TIDY BOXES

Table-tennis balls

Card

Sticky tape

Four table-tennis balls have been packed neatly into an open box.

Design and make some boxes to hold four table-tennis balls.

Try designing boxes to hold a different number of balls.

MATHS INVESTIGATIONS

CUBES

TIDY BOXES

51

Table-tennis balls
Card
Sticky tape

△ Boxes can be made using card or cartridge paper.

△ Some examples include:

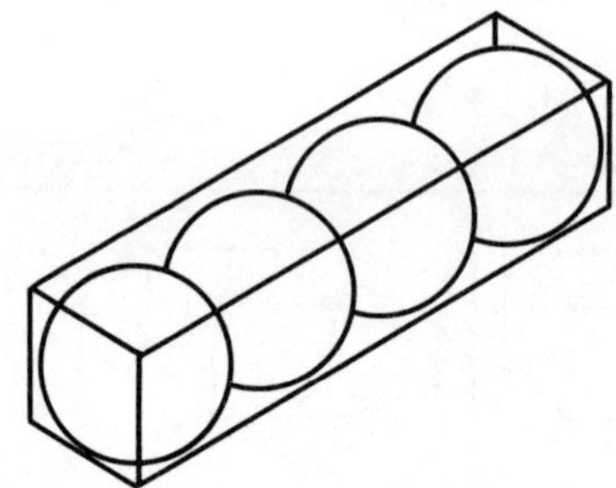

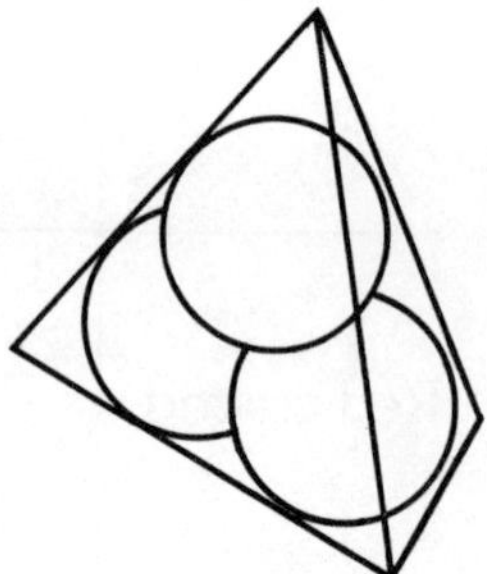

△ Which box uses the least amount of card?

△ What is the volume of each box?

△ An example of a six-ball box:

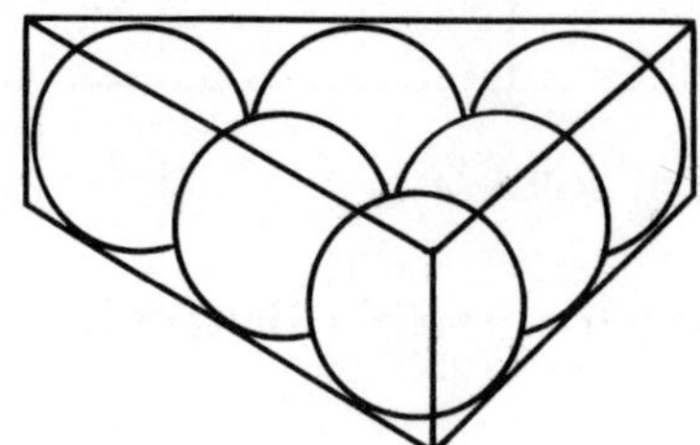

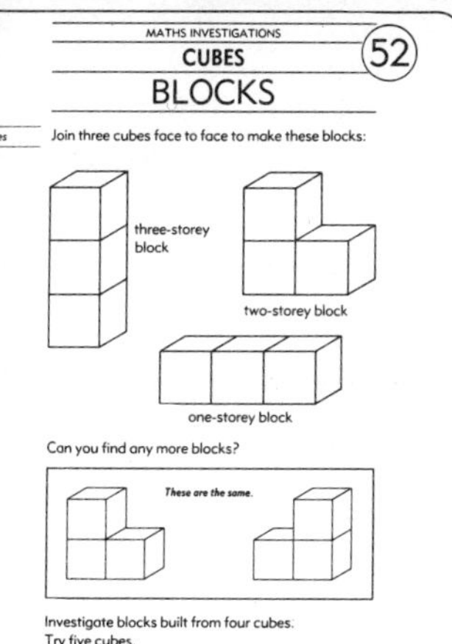
MATHS INVESTIGATIONS
CUBES
52
BLOCKS

Join three cubes face to face to make these blocks:

Can you find any more blocks?

Investigate blocks built from four cubes.
Try five cubes.

MATHS INVESTIGATIONS

CUBES 52

BLOCKS

Cubes

△ Ensure pupils realise that the blocks must be stable:

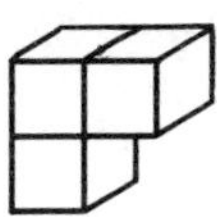

This will fall down.

△ With three cubes: one-storey 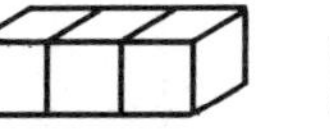 

two-storey 

three-storey 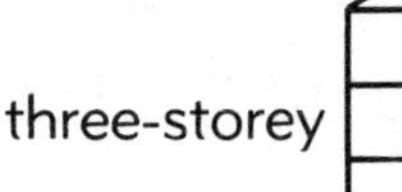

With four cubes: one-storey 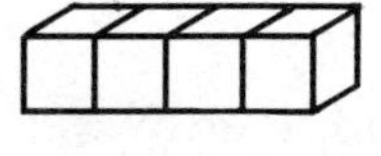 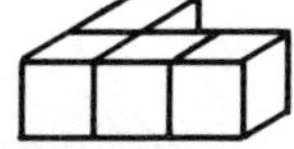 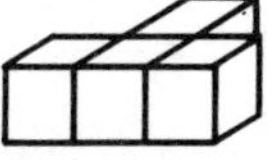

two-storey 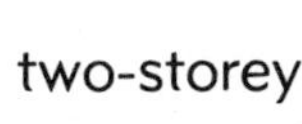 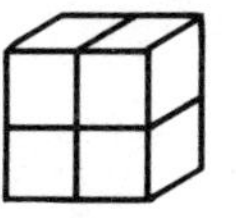 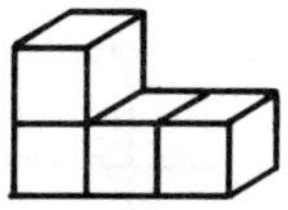 

three-storey 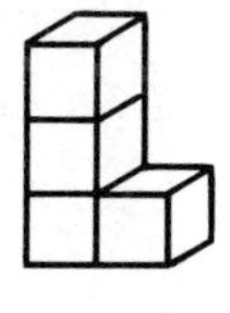

four-storey 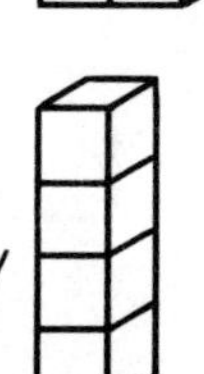

△ Suppose the blocks are to be painted. How many square walls will require paint for each block?

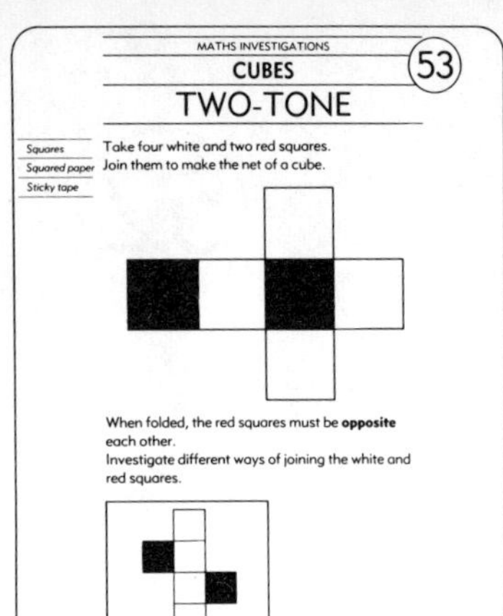
MATHS INVESTIGATIONS

CUBES

TWO-TONE

53

Squares
Squared paper
Sticky tape

Take four white and two red squares.
Join them to make the net of a cube.

When folded, the red squares must be **opposite** each other.
Investigate different ways of joining the white and red squares.

MATHS INVESTIGATIONS

CUBES

TWO-TONE

53

Squares
Squared paper
Sticky tape

△ Nets can be recorded on squared paper.

△ Possible nets include:

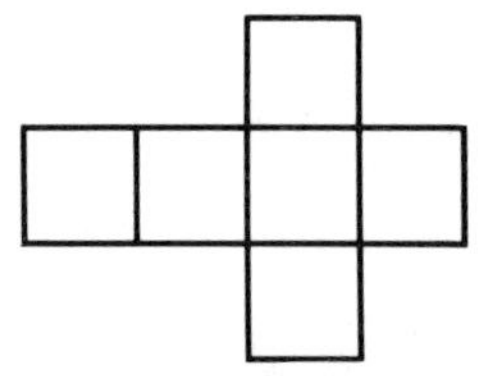 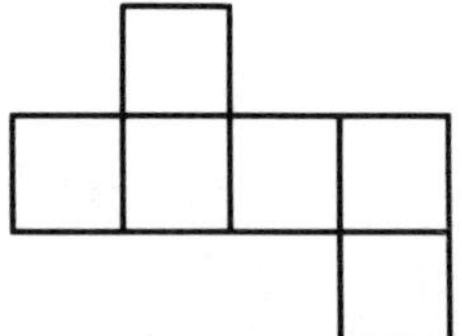 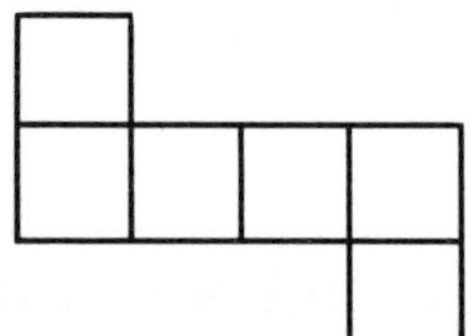 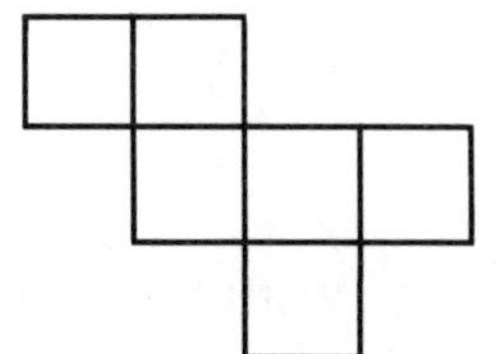

For each net several arrangements exist:

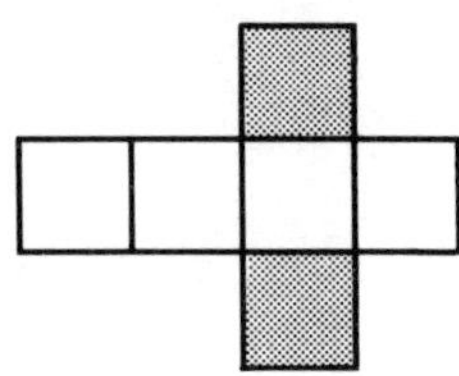 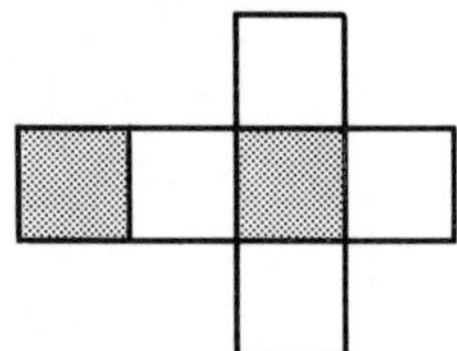 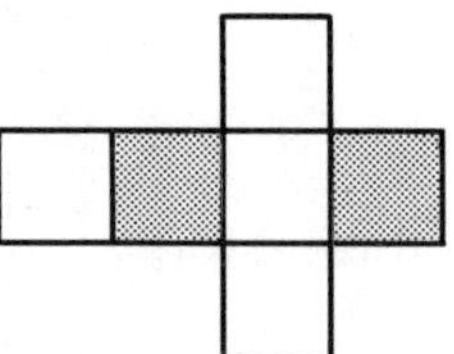

△ Suppose red faces must be adjacent when the net is folded:

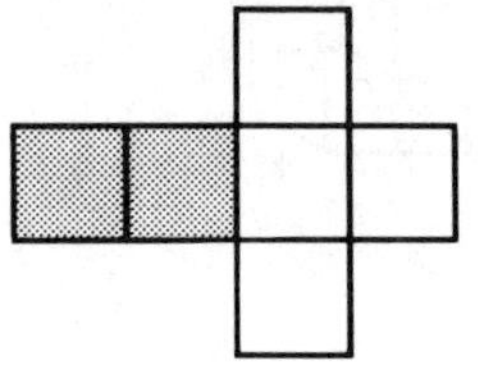 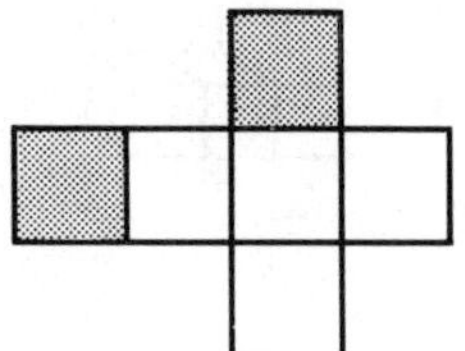 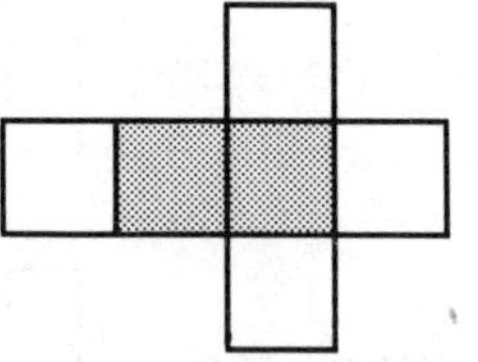 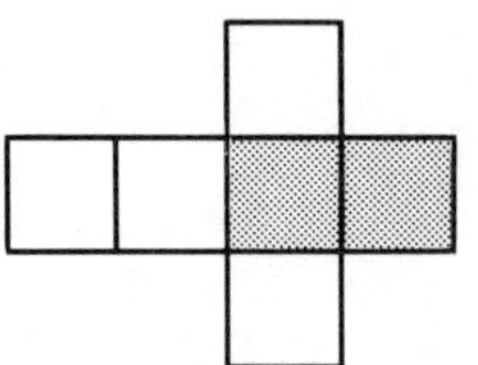

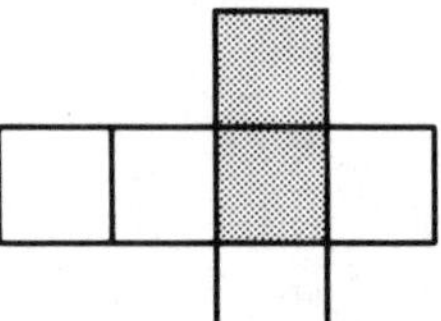 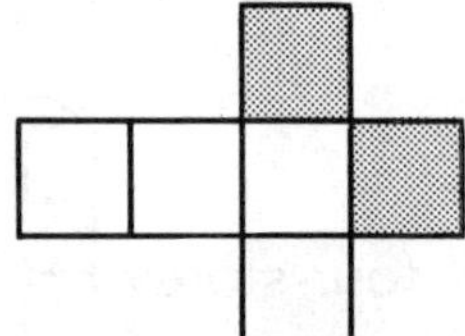

△ Try arrangements with three red squares.

MATHS INVESTIGATIONS

CUBES

NET DESIGNS

54

Card

Scissors

Glue

This net is designed to produce half a cube.

Draw two of the nets on card, add tabs, and make the models.

Place tabs on alternate edges.

Design some more nets which produce half cubes.

Make models from your designs.

MATHS INVESTIGATIONS

CUBES

NET DESIGNS

54

Card

Scissors

Glue

△ Care must be taken to ensure that the nets are not too small.
The measurements on the net design on the pupil's card could be doubled.
Pupils will need reminding to attach tabs to the nets, and some guidance about the positioning of the tabs (i.e. on any edge provided they alternate).
Quick drying glue, such as glue-stick, is most suitable.

△ Some other half-cubes include:

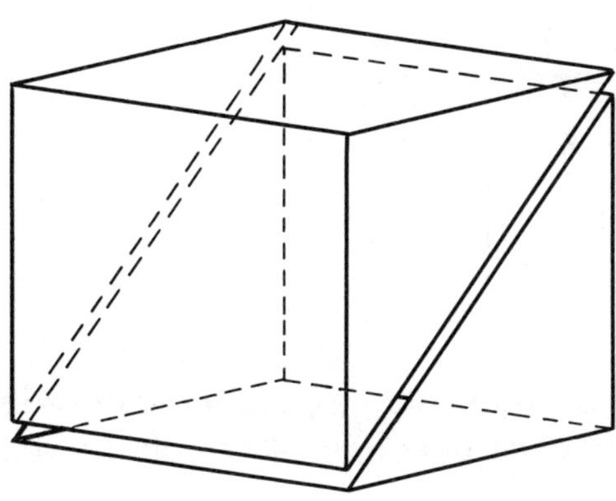

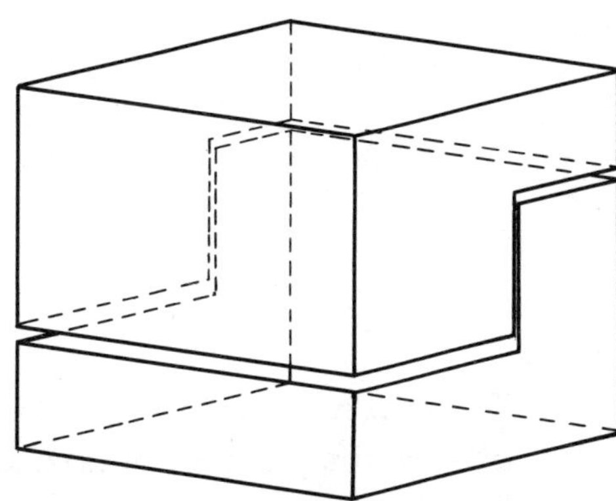

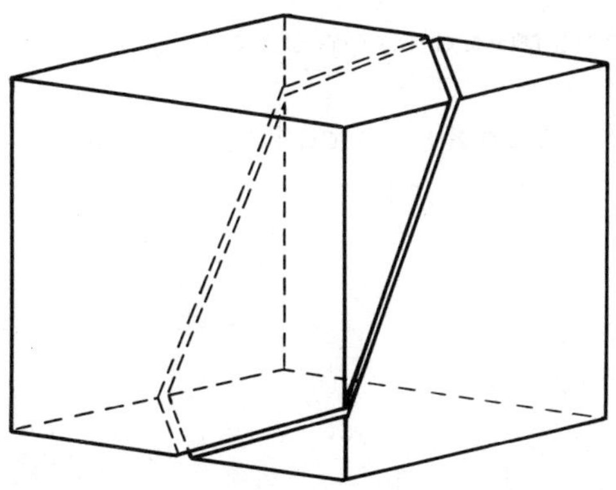

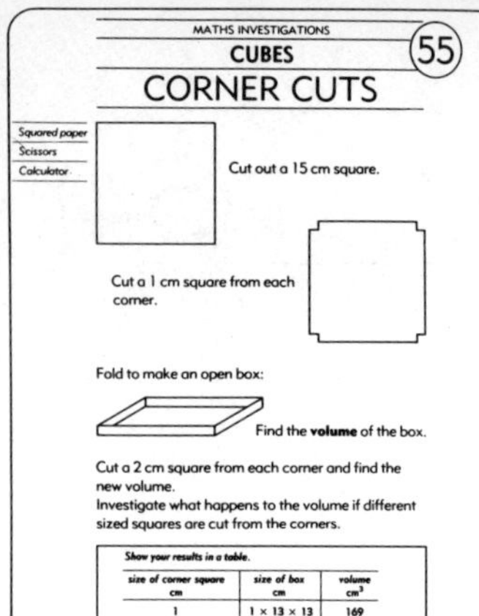

MATHS INVESTIGATIONS

CUBES (55)

CORNER CUTS

Squared paper
Scissors
Calculator

Cut out a 15 cm square.

Cut a 1 cm square from each corner.

Fold to make an open box:

Find the **volume** of the box.

Cut a 2 cm square from each corner and find the new volume.
Investigate what happens to the volume if different sized squares are cut from the corners.

Show your results in a table.

size of corner square cm	size of box cm	volume cm^3
1	1 × 13 × 13	169
2	2 × 11 × 11	242

MATHS INVESTIGATIONS

CUBES

CORNER CUTS

55

Squared paper
Scissors
Calculator

△ Assuming the edge of the cut squares are whole centimetres then the largest sized square which can be cut from each corner is 7 cm.

The table of results is:

Size of corner square cm	Size of box cm	Volume cm^3
1	1 × 13 × 13	169
2	2 × 11 × 11	242
3	3 × 9 × 9	243
4	4 × 7 × 7	196
5	5 × 5 × 5	125
6	6 × 3 × 3	54
7	7 × 1 × 1	7

The maximum volume is 243 cm^3. Can a larger volume be found if the edge of the cut square is not an exact number of centimetres? A calculator would be helpful.

△ Results can be graphed, with edge of cut square against volume.

△ Explore surface areas of the boxes.

△ Try cutting corners from a 16 cm square

△ Try starting with rectangles instead of squares.

MATHS INVESTIGATIONS

CUBES 56

FACES

MATHS INVESTIGATIONS
CUBES 56
FACES

Cubes

Take two cubes.
Write 1, 2, 3, 4, 5, 6 on the faces of one cube.
Write 3, 4, 5, 6, 7, 8 on the faces of the other.

[4] From above, one cube will show a **single-digit** number.

[4] [7] Two cubes can be placed to show a **two-digit** number.

Which single-digit and two-digit numbers can be shown?
How many **square numbers** can be shown?

Two other cubes have some **blank** faces.
One has 2, 3, 5, 6, 8 __ and the other has
1, 2, 4, __, __, 9
How many **square numbers** can be shown now?

Write numbers on the faces of two cubes to show **all** possible single-digit and two-digit square numbers.
Suppose some faces are blank. Can it be done now?

What happens if you want to show prime numbers instead of square numbers?
Investigate for other numbers.

△ (1, 2, 3, 4, 5, 6) and (3, 4, 5, 6, 7, 8)
single-digit numbers are 1, 2, 3, 4, 5, 6, 7, 8

Two-digit numbers are:

		31	41	51	61	71	81
		32	42	52	62	72	82
13	23	33	43	53	63	73	83
14	24	34	44	54	64	74	84
15	25	35	45	55	65	75	85
16	26	36	46	56	66	76	86
17	27	37	47	57	67		
18	28	38	48	58	68		

Pupils can make card cubes from nets to demonstrate solutions and to keep as a record.
Possible square numbers are 1, 4, 9, 16 ,25, 64, 81
– this leaves 9 and 49.

△ (2, 3, 5, 6, 8, __) and (1, 2, 4, __, __, 9)
Possible square numbers are 1, 4, 9, 16, 25, 64, 81
– leaving 36 and 49.

△ Several solutions exist to produce all square numbers.
e.g. (1, 2, 3, 4, 5, 6) and (4, 5, 6, 7, 8, 9,)
This solution has four blanks:
(1, 5, 3, 4, _, _) and (2, 6, 8, 9, _, _)
Can a solution be found with five blanks?

△ A solution revealing all 25 prime numbers is
(1, 2, 3, 6, 7, 9) and (1, 2, 4, 5, 7, 8)
Can a solution be found containing blanks?

△ Explore solutions for the triangular numbers:
1, 3, 6, 10, 15, 21, 28, 36, 45, 55, 66, 78, 91

DIGITS

	APPARATUS	ACTIVITY
57 One to Four		Inventing expressions using four digits
58 Row Sums	Number cards (0 – 9)	Arrangement of digits to produce different totals
59 Digital Sums		Pattern of digital sums for multiples
60 Square Card Sets	Number cards (0 – 9)	Arrangement of digits to produce square numbers
61 Equation Sets		Making equations using four digits
62 Odds and Evens	Number cards (1 – 9)	Arrangement of digits to produce odd and even totals
63 Grid Sums	Squared paper (1 cm)	Arranging digits to produce known totals
64 Target	Number cards (0 – 9): operation cards (+, −, ×, ÷)	Arranging digits and operations to produce given targets

Although most of these investigations can be performed with pencil and paper, the use of number cards makes the activities more dynamic. The cards provide pupils with the confidence and ease to experiment. Most of these ideas will help to develop pupils competence with the number operations.

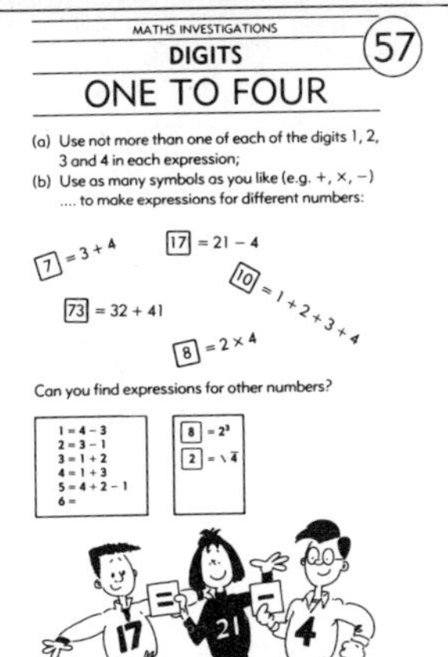

MATHS INVESTIGATIONS

DIGITS 57

ONE TO FOUR

(a) Use not more than one of each of the digits 1, 2, 3 and 4 in each expression;
(b) Use as many symbols as you like (e.g. +, ×, −) .... to make expressions for different numbers:

7 = 3 + 4
17 = 21 − 4
10 = 1 + 2 + 3 + 4
73 = 32 + 41
8 = 2 × 4

Can you find expressions for other numbers?

1 = 4 − 3
2 = 3 − 1
3 = 1 + 2
4 = 1 + 3
5 = 4 + 2 − 1
6 =

$8 = 2^3$
$2 = \sqrt{4}$

MATHS INVESTIGATIONS

DIGITS 57

ONE TO FOUR

△ One approach is to list the numbers in order, and then seek expressions for each in turn.

△
$1 = 4 - 3$
$2 = 3 - 1$
$3 = 1 + 2$
$4 = 1 + 3$
$5 = 4 + 2 - 1$
$6 = 2 \times 3$
$7 = (2 \times 4) - 1$
$8 = 2 \times 4$
$9 = 3^2$
$10 = 3^2 + 1$
$11 = 13 - 2$
$12 = 3 \times 4$
$13 = (3 \times 4) + 1$
$14 = (3 \times 4) + 2$
$15 = 13 + 2$
$16 = 4^2$
$17 = 4^2 + 1$
$18 = 3 \times (4 + 2)$
$19 = 4^2 + 3$
$20 = 4 \times (3 + 2)$

$21 = 24 - 3$
$22 = 23 - 1$
$23 = 24 - 1$
$24 = 23 + 1$
$25 = 24 + 1$
$26 = 13 \times 2$
$27 = 31 - 4$
$28 = 14 \times 2$
$29 = 31 - 2$
$30 = (13 \times 2) + 4$
$31 = (14 \times 2) + 3$
$32 = 34 - 2$
$33 = 34 - 1$
$34 = (14 + 3) \times 2$
$35 = 34 + 1$
$36 = 34 + 2$
$37 = 32 + 1 + 4$
$38 = 41 - 3$
$39 = 41 - 2$
$40 = 43 - 2 - 1$

△ Find numbers which have several expressions:

$$24 = 4 \times 3 \times 2$$
$$= 23 + 1$$
$$= 21 + 3$$
$$= 12 \times \sqrt{4}$$

△ Change the digits to 2, 3, 4 and 5 for example.

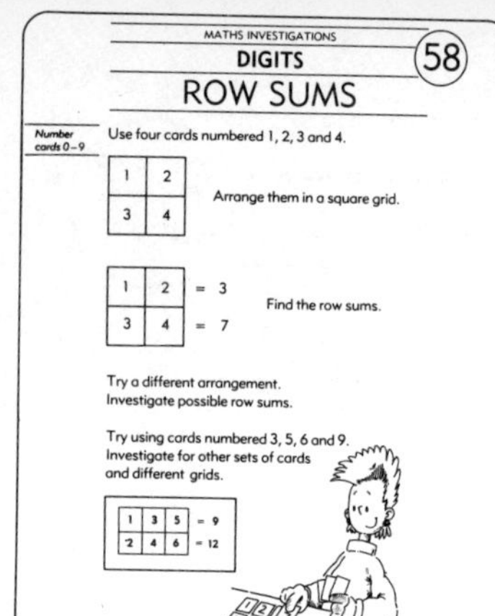
MATHS INVESTIGATIONS

DIGITS

ROW SUMS

58

Number cards 0–9

Use four cards numbered 1, 2, 3 and 4.

1	2
3	4

Arrange them in a square grid.

1	2	= 3
3	4	= 7

Find the row sums.

Try a different arrangement.
Investigate possible row sums.

Try using cards numbered 3, 5, 6 and 9.
Investigate for other sets of cards and different grids.

1	3	5	= 9
2	4	6	= 12

MATHS INVESTIGATIONS

DIGITS

ROW SUMS

58

Number cards 0–9

△ Using cards numbered 1, 2, 3 and 4, five different row sums are possible:

3, 4, 5, 6 and 7

△ With four consecutive numbers, only five different row sums are possible. Why?

△ Why is six the maximum number of row sums?
Which four digits will result in six row sums?

△ Try different grids:

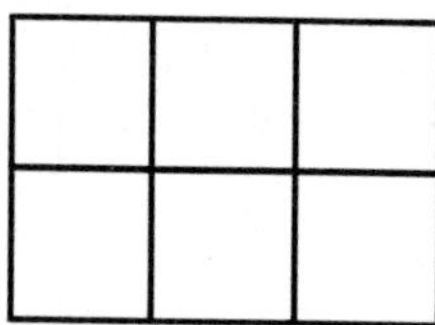

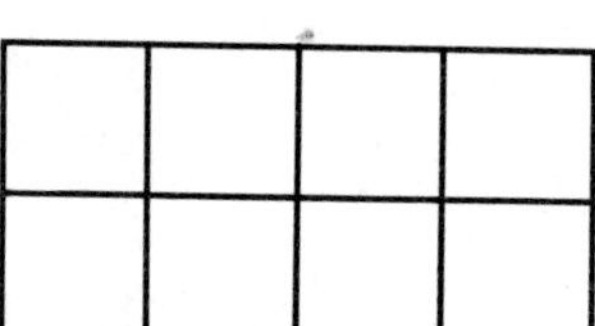

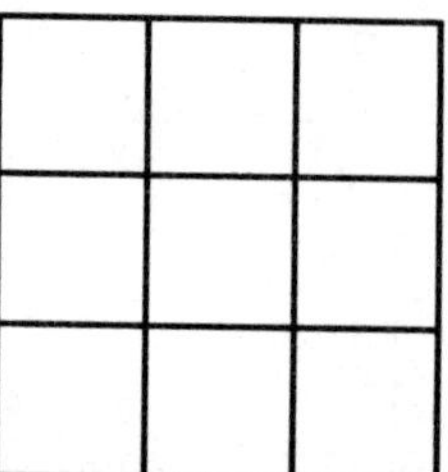

△ Try:

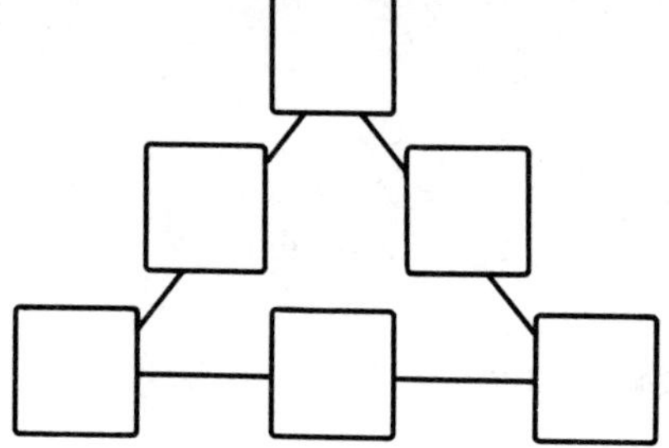

and consider the sums along the sides.

MATHS INVESTIGATIONS

DIGITS 59

DIGITAL SUMS

The **digital sum** of 23 is 5.

$23 \rightarrow 2 + 3 = 5$

The **digital sum** of 84 is 3.

$84 \rightarrow 8 + 4 = 12$
$12 \rightarrow 1 + 2 = 3$

The digital sums of multiples of 3 are:

multiples of three	digital sums
3	3
6	6
9	9
12	3
15	6
18	9
21	3
24	6
27	9

What do you notice?
Investigate digital sums of other multiples.

MATHS INVESTIGATIONS

DIGITS

DIGITAL SUMS

△ Pupils should be encouraged to try to explain the patterns or sequences.

△ Results may be tabulated:

	digital sums													
2 ×	2	4	6	8	1	3	5	7	9	2	4	6	8	1
3 ×	3	6	9	3	6	9	3	6	9	3	6	9	3	6
4 ×	4	8	3	7	2	6	1	5	9	4	8	3	7	2
5 ×	5	1	6	2	7	3	8	4	9	5	1	6	2	7
6 ×	6	3	9	6	3	9	6	3	9	6	3	9	6	3
7 ×	7	5	3	1	8	6	4	2	9	7	5	1	8	6

△ Patterns often exist in alternate digital sums:

e.g. 4 ×, 5 ×.

△ Investigate digital sums of special numbers
e.g. square numbers.

△ Investigate **final digits** for different multiples.

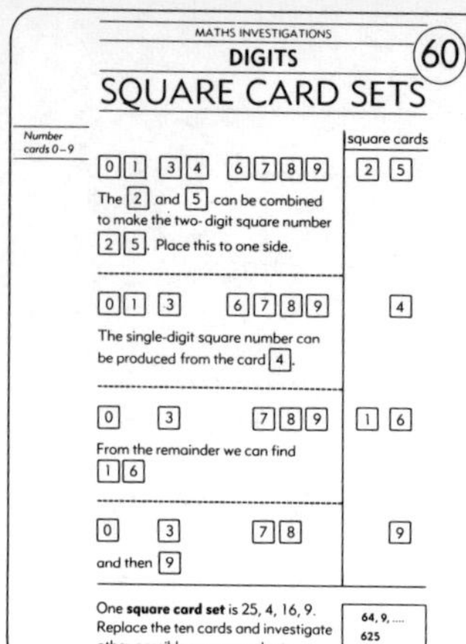
MATHS INVESTIGATIONS

DIGITS 60

SQUARE CARD SETS

Number cards 0–9

	square cards
[0][1] [3][4] [6][7][8][9] The [2] and [5] can be combined to make the two-digit square number [2][5]. Place this to one side.	[2] [5]
[0][1] [3] [6][7][8][9] The single-digit square number can be produced from the card [4].	[4]
[0] [3] [7][8][9] From the remainder we can find [1][6]	[1] [6]
[0] [3] [7][8] and then [9]	[9]

One **square card set** is 25, 4, 16, 9. Replace the ten cards and investigate other possible square card sets.

64, 9, ...
625

MATHS INVESTIGATIONS

DIGITS

SQUARE CARD SETS

60

Number cards 0 – 9

△ Pupils may need reminding that each digit can only be used **once** for each set of square numbers.

△ Which is the largest set of square numbers you can make?
1, 4, 9, 25, 36 – this set contains five square numbers.

△ Which set of square numbers requires most digits?
49, 36, 25, 81 – this set uses eight digits.
Can you find a set which uses more than eight?

△ How many three-digits square numbers can be made?

△ What is the largest square number that can be made?

△ Investigate sets of other numbers.

e.g. 'prime' card sets
'multiples of four' card sets.

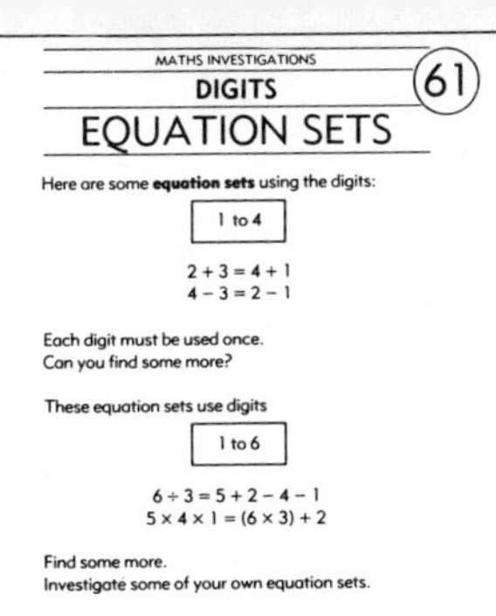
MATHS INVESTIGATIONS

DIGITS (61)

EQUATION SETS

Here are some **equation sets** using the digits:

1 to 4

$2 + 3 = 4 + 1$
$4 - 3 = 2 - 1$

Each digit must be used once.
Can you find some more?

These equation sets use digits

1 to 6

$6 \div 3 = 5 + 2 - 4 - 1$
$5 \times 4 \times 1 = (6 \times 3) + 2$

Find some more.
Investigate some of your own equation sets.

MATHS INVESTIGATIONS

DIGITS (61)

EQUATION SETS

△ Pupils should be encouraged to use a variety of symbols and operations:

e.g. $1^2 = 4 - 3$ $\qquad \frac{(3 + 1)}{2} = \sqrt{4}$ $\qquad 3 + 1 = 2 + \sqrt{4}$

$1 = (2 + 3) - 4$ $\qquad 3! = \frac{(4 + 2)}{1}$

△ The use of six digits raises the potential for utilising a wider range of operations.

△ Investigate equation sets using a random set of digits:

1, 3, 5, 6,
2, 4, 5, 8, 9

△ Use decimal points:
for 1, 2, 3, 5, 6
$3.2 \times 5 = 16$

MATHS INVESTIGATIONS

DIGITS (62)

ODDS AND EVENS

Number cards 1–9

Choose cards 2, 3, 4 and 7.
Arrange them on a 2 × 2 grid.
Find the **row sums**.

		row sums
2	3	5
4	7	11

Repeat using a different arrangement of 2, 3, 4 and 7.

Investigate arrangements which give:
(a) even row sums
(b) odd row sums
(c) an even and an odd row sum.

Find the column sums.

	2	3
	4	7
column sums	6	10

Investigate arrangements which give:
(a) even row and odd column sums
(b) odd row and odd column sums

Investigate for a different set of cards.

MATHS INVESTIGATIONS

DIGITS (62)

ODDS AND EVENS

Number cards 1 – 9

△ Pupils can eventually be encouraged to look at the combinations of odd and even numbers, rather than the value of the cards themselves (i.e. choosing 1, 3, 5 and 7 is no different from choosing 3, 5, 7 and 9 as they both consist of four odd numbers).

△ For four cards the possible combinations are:

4 evens; 4 odds; 2 odds and 2 evens; 3 odds and 1 even; 3 evens and 1 odd.

The arrangements are:

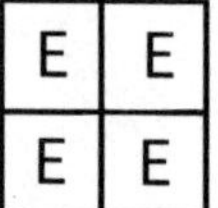

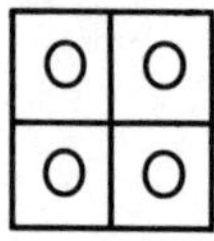

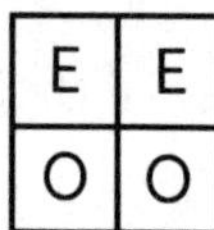

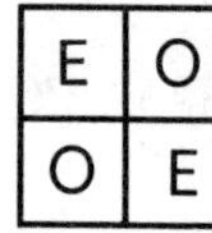

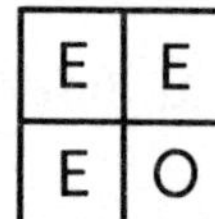

From this the possible combinations of odd and even row and column sums can clearly be seen.

△ What happens if you allow a repeated digit?

△ Extend to grids of other sizes (e.g. 3 × 3).

MATHS INVESTIGATIONS

DIGITS

GRID SUMS

63

Squared paper

Copy this grid in which row sums and column sums are given.

		8
		6
10	4	

Write numbers on the grid which produce these row and column sums.

Here is a solution:

7	1	8
3	3	6
10	4	

How many other solutions can you find?

Investigate different solutions to these:

		9
		7
8	8	

		8
		6
9	5	

Invent some grids of your own and investigate possible solutions.

MATHS INVESTIGATIONS

DIGITS

GRID SUMS

63

1 cm Squared paper

△ Pupils may need reminding that:

(a) a number can be used more than once.
(b) a zero can be used.

△ Is there a relationship between the number of solutions and the given sums? The number of solutions will be decided by the smallest given sum.

△ Why won't this work?

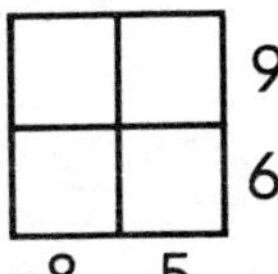

The row sum total and column sum total must be the same.

△ Can you find a grid which has nine solutions?

△ Extend the size of the grid.

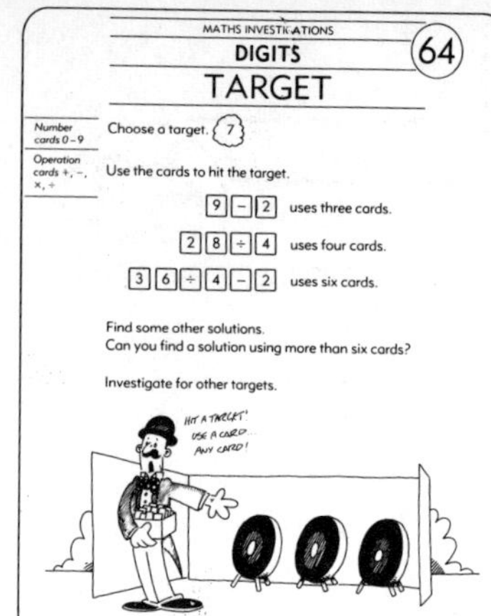

MATHS INVESTIGATIONS

DIGITS

TARGET

(64)

Number cards 0 – 9

Operation cards +, −, ×, ÷

Choose a target. (7)

Use the cards to hit the target.

[9] [−] [2] uses three cards.

[2] [8] [÷] [4] uses four cards.

[3] [6] [÷] [4] [−] [2] uses six cards.

Find some other solutions.
Can you find a solution using more than six cards?

Investigate for other targets.

MATHS INVESTIGATIONS

DIGITS

TARGET

(64)

Number cards 0 – 9

Operation cards
+, −, ×, ÷

△ Some discussion may be necessary about the order in which the operations are carried out. Bracket cards can be used.

△ Investigate solutions involving the largest possible number of cards (maximum 14).

△ Investigate six-card solutions. For example, which targets can be hit using six cards?

△ Remove the card [0]. Investigate different ways of achieving the target, 0.

six-card solution [4] [2] [÷] [6] [−] [7]

seven-card solution [2] [÷] [1] [×] [4] [−] [8]

eight-card solution [4] [3] [+] [2] [5] [−] [6] [8]

CALCULATOR

	APPARATUS	ACTIVITY
65 Switch	Calculator	Relationship between digits to produce equal product
66 Two Digits	Calculator	Fewest keystrokes to produce a target with two digits
67 Light Bars	Calculator	Number of display light bars needed to produce numbers
68 Forbidden Keys	Calculator	Operations avoiding the use of certain keys
69 Finger Tapping	Calculator	Patterns of touching keys to produce certain totals
70 Repeats	Calculator	Divisions which produce a pattern of repeating digits
71 Solitaire	Calculator	One digit and several functions to produce known answers
72 Big Times	Calculator: card squares	Arranging digits to produce largest product

Although the theme of these investigations is **calculators** it is expected that calculators may be used in other sections of the investigations.

Several of these activities consolidate the use of number operations in general and multiplication and division in particular.

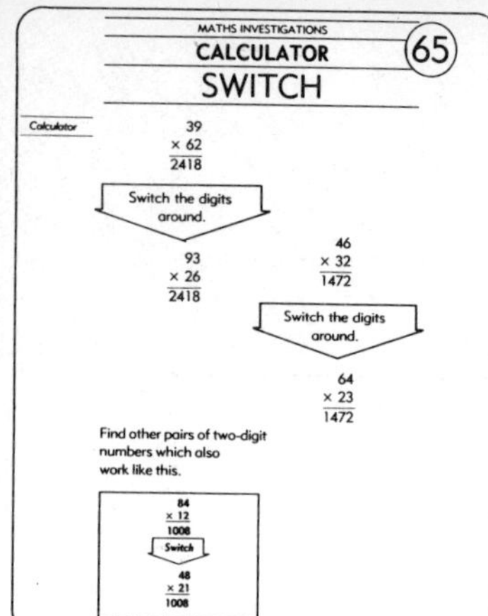
MATHS INVESTIGATIONS
CALCULATOR
SWITCH
65

Calculator

39
× 62
2418

Switch the digits around.

93
× 26
2418

46
× 32
1472

Switch the digits around.

64
× 23
1472

Find other pairs of two-digit numbers which also work like this.

84
× 12
1008

Switch

48
× 21
1008

MATHS INVESTIGATIONS

CALCULATOR SWITCH

65

Caluclator

△ Pairs of numbers may well be found by trial and error initially.

△ Several solutions exist:

repeated digits:

$$\begin{array}{r} 22 \\ \times\ 44 \\ \hline 968 \end{array} \qquad \begin{array}{r} 22 \\ \times\ 44 \\ \hline 968 \end{array}$$

one number the reverse of the other:

$$\begin{array}{r} 21 \\ \times\ 12 \\ \hline 252 \end{array} \qquad \begin{array}{r} 12 \\ \times\ 21 \\ \hline 252 \end{array}$$

tens digit product equals units digit product:

$$\begin{array}{r} 39 \\ \times\ 62 \\ \hline 2418 \end{array} \qquad \begin{array}{r} 93 \\ \times\ 26 \\ \hline 2418 \end{array}$$

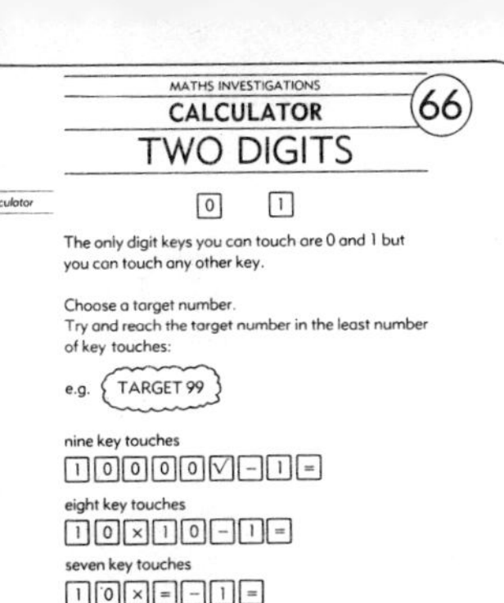

MATHS INVESTIGATIONS

CALCULATOR (66)

TWO DIGITS

[0] [1]

The only digit keys you can touch are 0 and 1 but you can touch any other key.

Choose a target number.
Try and reach the target number in the least number of key touches:

e.g. TARGET 99

nine key touches
[1] [0] [0] [0] [0] [√] [−] [1] [=]

eight key touches
[1] [0] [×] [1] [0] [−] [1] [=]

seven key touches
[1] [0] [×] [=] [−] [1] [=]

six key touches
[1] [0] [0] [−] [1] [%]

Try different target numbers.

MATHS INVESTIGATIONS

CALCULATOR (66)

TWO-DIGITS

Calculator

△ Encourage pupils to select a wide range of target numbers and experiment with all the calculator functions :

[√] [M] [M+] [%] etc.

△ If six key-touches is the minimum, investigate other solutions involving six key-touches:

[1] [0] [0] [−] [1] [=]

[1] [0] [0] [−] [1] [%]

△ Investigate solutions using a different pair of digit keys: e.g.

[0] [3]

MATHS INVESTIGATIONS

CALCULATOR (67)

LIGHT BARS

Calculator

△ It is worthwhile recording the light bars needed for each digit.

digit	0	1	2	3	4	5	6	7	8	9
light bars	6	2	5	5	4	5	6	4	7	6

Several lines of investigation can be followed

e.g. With 10 light bars:

- Which is the largest/smallest number that can be displayed?
- How many different numbers can be displayed?
- Try with different numbers of light bars.
- How many light bars are needed to make special numbers (square, prime, etc.)?

△ Which two digit numbers use most (or fewest) light bars?

MATHS INVESTIGATIONS
CALCULATOR
FORBIDDEN KEYS 68

ulator

Calculate: 78×36

[7] [8] [×] [3] [6] [=]

Suppose you are forbidden to use the [6] key.

Here are two ways of obtaining the answer:

[7] [8] [×] [3] [5] [+] [7] [8] [=]

[7] [8] [×] [7] [2] [÷] [2] [=]

[7] [×]

Find some other ways.
Choose other forbidden keys and investigate different ways of obtaining the answer.

Calculate: $1728 \div 36$

[1] [7] [2] [8] [÷] [3] [6] [=]

Choose a forbidden key.
Investigate different ways of obtaining the answer.

MATHS INVESTIGATIONS

CALCULATOR

FORBIDDEN KEYS

68

Calculator

- △ Encourage pupils to find several ways for each forbidden key.

- △ Some solutions for the forbidden key, [7] :

  $2 \times 39 \times 36$
  $(80 \times 36) - (2 \times 36)$
  $6 \times 13 \times 36$
  $(68 \times 36) + (10 \times 36)$
  $(156 \div 2) \times 36$

  Some of these solutions require the use of a memory key.

- △ Some solutions for the forbidden key, [X] :

  $780 + 780 + 780 + 78 + 78 + 78 + 78 + 78 + 78$
  $780 + 780 + 780 + 780 - 78 - 78 - 78 - 78$

- △ Consider calculations involving other operations e.g. +, −

- △ Suppose more than one key is forbidden.

- △ Investigate calculations involving decimals e.g. 3.4×5.6

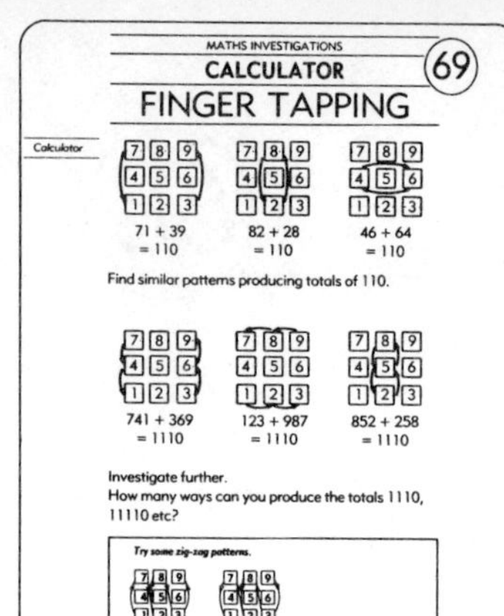
MATHS INVESTIGATIONS
CALCULATOR 69
FINGER TAPPING

Calculator

71 + 39 = 110
82 + 28 = 110
46 + 64 = 110

Find similar patterns producing totals of 110.

741 + 369 = 1110
123 + 987 = 1110
852 + 258 = 1110

Investigate further.
How many ways can you produce the totals 1110, 11110 etc?

Try some zig-zag patterns.

172839 + 938271 = 1111110

MATHS INVESTIGATIONS

CALCULATOR

FINGER TAPPING

69

Calculator

△ There are at least eight ways of producing 110:

19 + 91   17 + 93   71 + 39   28 + 82

79 + 31   97 + 13   73 + 37   46 + 64

△ There are at least 20 ways of producing 1110:

147 + 963   741 + 369   174 + 936   714 + 396.

What about producing 11110; 111110, etc?

△ It may be noticed that:

hundreds digits total 10
tens digits total 10
units digits total 10

e.g. 147 + 963

△ What about producing 1111111110?
(This is too large for the display.)

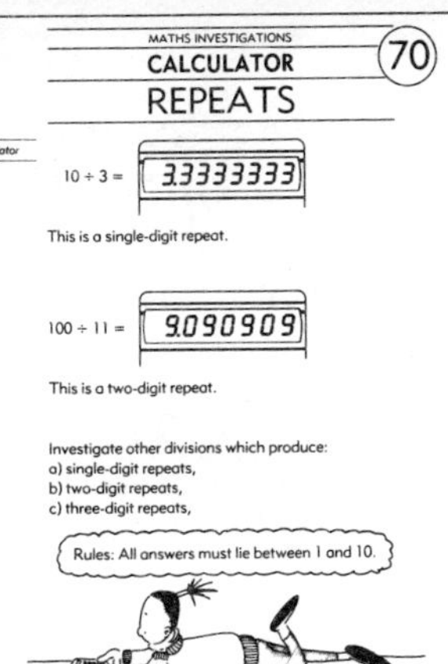

MATHS INVESTIGATIONS

CALCULATOR REPEATS

70

Calculator

△ Single digit repeats up to 8.8888888 can be obtained by dividing multiples of 10 by 9. Producing 9.9999999 causes problems, either 9.9999998 or 0.9999999 are possible.

△ Two-digit repeats can be obtained in many ways:

dividing by 99
dividing by some multiples of 11
dividing multiples of 10 by 11

△ Will you allow 83.333333 ÷ 11 or 100 ÷ 27.5?

△ Which divisions will produce 3.6363636; 1.2121212? etc.

△ Three-digit repeats can be quite challenging. Because only eight digits can be displayed some discussion may be necessary as to the 'missing numbers'. It is worth considering the introduction of notation for repeated digits e.g. 100 ÷ 54 = 1.8518518 or $1.\overline{851}$

△ Dividing by 999 will produce three nominated repeating digits (e.g. If you want 347 to repeat, then divide 3470 by 999).

△ Dividing multiples of 100 by 101 produces a four-digit repeat.

△ Explore repeating digits obtained by multiplication:

37037 × 3 ... and other multiples of 3.
15873 × 7 ... and other multiples of 7.

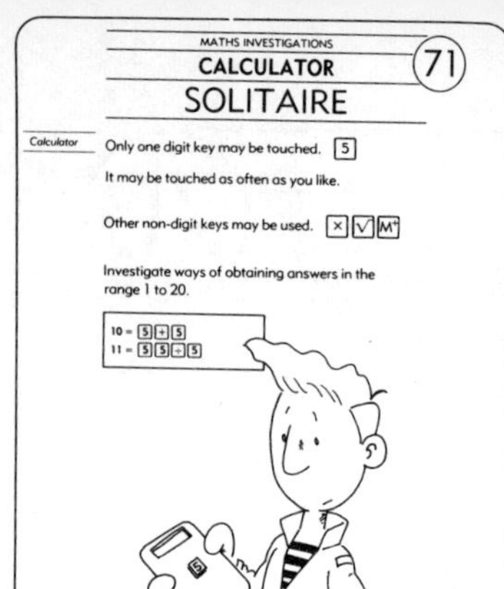

MATHS INVESTIGATIONS

CALCULATOR SOLITAIRE

71

Calculator

△ It helps if pupils are familiar with, and can use, the memory function.

△ Encourage the use of 'fewest key strokes'.

△ Pupils are more likely to see patterns if the numbers 1 to 20 are listed.

△ Some solutions for 5:

1) $5 \div 5$
2) $(5 + 5) + 5$
3) $(5 + 5 + 5) \div 5$
4) $(5 + 5 + 5 + 5) \div 5$
5) 5
6) $5 + (5 \div 5)$
7) $(5 + 5) \div 5 + 5$
8) $(5 + 5 + 5) \div 5 + 5$
9) $(5 + 5 + 5 + 5) \div 5 + 5$
10) $5 + 5$
11) $55 \div 5$
12) $(5 + 5) \div 5 + (5 + 5)$
13) $(5 + 5 + 5) \div 5 + (5 + 5)$
14) $(5 + 5 + 5) \div 5 + (55 \div 5)$
15) $5 + 5 + 5$

△ How many different solutions can you find for 7; 11?

△ Try with different digits e.g. 8.

△ Choose a target number e.g. 137, and reach it with a range of single-digits. Which digit requires fewest key strokes?

MATHS INVESTIGATIONS

CALCULATOR

BIG TIMES

Use 2,3,4 and the multiplication sign.
Find the largest possible product.

Investigate largest products with other sets of three digits.

Investigate largest products with four and five digits.

Write the digits on card.

MATHS INVESTIGATIONS

CALCULATOR

BIG TIMES

Calculator

Number cards 0 – 9

△ Pupils may gain confidence from the use of digit cards.

△ With three digits *a*, *b*, *c* where *a*, *b*, and *c* are in ascending order of value, the largest product will be:

$$\begin{array}{r} bc \\ \times\ a \\ \hline \end{array} \qquad \text{e.g.} \quad \begin{array}{r} 32 \\ \times\ 4 \\ \hline 128 \end{array}$$

△ With four digits the largest product will be:

$$\begin{array}{r} bc \\ \times\ ad \\ \hline \end{array} \qquad \text{e.g.} \quad \begin{array}{r} 43 \\ \times\ 52 \\ \hline 2236 \end{array}$$

△ With five digits the largest product will be:

$$\begin{array}{r} bce \\ \times\ ad \\ \hline \end{array} \qquad \text{e.g.} \quad \begin{array}{r} 542 \\ \times\ 63 \\ \hline 34146 \end{array}$$

△ What happens if a digit can be repeated?

△ Investigate smallest products.

△ Investigate for other number operations.

NUMBER II

	APPARATUS	ACTIVITY
73 Spike	Counters	Arrangement of counters to produce different numbers
74 Balance	(Equaliser – optional)	Combination of numbers which sum to equal totals
75 Three Card Tricks	Number cards (0 – 20)	Summing numbers to a given total
76 Darts		Different scores on a dart board
77 Tidy Numbers	Squared paper (1 cm)	Arrangement of squares inside a square
78 Neighbours		Expression of numbers in terms of consecutive whole numbers
79 Threes and Fives		Expression of numbers using multiples of two numbers
80 Palindromes		Algorithms on numbers to form palindromes

- Calculators could be used during these activities.

These ideas employ a variety of pieces of apparatus eg. cards, dartboards, abacii, balances etc.

A theme common to several of the investigations is the search for different expressions for numbers based on sums of other numbers.

All of the activities provide opportunities for pupils to:

develop a general facility with number,

practice operations on numbers and associated ideas eg. doubles, trebles, multiples.

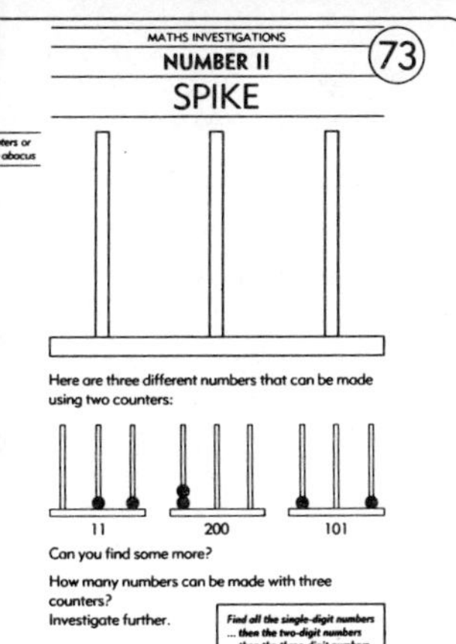

MATHS INVESTIGATIONS

NUMBER II

SPIKE

(73)

Counters or
Spike abacus

△ Pupils can (a) use an abacus with beads, or (b) use the pupil card with counters.

△ Two-counter solutions: 2, 11, 20, 101, 110, 200.
Three-counter solutions: 3, 12, 21, 30, 102, 111, 120, 201, 210, 300.

△ Pupils should be encouraged to be systematic

e.g. find all single-digit numbers
then two-digit numbers
then three-digit numbers

Are there any patterns in the lists of numbers?
The results may be tabulated:

e.g. Four counters

single-digit	two-digit	three-digit
4	13	103
	22	112
	31	121
	40	130
		202
		211
		301
		310
		400

number of counters	total different numbers
2	6
3	10
4	15
5	21

△ Is it possible to predict how many two-digit and three-digit numbers can be created from (a) four counters (b) five counters (c) ten counters?
Is it possible to predict the total possible numbers for (a) five counters (b) six counters (c) ten counters?

△ Investigate the patterns obtained using an abacus with four spikes.

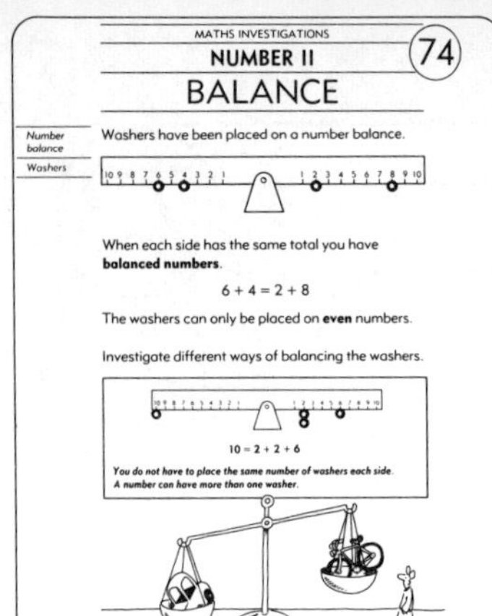
MATHS INVESTIGATIONS

NUMBER II

BALANCE

74

Number balance

Washers

Washers have been placed on a number balance.

When each side has the same total you have **balanced numbers**.

6 + 4 = 2 + 8

The washers can only be placed on **even** numbers.

Investigate different ways of balancing the washers.

10 = 2 + 2 + 6

You do not have to place the same number of washers each side.
A number can have more than one washer.

MATHS INVESTIGATIONS

NUMBER II

BALANCE

74

Number balance

Washers

△ A number balance may be used for this activity.

△ Pupils can be systematic by considering –
(a) one washer balancing three.
(b) two washers balancing two.

△ The table below ignores simple cases in which washers are placed on the same numbers each side.

L.H.S.	R.H.S.
one washer	three washers
(6) (8) (10)	(2, 2, 2) (2, 2, 4) (2, 2, 6) (2, 4, 4)
two washers	two washers
(2, 6) (2, 8) (2, 10) (4, 10) (6, 10)	(4, 4) (4, 6) (6, 6) (4, 8) (6, 8) (8, 8)

△ What happens if only odd numbers can be used?

△ What happens if you are restricted to odd numbers on one side and even numbers on the other?

△ Investigate the situation when more washers are available.

△ Investigate the situation if the balance is extended to 20.

MATHS INVESTIGATIONS

NUMBER II 75

THREE CARD TRICKS

0 1 2 3 4 5 6 7 8 9 10

11 12 13 14 15 16 17 18 19 20

Make a set of **three-card tricks** from the 21 cards. The total for each trick must be 17.

Here is a set:

3 4 10   12 5 0   1 2 14

Investigate other sets for which each trick totals 17.

Investigate sets for other totals.

MATHS INVESTIGATIONS

NUMBER II

THREE CARD TRICKS

75

Number cards 0 – 20

△ The activity can be performed as a paper and pencil exercise. Most pupils, however, will perform better by using the cards.

△ Three-card tricks – total 17
Here are two more possible sets, each containing three tricks:

How many other different sets can be found?

A set containing four tricks.

1 7 9   2 5 10   3 6 8   4 0 13

How many other four trick sets are possible?

△ Change the target total. Here is a set of four tricks – total 20

1 5 14   2 6 12   3 7 10   0 4 16

For which totals is it possible to find a set of four tricks?

△ Change the number of cards in a trick (e.g. two/four card tricks.)

△ Change the numbers on the pack of cards (e.g. 0 to 30, 10 to 30).

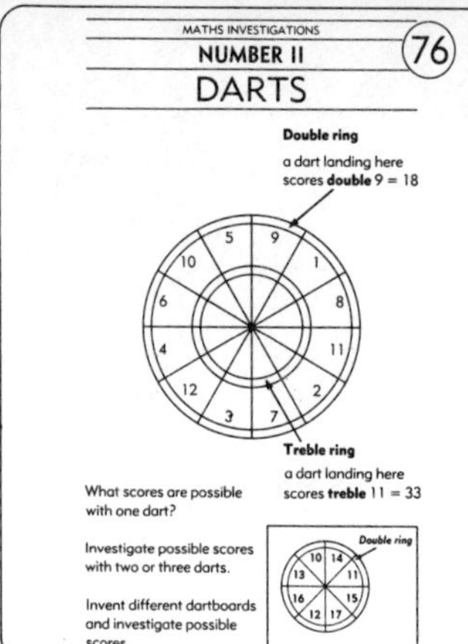

MATHS INVESTIGATIONS

NUMBER II

DARTS

(76)

△ Pupils can be systematic by constructing a table.

One dart scores

S singles	1	2	3	4	5	6	7	8	9	10	11	12
D doubles	2	4	6	8	10	12	14	16	18	20	22	24
T trebles	3	6	9	12	15	18	21	24	27	30	33	36

All scores in the range 1 to 36 are possible except:
13, 17, 19, 23, 25, 26, 28, 29, 31, 32, 34, 35.

△ Which scores are possible in two different ways?

2 (2, D1) 4 (4, D2) 8 (8, D4) 10 (10, D5) 18 (D9, T6) 24 (D12, T8)

△ Which scores are possible in three different ways?

6 (6, D3, T2) 12 (12, D6, T4)

△ **Two-dart scores**
An addition table can be constructed to find all possible scores:

	1	2	3	4.....	36
1	2	3	4	5.....	37
2	3	4	5	6.....	38
3	4	5	6	7.....	39
⋮					
36	37				

What scores are possible if: (a) both darts score single numbers?
(b) one dart scores single and the other double?
(c) one dart scores single and the other treble?
(d) both darts score double?

△ Try different dartboards:

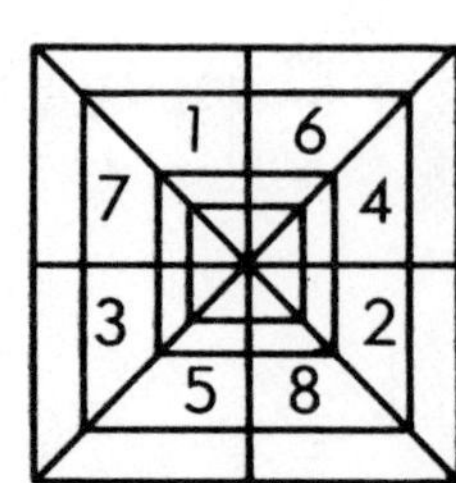

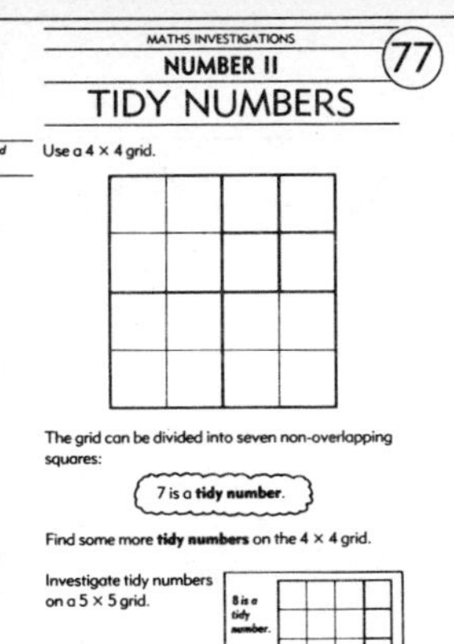

MATHS INVESTIGATIONS
NUMBER II
TIDY NUMBERS
77

Use a 4 × 4 grid.

The grid can be divided into seven non-overlapping squares:

7 is a **tidy number**.

Find some more **tidy numbers** on the 4 × 4 grid.

Investigate tidy numbers on a 5 × 5 grid.

8 is a tidy number.

MATHS INVESTIGATIONS

NUMBERS II

TIDY NUMBERS

77

1 cm Squared paper

△ 4 × 4 grid.

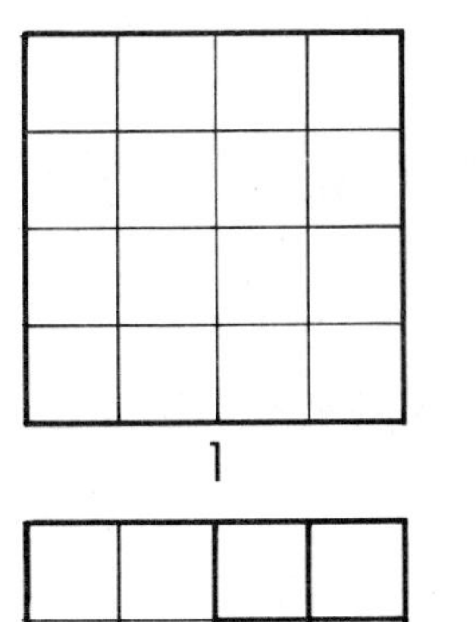

1

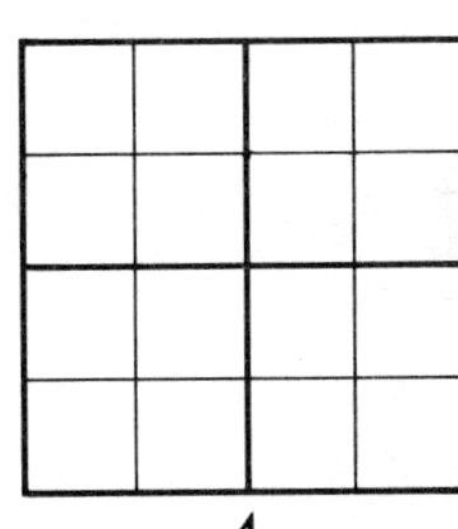

4

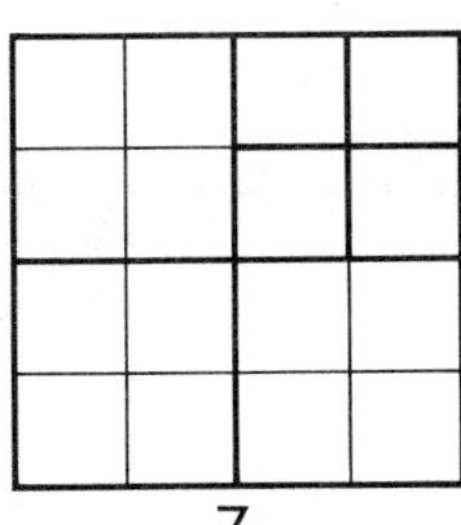

7

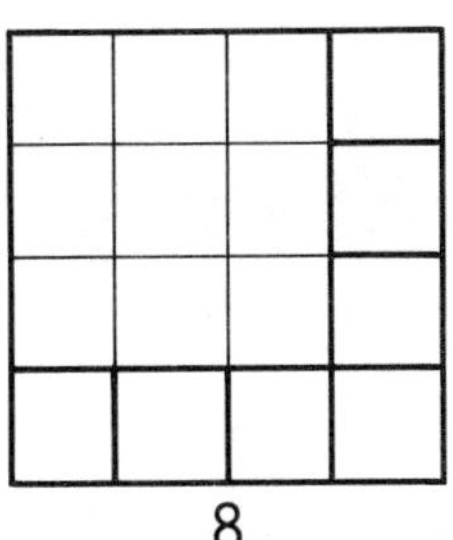

8

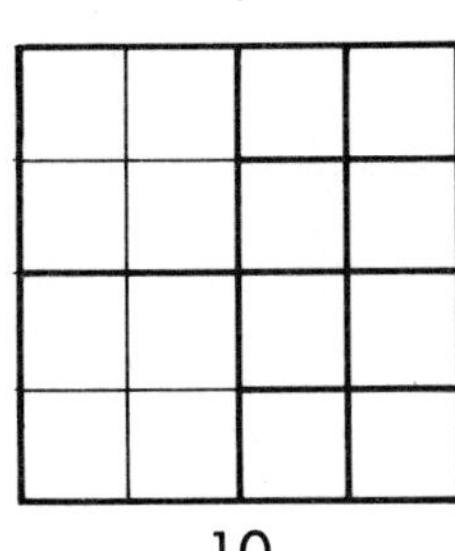

10

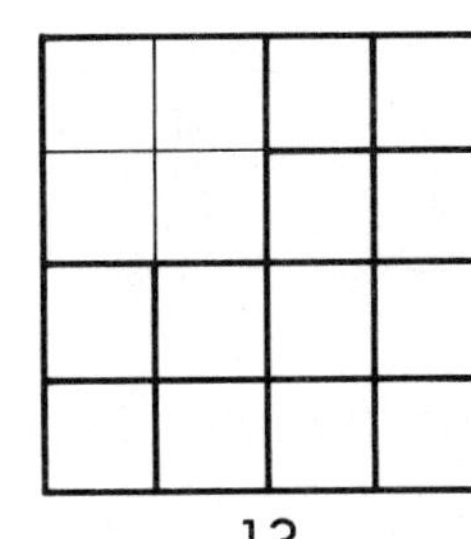

13

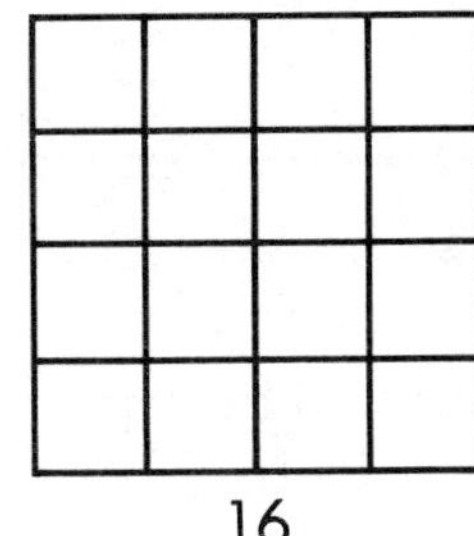

16

△ A 5 × 5 grid produces eleven tidy numbers:
1, 8, 10, 11, 13, 14, 16, 17, 19, 22 and 25

△ Try different sized square grids.

△ Try rectangular grids.

△ Consider dividing triangular grids into triangles.

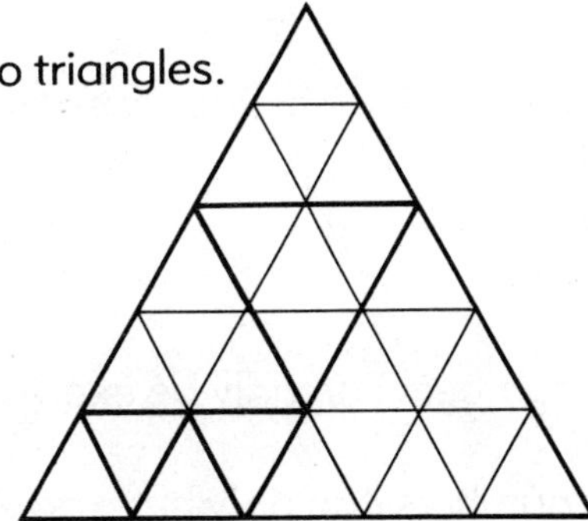

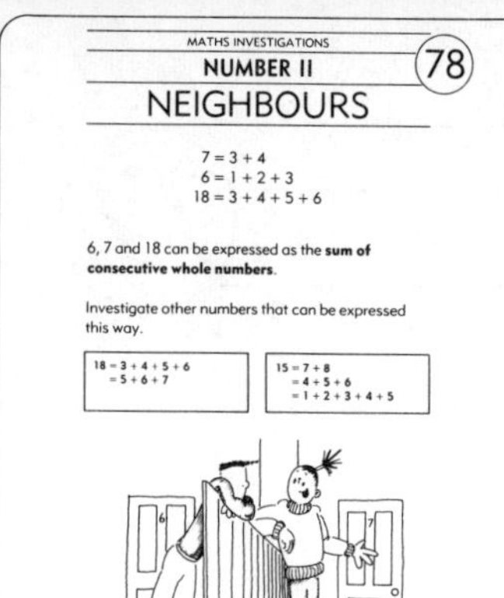

MATHS INVESTIGATIONS

NUMBER II (78)

NEIGHBOURS

△ Pupils can begin by making a list of counting numbers.
Expressions can then be sought at random until some patterns appear.

△

1		11	5 + 6
2		12	3 + 4 + 5
3	1 + 2	13	6 + 7
4		14	2 + 3 + 4 + 5
5	2 + 3	15	7 + 8, 4 + 5 + 6, 1 + 2 + 3 + 4 + 5
6	1 + 2 + 3	16	
7	3 + 4	17	8 + 9
8		18	5 + 6 + 7, 3 + 4 + 5 + 6
9	4 + 5, 2 + 3 + 4	19	9 + 10
10	1 + 2 + 3 + 4	20	2 + 3 + 4 + 5 + 6

△ Numbers which cannot be expressed as the sum of consecutive whole numbers are 1, 2, 4, 8, 16, 32 ... the powers of 2.

△ All odd numbers, except 1, can be expressed as the sum of two consecutive numbers.

△ All multiples of 3, except 3, can be expressed as the sum of three consecutive numbers.

△ All multiples of 5, except 5, can be expressed as the sum of five consecutive numbers.

△ Some numbers can be expressed in more than one way:

$$\begin{aligned} 30 &= 9 + 10 + 11 \\ &= 6 + 7 + 8 + 9 \\ &= 4 + 5 + 6 + 7 + 8 \end{aligned}$$

△ Which numbers can only be expressed in one way?

△ Which numbers can only be expressed in two ways?

MATHS INVESTIGATIONS

NUMBER II (79)

THREES AND FIVES

6, 8, 11 and 15 can be obtained by adding **threes** and **fives**:

6 = 3 + 3
8 = 3 + 5
11 = 3 + 3 + 5
15 = 5 + 5 + 5

Investigate other totals obtained by adding **threes** and **fives**.
Try adding twos and sevens.

Investigate for other pairs of numbers.

15 = 5 + 5 + 5
= 3 + 3 + 3 + 3 + 3

MATHS INVESTIGATIONS

NUMBERS II (79)

THREES AND FIVES

△ Pupils can start by writing a list of counting numbers and then write in the totals of threes and fives:

1		9	3 + 3 + 3
2		10	5 + 5
3	3	11	3 + 3 + 5
4		12	3 + 3 + 3 + 3
5	5	13	3 + 5 + 5
6	3 + 3	14	3 + 3 + 3 + 5
7		15	5 + 5 + 5
8	3 + 5	16	3 + 3 + 5 + 5

△ Which numbers cannot be found by adding threes and fives?

△ Which numbers can be expressed in more than one way?

△ Another approach is to make an addition table:

		0 × 3	1 × 3	2 × 3	3 × 3	4 × 3	5 × 3....
		0	3	6	9	12	15 ......
0 × 5	0	0	3	6	9	12	15
1 × 5	5	5	8	11	14	17	20
2 × 5	10	10	13	16	19	22	25
3 × 5	15	15	18	21	24	27	30
4 × 5	20	20	23	26	29	32	35
5 × 5	25	25	28	31	34	37	40
⋮	⋮						

△ What happens if the two numbers are both even?

△ What happens if one number is a multiple of the other?

△ What happens if the two numbers have no common divisor?

△ Suppose three numbers are available (e.g. 3, 4 and 8)?

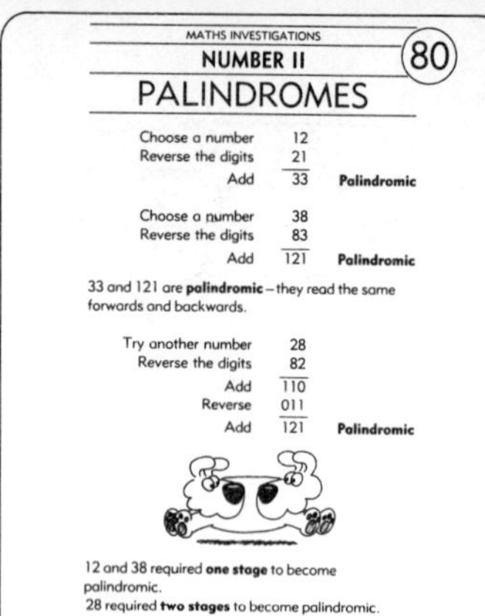

MATHS INVESTIGATIONS

NUMBER II (80)

PALINDROMES

Choose a number 12
Reverse the digits 21
Add 33 **Palindromic**

Choose a number 38
Reverse the digits 83
Add 121 **Palindromic**

33 and 121 are **palindromic** – they read the same forwards and backwards.

Try another number 28
Reverse the digits 82
Add 110
Reverse 011
Add 121 **Palindromic**

12 and 38 required **one stage** to become palindromic.
28 required **two stages** to become palindromic.

Investigate other numbers.

MATHS INVESTIGATIONS

NUMBER II (80)

PALINDROMES

△ Pupils can start by trying numbers at random.
Obviously a number and its reverse, e.g. 38 and 83, can become palindromic in the same number of stages.

△ three-stage palindrome

86
68
154
451
605
506
1111

four-stage palindrome

78
87
165
561
726
627
1353
3531
4884

△ Do all numbers result in a palindromic number?
Which numbers less than 100, become palindromic in one-stage, two-stages, three-stages ...?
Which number requires the most stages?

△

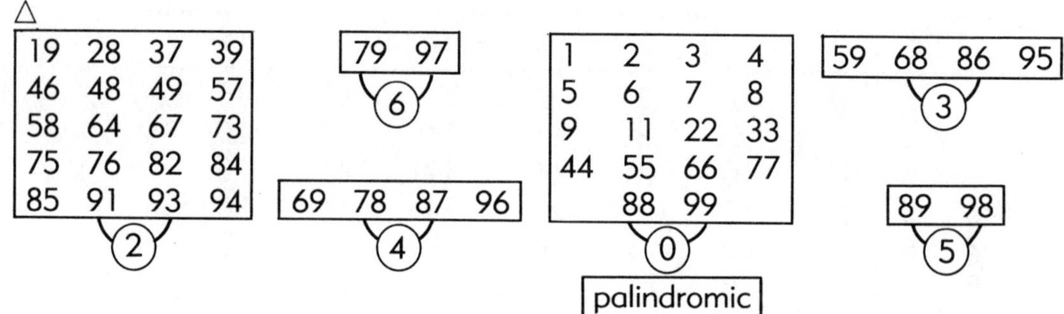

△ How many palindromic numbers can be found between 100 and 1000?

△ Find some palindromic words (e.g. noon, radar).

△ Find palindromic square numbers.

△ Find palindromic prime numbers.

Letts study aids

Revise Religious Studies

A complete revision course for O level and CSE

Reverend John P Mackney MA

Formerly Head of Religious Studies, Theale Green School, Berkshire

Charles Letts & Co Ltd
London, Edinburgh & New York

First published 1984
by Charles Letts & Co Ltd
Diary House, Borough Road, London SE1 1DW

Editor: Liz Davies
Design: Ben Sands & Anne Davison
Illustrations: Tek-Art

ISBN 0 85097 612 X

Printed by Charles Letts (Scotland) Limited

Preface

This book is designed to meet the needs of students taking a course of Religious Studies for O Level, GCE, CSE or Scottish O-Grade examinations. It is intended to be a commentary for constant reference and above all to be of particular use during revision.

Examination candidates should note that in both the Old Testament and New Testament papers, detailed knowledge of the text is of paramount importance and this can be acquired only by frequent reading; no commentary should be considered a substitute for this. The aim of this book is therefore to complement and illuminate the reader's existing knowledge and to provide those deeper insights into the meaning of the prescribed texts which are demanded of good candidates. Textual references throughout this commentary are from the New English Bible unless otherwise stated. (New English Bible, © 1970 by permission of Oxford and Cambridge University Presses.)

The section on Christian Social Responsibility is designed to introduce students to some of the basic facts and principles involved in the more common social problems of today, in the hope of stimulating thought and enabling the student to express an informed and balanced personal opinion on certain controversial social and moral problems.

The section on World Religions is intended to provide essential information about the tenets of belief and practices of each of the five religions covered.

Many teachers now find that they have to diversify and teach subjects in which they have no specialism. It is therefore hoped that non-specialist teachers, with little background knowledge of Religious Studies, who, for various reasons are required to teach the subject, will find in this commentary a useful source of information for the planning of interesting and relevant lessons.

In preparing this book I have received indispensable help and advice from Mr Roger Bates, Dr J G Harris and Mr Bruce Wallace and it is impossible for me adequately to express my gratitude to them. I am further indebted to Mrs Pat Rowlinson, Mrs Karen Sparrock and the editorial team of Charles Letts and Co Ltd for their encouragement and expertise.

I am grateful to the following examination boards for their permission to use questions from past examination papers:

Associated Examining Board (AEB)
Associated Lancashire Schools Examining Board (ALSEB)
East Anglian Examinations Board (EAEB)
East Midland Regional Examinations Board (EMREB)
Joint Matriculation Board (JMB)
London Regional Examinations Board (LREB)
Oxford Delegacy of Local Examinations (Oxford)
Oxford and Cambridge Schools Examination Board (O & C)
North West Regional Examinations Board (NWREB)
Northern Ireland Schools GCE Examinations Council (NIEC)
Northern Ireland CSE Examinations Board (NIEB)
Scottish Examination Board (SEB)
South East Regional Examinations Board (SEREB)
Southern Universities' Joint Board for School Examinations (SUJB)
University of Cambridge Local Examinations Syndicate (Cambridge)
University of London University Entrance and School Examinations Council (London)
Welsh Joint Education Committee (WJEC)
Yorkshire Regional Examinations Board (YREB)

John Mackney 1984

Contents

Section I
Introduction and guide to using this book

This book has been written for O-Level and CSE candidates to help them gain the knowledge and understanding which will enable them to approach the examination with greater confidence and to give their best performance.

1 Using the table of analysis of examination syllabuses

The table of analysis which can be found on page x shows how the core material relates to the specific syllabuses set by the examination boards.

Many syllabuses contain options so that candidates have a choice of topics; all the options have been included in the chart. Consequently you should check which options your school expects you to take and concentrate your work on these topics.

Syllabuses change from time to time so that the information provided may not remain accurate; if in doubt, consult your teacher or write to the appropriate examination board, the address of which you will find on page xii. The syllabus analysis chart has been divided into the sections of this book and the number of papers shown relates to the total number of papers for each examination.

2 Revision

(a) Remember that an indispensable requirement for the Bible papers is close knowledge of your prescribed texts, and that a major part of your revision for these papers must be regular reading of the relevant scriptures. You should be familiarizing yourself with this material from the beginning of the course.
(b) In conjunction with the text, study the explanatory notes in the relevant section of this commentary. Then, referring to the appropriate questions in Section 3 of the book, test your own knowledge by writing answers to the specimen questions on the topic you have been revising.
(c) The importance of beginning revision well in advance of your examination cannot be over-emphasized. The initial memory retention of new material is very short but can be dramatically increased by regular and systematic revision. It is essential to have a plan of revision to take account of the large amount of material that you will have to assimilate.
(d) A valuable revision technique is to write timed essays under simulated examination conditions.

3 Examination technique

(a) Read the instructions at the top of the paper carefully and ascertain the number of questions to be answered, whether one or more is compulsory and whether there is a limit to the number of questions to be answered in any one section of the paper.

(b) Read all the questions before deciding which to answer; then re-read them, ticking off those you intend to attempt.

(c) Plan each answer before you begin, and before leaving each essay answer, check the question to ensure that each part has been dealt with.

(d) The golden rule is to be relevant. You will be given no extra marks for displaying knowledge which is not strictly necessary.

(e) Apportion roughly equal periods of time to each question since you cannot do justice to yourself if you fail to complete the paper.

(f) Try to complete your paper so as to allow sufficient time for reading it through to check spelling, punctuation, grammar and legibility. Badly written and untidy answers make unnecessary work for examiners and give a poor impression of your work.

Table of analysis of examination syllabuses

So many options are offered by the various examination boards that it is important for students to consult the syllabus of any examination they are entering.

	AEB Syllabus 1 Section A	UCLES	JMB Syllabus C	London Syllabus 560 Section 2	Oxford Syllabus 2841/3	Oxford and Cambridge Syllabus 2600/1	SUJB Section B1	SEB Section A	NIGCE Section 1
Level	O	O	O	O	O	O	O	O	O
Theory Papers	2	1	1	1	1	1	1	1	2
No. Hours	40 + 2	2½	2½	2½	2½	2½	2½	2½	2 + 2
Teacher Assessment	Nil	Nil	Nil	Nil	Nil	Nil	Nil	25%	Nil
Part I The Synoptic Gospels									
1 The Background to the Gospels	●	●	●	●	●	●	●	●	●
2 The Background to the Synoptic Gospels	●	●	●	●	●	●	●	●	●
3 The Characteristics of the Gospels	●	(●)	(●)	(●)	(●)	(●)	(●)	(●)	●
4 The Nativity	●	●	●	●	●	●	●	●	●
5 John the Baptist and Jesus	●	●	●	●	●	●	●	●	●
6 The Ministry in Galilee	●	●	●	●	●	●	●	●	●
7 The Parables	●	●	●	●	●	●	●	●	●
8 The Miracles	●	●	●	●	●	●	●	●	●
9 Important Events in the Life of Jesus	●	●	●	●	●	●	●	●	●
10 The Passion Narrative	●	●	●	●	●	●	●	●	●
11 A Selection of Important Teachings	●	●	●	●	●	●	●	●	●

	AEB Paper 2	UCLES Options 2041/4,5,6			JMB Syllabus B	London Section 1 Option A or B		Oxford Options 1 & 2		Oxford and Cambridge Paper 4	SUJB Section A	SEB Section A!	NIGCE Paper 1 Sections 2 & 3	
Level	O	O			O	O		O		O	O	O	O	
Theory Papers	1	1			1	1 (2 of 7 options)		1		1	1 (2 Sections)	1	2	
No. Hours	2	1½			2½	2½		2½		2½	2½	2½	2+2	
Teacher Assessment	Nil	Nil			Nil	Nil		Nil		Nil	Nil	25%	Nil	
Part II The Old Testament		(4)	(5)	(6)		A	B	1	2				2	3
12 Genesis		●			●	●		●					●	
13 The Exodus	(●)	●			●	●		●			●	●	●	
14 The Settlement in Canaan	(●)	●			●	●		●		(●)	●			
15 The Book of Samuel	●	●			●	●		●		●	●			●
16 The Books of Kings	●	●	●		●	●	●	●	●	●	●			●
17 The Minor Prophets	●		●		●		●		●	●	●	●		●
18 The Major Prophets, Isaiah	●		●	●	●		●		●	●		●		●
19 The Major Prophets, Jeremiah	●		●	●	●		●		●		●	●		
20 The Major Prophets, Ezekiel	●			●					●		●			
21 The Second or Deutero-Isaiah	●				●							●		

	AEB Paper 2 Section B3	UCLES	JMB Syllabus A	London Syllabus 856 Part B	Oxford Alternative Paper 6	Oxford and Cambridge	SUJB Section C	SEB Section Ac	NIGCE Section B
Level	O		O	O	O		O	O	O
Theory Papers	2		1	1	1		1	1	2
No. Hours	⅔ + 2		2½	3	2½		2½	2½	2+2
Teacher Assessment	Nil		Nil	Nil	Nil		Nil	25%	Nil
Part III Christian Social Responsibility									
22 Introduction	●		●	●	●		●	●	●
23 The Family	●		●	●	●		●	●	●
24 A Christian View of Work	●		●	●	●		●	●	●
25 Money	●		●	●	●		●	●	●
26 The Christian and the Community	●		●	●	●		●	●	●
27 Discrimination	●		●	●	●		●	●	●
28 Crime and Punishment	●		●	●	●		●	●	●
29 Drug Abuse	●		●	●	●		●	●	●
30 Medical Ethics	●		●	●	●		●	●	●
31 The Population Explosion	●		●	●	●		●	●	●

	AEB Syllabus 11	UCLES Paper 2042/4	JMB	London 560 Section 7 and 856	Oxford	Oxford and Cambridge	SUJB	SEB Section B	NIGCE Paper 2 Section A Theme 6
Level	O	O		O				O	O
Theory Papers	1	1		1				1	2
No. Hours	2½	2½		2½				2½	2+2
Teacher Assessment	25%	Nil		Nil				25%	Nil
Part IV World Religions									
32 Judaism	●	●		●				●	●
33 Islam	●	●		●				●	●
34 Hinduism	●	●		●				●	
35 Sikhism	●	●		●					
36 Buddhism		●		●					●

NICSE Paper 2	WJEC (GCE) Paper Od	WJEC (CSE)	ALSEB	EAEB	EMREB	LREB Syllabus A	NREB	NWREB	SREB	SEREB	SWEB	WMEB	YHREB YREB	WY&LREB	
CSE	O	CSE	CSE	CSE	CSE	CSE	CSE	CSE	CSE	CSE	CSE	CSE	CSE	CSE	
2	2	1	1	2	2	1	2	2	1	2	2	2	1	2	
2 + 2	2 + 2	2¼	2½	2 + 2	1½ + 1	2½	1¾ + 1¼	1¾ + 1¾	2	¾ + 1½	2 + 1¼	2¼ +2¼	2½	2 + 2	
Nil	Nil	Optional Project	20%	Optional Project	30%	Nil	20%	Nil	30%	20%	Nil	Optional special study 50%	Nil	Nil	
															I
•	•	•	•	•	•	•	•	•	•	•	•	•	•	•	1
	•	•	•		•	•	•	•	•	•	•	•	•	•	2
	•						•								3
	•	•	•	•	•	•	•	•	•	•	•	•	•	•	4
•	•	•	•	•	•	•	•	•	•	•	•	•	•	•	5
•	•	•	•	•	•	•	•	•	•	•	•	•	•	•	6
•	•	•	•	•	•	•	•	•	•	•	•	•	•	•	7
•	•	•	•	•	•	•	•	•	•	•	•	•	•	•	8
•	•	•	•	•	•	•	•	•	•	•	•	•	•	•	9
•	•	•	•	•	•	•	•	•	•	•	•	•	•	•	10
•	•	•	(•)	(•)	(•)	(•)	(•)	(•)	(•)	(•)	(•)	(•)	(•)	(•)	11

NICSE Paper 1	WJEC(GCE) Paper 2 Options a,b&c	WJEC (CSE) Section B Options 1&2	ALSEB	EAEB Paper 2 Option D	EMREB Paper 2 Option 2	LREB Syllabus A	NREB Paper 2A	NWREB	SREB	SEREB	SWEB	WMEB Paper 2 Option B	YHREB YREB	WY&LREB Paper 2 Option B	
CSE	O	CSE		CSE	CSE	CSE	CSE					CSE		CSE	
2	1	1		2	2	1	2					2		2	
2+2	2	2¼		2 + 2	1½ + 1	2½	1¾ + 1¼					2¼ + 2¼		2 + 2	
Nil	Nil	Nil		Nil	30%	Nil	20%					Optional to Paper 2 Special Study 50%		Nil	
	a b c	1 2													II
•	•	•		•	(•)	•						•		•	12
•	•	•		•	(•)	•	•					•		•	13
•	•	•												•	14
•	•	•		•		•						•		•	15
•	• •	• •		•		•						•		•	16
	•	•		•		•	•					•		•	17
	•	•		•		•	•					•		•	18
	• •	•		•	(•)	•	•					•			19
	•													•	20
	•			•		•	•					•			21

NICSE Paper 2 Topics	WJEC (GCE)	WJEC (CSE)	ALSEB Section B	EAEB Paper 2 Option F	EMREB Paper 2 Option 1	LREB Syllabus B	NREB Paper 1 Section B	NWREB Syllabus B Option 1	SREB Alternative 2 & 3	SEREB Section 1	SWEB Paper 2 Part A	WMEB Paper 2A	YHREB YREB	WY&LREB Paper 2 Theme A	
CSE			CSE	CSE	CSE	CSE	CSE	CSE	CSE	CSE	CSE	CSE	CSE	CSE	
2			1	2	2	1	2	2	1	1	2	2	1	2	
2+2			2½	2 + 2	1½ + 1	2½	1¾ + 1¼	1¾ + 1¾	2	2¼	2 + 1¼	2¼ + 2¼	2½	2 + 2	
Nil			Nil	Nil	Nil	Nil	Nil	Nil	30%	50%	Nil	50% Optional to Paper 2	50%	Nil	
															III
•			•	•	•	•	•	•	•	•	•	•	•	•	22
•			•	•	•	•	•	•	•	•	•	•	•	•	23
•			•	•	•	•	•	•	•	•	•	•	•	•	24
•			•	•	•	•	•	•	•	•	•	•	•	•	25
•			•	•	•	•	•	•	•	•	•	•	•	•	26
•			•	•	•	•	•	•	•	•	•	•	•	•	27
•			•	•	•	•	•	•	•	•	•	•	•	•	28
•			•	•	•	•	•	•	•	•	•	•	•	•	29
•			•	•	•	•	(•)	•	•	•	•	•	•	•	30
•			•	•	•	•		(•)	•	•	•	•	•	•	31

NICSE	WJEC (GCE)	WJEC (CSE)	ALSEB Section B Part 3	EAEB Option G	EMREB Option 3	LREB Syllabus C	NREB Paper 2C	NWREB Syllabus B Option 2	SREB	SEREB Option B	SWEB Paper 2 Part B	WMEB Syllabus B	YHREB YREB	WY&LREB Paper 2C	
			CSE	CSE	CSE	CSE	CSE	CSE		CSE	CSE	CSE		CSE	
			2	2	2	1	2	1		2	2	2		2	
			¾ + 1¾	2 + 2	1½ + 1	2½	1¾ + 1¼	1¾		¾ +1½	2 + 1¼	2¼ +2¼		2+	
			20%	Nil	30%	Nil	20%	50%		20%	Nil	Nil		Nil	
															IV
			•	•	•	•		•		•	•	•		•	32
			•	•	•	•	•	•		•	•	•		•	33
			•	•	•	•		•		•	•	•			34
				•	•	•				•	•	•			35
					•			•		•					36

Examining Boards: Addresses

General Certificate of Education–Ordinary Level (GCE)

AEB	Associated Examining Board Wellington House, Aldershot, Hampshire GU11 1BQ
Cambridge	University of Cambridge Local Examinations Syndicate Syndicate Buildings, 17 Harvey Road, Cambridge CB1 2EU
JMB	Joint Matriculation Board Manchester M15 6EU
London	University of London, School Examinations Department 66–72 Gower Street, London WC1E 6EE
NIEC	Northern Ireland Schools GCE Examinations Council Beechill House, 42 Beechill Road, Belfast BT8 4RS
Oxford	Oxford Delegacy of Local Examinations Ewert Place, Summertown, Oxford OX2 7BX
O and C	Oxford and Cambridge Schools Examination Board 10 Trumpington Street, Cambridge; *and* Elsfield Way, Oxford OX2 8EP
SUJB	Southern Universities' Joint Board for School Examinations Cotham Road, Bristol BS6 6DD
WJEC	Welsh Joint Education Committee 245 Western Avenue, Cardiff CF5 2YX

Certificate of Secondary Education

ALSEB	Associated Lancashire Schools Examining Board 77 Whitworth Street, Manchester M1 6HA
EAEB	East Anglian Examinations Board The Lindens, Lexden Road, Colchester, Essex CO3 3RL
EMREB	East Midland Regional Examinations Board Robins Wood House, Robins Wood Road, Apsley, Nottingham NG8 3RL
LREB	London Regional Examinations Board (*formerly:* MREB Middlesex Regional Examinations Board) Lyon House, 104 Wandsworth High Street, London SW18 4LF
NIEB	Northern Ireland CSE Examinations Board Beechill House, 42 Beechill Road, Belfast BT8 4RS
NREB	North Regional Examinations Board Wheatfield Road, Westerhope, Newcastle upon Tyne NE5 5JZ
NWREB	North West Regional Examinations Board Orbit House, Albert Street, Eccles, Manchester M30 0WL
SEREB	South East Regional Examinations Board Beloe House, 2–4 Mount Ephraim Road, Royal Tunbridge Wells, Kent TN1 1EU
SREB	Southern Regional Examinations Board 53 London Road, Southampton SO9 4YI
SWEB	South Western Examinations Board 23–29 Marsh Street, Bristol BS1 4BP
WJEC	Welsh Joint Education Committee 245 Western Avenue, Cardiff CF5 2YX
WMEB	West Midland Examination Board Norfolk House, Smallbrook Queensway, Birmingham B5 4NJ
*WYLREB**	West Yorkshire and Lindsey Regional Examining Board Scarsdale House, 136 Derbyshire Lane, Sheffield S8 8SE
*YREB**	Yorkshire Regional Examinations Board 31–33 Springfield Avenue, Harrogate, North Yorkshire HG1 2HW

*Yorkshire and Humberside Regional Examinations Board, at the *YREB* address, now embraces *WYLREB* and *YREB*

Section II Core units 1–36

Part I The Synoptic Gospels

1 The Background to the Gospels

1.1 The Political Background

Herod the Great

At the time of Jesus' birth, Palestine had been governed for some thirty years as a Roman protectorate. Its ruler, Herod the Great (37–4 BC), was a tribute-paying vassal of Rome. It was his policy ruthlessly to preserve his throne against all comers, and to this end to be on good terms with whoever ruled in Rome. A considerable proportion of Jewish opinion was hostile to him because:

1 He was an Idumean and had no Jewish blood. According to the Jewish Law he therefore had no claim to the royal throne, for Deuteronomy 17:15 explicitly states that it was forbidden to accept as king anyone who was not of pure Jewish blood. Herod was therefore regarded as an illegitimate usurper. This applied also to his heirs.

2 He had been thrust upon the nation by the Roman conqueror.

3 His observance of the Law of Moses was perfunctory and he himself claimed to be 'more a Greek than a Jew'. He antagonized the people by promoting Greek culture and building a theatre and amphitheatre close to the Holy City.

Herod was a very able man, but history condemns him as a cruel tyrant, and the story of the massacre of the children of Bethlehem is quite typical of his conduct.

After Herod the Great

On the death of Herod, the Emperor Augustus divided the kingdom among Herod's sons. They were not, however, granted the title of king which their father had enjoyed as a mark of the special favour of Rome: Archelaus was an ethnarch or ruler of the nation, while Philip and Antipas were tetrarchs or rulers of a fourth part. (In Chapter 6:14 of his Gospel, Mark mistakenly refers to Herod Antipas as 'King Herod'.) Archelaus was deposed after ten years and his territory was made into a province administered by a Roman procurator. Philip was established in the hill country of the north and took no active part in Palestinian affairs, so that at the time of the ministry of Jesus, Palestine was effectively governed by two men, the tetrarch Herod Antipas in Galilee and a Roman procurator in Judaea and Samaria. In 26 AD the proconsul was Pontius Pilate.

Herod Antipas

It is Herod Antipas to whom the Gospel refers as the one who provoked the public denunciation by John the Baptist for his marriage to Herodias, the divorced wife of his half-brother. Consequently he had John imprisoned and later beheaded. This was the Herod to whom Pilate sent the accused Jesus during his trial, on the pretext that because Jesus was a Galilean, jurisdiction belonged to the tetrarch of Galilee.

Tax collecting

The Roman system of taxation was established in Palestine in 6 AD and under it the right to impose taxes in a given area was farmed out to the highest bidder, who agreed to pay the state a fixed sum for the privilege. The right was further contracted out, at a profit, to chief tax gatherers, and finally to an army of local men who actually collected the tax, so that the total sum collected was much greater than the original Roman demand. Zacchaeus was a chief tax

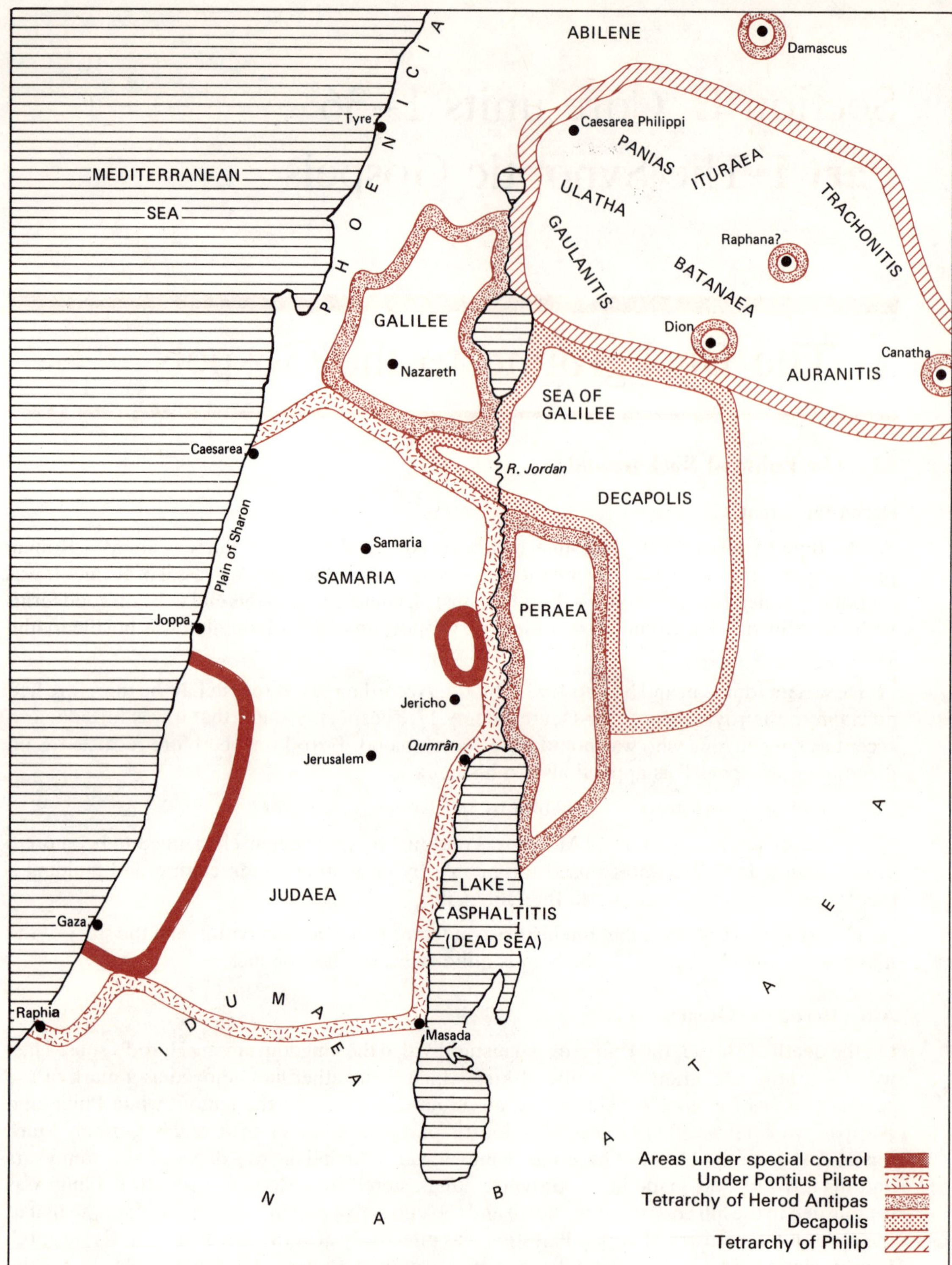

Fig. 1.1 Political map of Palestine during the time of Jesus

gatherer, for he is described in Luke 19:2 as 'superintendent of taxes, and very rich', while Matthew appears to have been a local tax gatherer, for he is described as being 'at his seat in the custom-house' (Matthew 9:9) when Jesus called him. The system was convenient to Rome but extremely inefficient because it made possible every form of dishonesty and extortion. **The publicans** were not only hated for their greed, but also because they served the conquerors. The Gospels reveal that the Jews associated the publicans with the worst kind of sinners.

The temper of the time

From the days of Herod the Great to the destruction of Jerusalem in 70 AD every procurator in succession had to deal with chronic agitation: violent uprisings, each in turn harshly put down, subsided, only to erupt again. **Palestine was a troubled province**, at best enjoying only an uneasy peace. The Jews were conscious of their unique status as the people of God, and this made the rule of foreigners unbearable to them. The Romans were not only conquerors but abominable heathens, and according to the Law, unclean. It is to be noted that at the trial of Jesus, Pilate had to come out of the Praetorium to speak to the Jews, for they themselves 'stayed outside the headquarters to avoid defilement, so that they could eat the Passover meal' (John 18:28).

1.2 The Religious Background

The Sanhedrin

Subject to the overall authority of the Romans or the reigning Herod, the Jews were free to

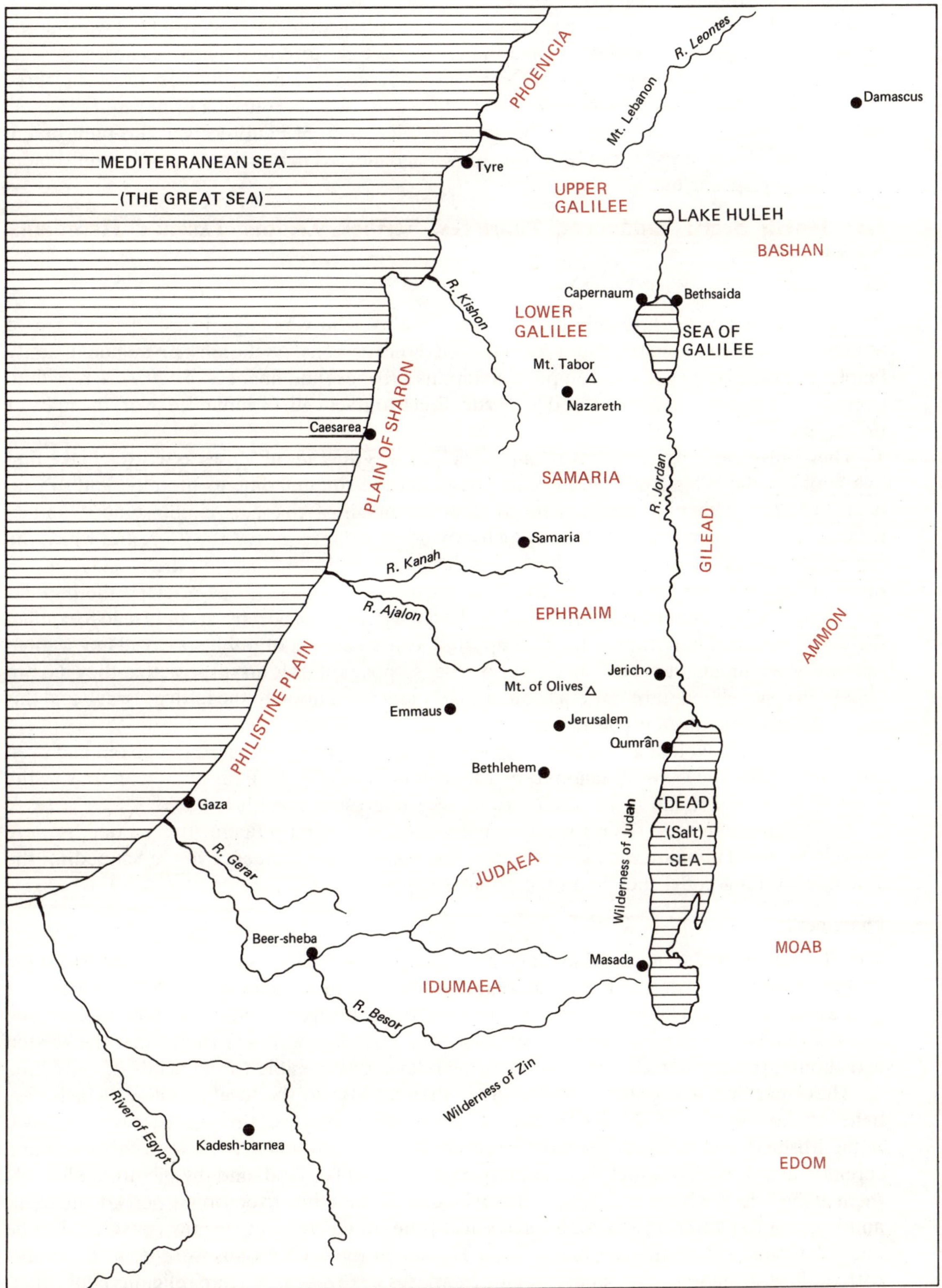

Fig. 1.2 Geographical map of Palestine during the time of Jesus

administer their own religious affairs, and did so under the High Priest and the Sanhedrin or Great Council, **the supreme Jewish court of law**. It dealt with cases of the utmost gravity and was also the final court of appeal: it was the supreme political and theological authority. The Jews believed that God alone ruled the nation and did so through His representatives on earth, mainly the Sanhedrin and the High Priest.

The Sanhedrin traced its origin to the seventy men Moses appointed to assist him in judging the people, and in keeping with the tradition there were seventy members (all co-opted) plus the High Priest. Membership was drawn mainly from the two opposing parties, **Sadducees** and **Pharisees**. Everything to do with religion came under the control of the Sanhedrin, which is the same as saying that for the Jewish communities it ruled everything. It made the laws and its own police enforced them.

In criminal cases it could pass a sentence of death but it could not be carried out without the consent of the Roman governor. It had the exclusive right to judge religious causes.

Every Jewish community had its own 'little Sanhedrin' normally consisting of three members. The most severe punishment it could impose was a scourging of not more than 39 strokes; to award more was regarded as liable to kill a man. **St Paul** received this punishment several times. It could also expel Jews from the synagogue, a very much more severe punishment than it might appear today.

1.3 Jewish Sects: Sadducees, Pharisees, Scribes, Zealots, Essenes, Herodians. Samaritans

Sadducees

The name Sadducee is probably derived from Zadok who was High Priest in the reign of Solomon, and from whom the Sadducees claimed direct descent. As **the most prestigious priestly family**, they had control of the Temple worship, its sacrifices and finances. They were recruited from the rich priestly aristocracy and had little direct contact with or influence over, the mass of the people.

Their only rule of religious, moral and social life was the **Law of Moses** as written in the first five books of the Bible (the **Pentateuch**). They took the commandments literally, and saw no need to interpret them or adapt them to meet the needs of changing circumstances, and so refused to be bound by the 'oral law' which expanded and interpreted the Law, and which the Pharisees believed was also given by Moses. They laid little emphasis on the message of the prophets. They were **ultra conservative** and distrusted new ideas. They rejected the popular belief in a Messiah whom God would send to liberate His people, because it could lead to conflict with the Romans whom they actively supported as the keepers of public order. They wanted nothing which might pose a threat to their privileged position. It is no surprise that they did not believe in resurrection since it was a comparatively recent addition to the faith of Israel, and this is a good example of their conservatism.

It is understandable that they were opposed to Jesus for he had dared to expand and even amend the Law; they saw his claim to be Messiah as both blasphemous and dangerous to the security of the state, and it was therefore in their interest to have him killed.

Unlike the Pharisees, whose whole activity was centred on religion, the Sadducees were actively involved in politics, and though they had considerable influence in the Sanhedrin, the Jewish governing body, they had little popular support.

Pharisees

The Pharisees were far more influential among the people than the Sadducees, and formed the **progressive wing of Judaism**. For them, religious life centred upon the study and observance of the Law of Moses, and they were closely connected to the synagogue. Since the Law could not be expected to make provision for every situation, the Pharisees were convinced that the written text should, by careful rethinking, be made as relevant as possible to the needs of daily life.

They appealed, not only to the Law (Torah), but also to the **'oral tradition'** which they believed was also given by Moses and was equally binding. This tradition was later written down in the Mishnah and Talmud. They observed no distinction in practice between the moral and ceremonial law, both of which were held to express the will of God, and the picture we have of them in the New Testament is of hypocrites who devote their time to enforcing petty distinctions and keeping the letter, rather than the spirit, of the Law. This view is now considered to be one-sided, and reflects the fact that in New Testament times Christians were in bitter conflict with the Jewish leaders. The majority of the Pharisees were men of sincere religious conviction, and it must not be forgotten that **modern Judaism owes almost everything to Pharisaism**.

The ideal of Pharisaism was to 'build a fence around the Law', to hedge the commandments with a multiplicity of minor related laws derived from the main principles, and so to protect them. They looked forward to the coming of the Messiah whom God would send to set up His Kingdom on earth and give His 'chosen people' their rightful place among the nations. The coming would be in God's own time, but the day would be, to a large extent, the reward for Israel's faithfulness in keeping the Law; the long-awaited event could be hastened by observing the Law in all its parts, and encouraging their fellow Jews to do the same. They separated themselves from all who did not keep the Law, unfaithful Jews (sinners) as well as gentiles.

Some leading Pharisees believed that even the heathens were being called by God to accept the Law, and it is clear from Matthew 23:15 that the Pharisees were actively engaged in missionary work among the gentiles.

Unlike the Sadducees, they believed in resurrection, and Paul was able to exploit this divergence of belief at his trial before the Sanhedrin (Acts 23:6). A good example of Jesus' view of the Sadducees can be found in Luke 18:10.

Scribes

The Scribes belonged mainly, though not exclusively, to the Pharisees. As the name suggests, they were originally responsible for making faithful copies of the Scriptures, and for guarding the text from corruption. As a result they became lawyers and authorities on the Scriptures. In the Gospels the terms 'Scribe' and 'lawyer' are synonymous. As administrators of the Law they were represented in the Sanhedrin. The parable of the Good Samaritan (Luke 10:25) is Jesus' answer to a lawyer's question.

Zealots

Sometimes described as the extreme left wing of Pharisaism, the Zealots differed from all other Jewish sects by their **violent and uncompromising nationalism**. Whereas each sect believed that the true ruler of Israel was God Himself, the Sadducees could compromise with and accommodate the Romans, the Pharisees could ignore them, but the Zealots openly hated them and believed that national liberation could only be achieved by violence. They are known as **'Sicarii', dagger-men** (hence possibly Judas 'Iscariot'), and were involved in terrorist activity against the Romans and those who co-operated with them. They were fanatically brave and were mainly responsible for the atmosphere of deep-seated discontent and agitation which finally led to the great revolt of 66 AD and the destruction of Jerusalem in 70 AD. They were religiously minded fanatical patriots. One of the disciples of Jesus, Simon (not Peter), was formerly a Zealot (Mark 3:18).

Essenes

This sect is not mentioned in the New Testament and until the discovery of the **Dead Sea Scrolls** and the Qumrân monastery little detailed information about them was available. They took their inspiration from the 'Teacher of Righteousness', a religious leader of the second century BC who believed that God had revealed to him the mysteries of Old Testament prophecy and given him the secrets of the 'Divine Plan' for the world. He founded a religious community in the desert, the purpose of which was to make it possible to follow the revealed will of God. The Essenes were **'volunteers for holiness'** who led lives of rigorous discipline, far outstripping even those of the Pharisees, in order to prepare for the coming of the Lord. This great event would take place as a result of their diligent study and observance of the Law.

The life of the community was to be an offering to God to make amends for the wrong-doings of the nation; they believed that by the purity of their lives, their sufferings, their submission to severe discipline, and their single-minded devotion to the Law, their offering would be acceptable to God as an atonement (at-one-ment).

Since the discovery of the Scrolls there has been a great deal of speculation among scholars about the possible influence of the Essenes on John the Baptist and even upon Jesus himself. There are, however, fundamental differences to be accounted for in each case. Some scholars hold that it was from this particular sect of Judaism that Christianity evolved.

Herodians

The Herodians were members of a very small party who saw hope for the future of Israel in the return of the rule of the Herodian family to all the Jewish territories. Mark 3:6 and 12:13 record that they joined with the Pharisees in attacks on Jesus, a very surprising alliance since they hated each other.

The sects were in theory, exclusive of each other, but they differed only within Judaism itself which could accommodate such widely differing beliefs because Jewish orthodoxy was largely a matter of practice rather than belief. Loyalty to the Law was the principle that firmly held together the diverse sects within Judaism.

Samaritans

Since the Samaritans were not recognized by the Jews as true Israelites, they cannot be regarded as a sect of Judaism.

The **enmity between Jews and Samaritans** was many centuries old. On the death of Solomon, circa 935 BC his kingdom was divided into two parts, the tribes of Judah and Benjamin in the south adhering to Rehoboam, Solomon's heir, while the ten northern tribes made Jeroboam, who was not descended from David, their king. The two factions were often at war. After the city of Samaria was built by King Omri, circa 880 BC, it was always considered to be a rival to the southern capital of Jerusalem.

In 721 BC the northern kingdom was conquered by the Assyrians who colonized it with pagans, and as a result of inter-marriage, a population of mixed race was produced. At first, heathen elements influenced worship, but eventually the Samaritans became as opposed to idolatry as the Jews, and accepted the five books of Moses (the Pentateuch) as their Scripture. They rejected the other Jewish books.

During the reign of Alexander the Great, circa 333 BC, a **Temple** was begun on Mount Gerizim to rival that at Jerusalem. The head of the Samaritan community, as of the Jewish, was a High Priest claiming descent from Aaron, the first High Priest. Sacrifices were made, and prayer was offered, to Yahweh, the one true God. As the woman of Samaria in John 4:19 implies, the Samaritans believed that their Temple was the only legitimate one.

At the time of Jesus there was bitter hatred between the groups. An ancient Jewish writing states that 'a piece of bread given by a Samaritan is more unclean than swine's flesh'. Against this background of hatred, Jesus dared to speak of the grateful Samaritan leper (Luke 17: 11-19) and the 'Good Samaritan' (Luke 10:30-37).

1.4 The Temple

The Temple, according to Jewish tradition, originated in the 'Tabernacle' or 'Tent of Meeting', the portable shrine said to have been made under the direction of Moses and which accompanied Israel in their forty years of wandering in the wilderness. It **symbolized the presence of God in the midst of His people**.

King David was responsible for the idea of a permanent building, though the first Temple dates from the reign of his son Solomon, circa 970–930 BC. This Temple in Jerusalem became the central shrine of the Jewish religion, and there alone could sacrificial worship be offered. The Babylonians destroyed the first Temple in 586 BC, but it was rebuilt on the return of the Jews from exile in 520 BC. After suffering various vicissitudes, including desecration at the hands of the Greek King Antiochus Epiphanes in 167 BC, a new and more splendid building was begun by Herod the Great in 20 BC. This was the Temple Jesus knew, and it was not completed until 64 AD, about 30 years after his death.

The Temple areas

Herod greatly increased the Temple area by enlarging the platform on which it was built. There were several open spaces or courts of increasing exclusiveness as the Temple building was approached. The outer area was the **Court of the Gentiles**, which, as the name implies, was a public place and a market where traders sold the sacrificial doves and animals of approved quality. Here also, pilgrims could change the many ceremonially 'unclean' foreign currencies into the sacred Temple coinage for the payment of Temple dues.

Next came the **Court of the Israelites** where Jews only were allowed, and there was an inscription in Latin and Greek forbidding the entrance of gentiles on pain of death. (This still exists.) There was a separate **Court of the Women** on this level.

The most exclusive area was the **Court of the Priests**: here was the altar, a huge block of stone on which animals were sacrificed.

On the highest level was the Temple building. It was not built to accommodate the worshippers, who gathered outside in the various courtyards, but rather as a dwelling of God. A

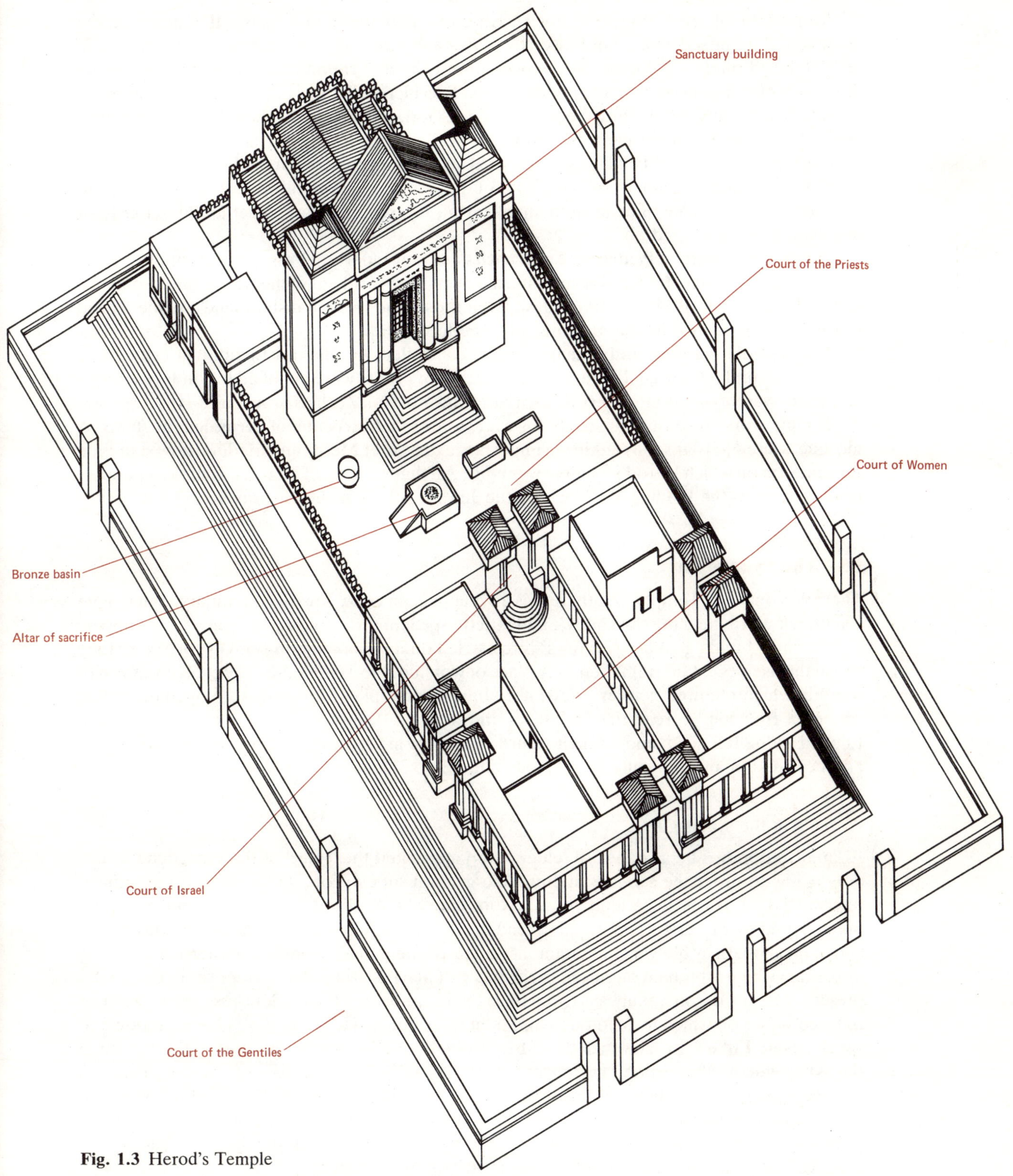

Fig. 1.3 Herod's Temple

porch and doorway led to the sanctuary which only the officiating priests had the right to enter. There stood the table for the Shew-Bread, the Seven Branched Candlestick and the Altar of Incense. A curtain covered the entrance to the innermost shrine, the **Holy of Holies**, which was completely empty, with no statue, no symbol, only the bare rock on which the High Priest stood on a single day in the year, the Day of Atonement.

Worship in the Temple

The importance of the Temple for Jews can hardly be overestimated. Central to its worship was sacrifice. Of special importance was the daily morning and evening sacrifice, both of which took the form of the offering of incense, the sacrifice of a lamb without blemish, the meal offering made with flour mixed with oil, followed by prayer and praise. Private offerings were also made daily, (Herod offered 300 oxen on the occasion of the dedication of his Temple), and many thousands of offerings were made on feast days. **Sacrifices**, whether bloodless or bloody, were intended as a means of obtaining the forgiveness of God, though ideally they should have been a means by which the sinner expressed in a practical way his sorrow for his sins.

The climax of the sacrificial system was the **Day of Atonement** when the High Priest entered the Holy of Holies, the innermost sanctuary of the shrine, the symbolical dwelling of God, and offered the blood of sacrifice as atonement for all the sins of Israel, whether consciously or unconsciously committed.

Three times a year, at **Pentecost, Tabernacles** and particularly at **Passover**, vast multitudes of Jewish pilgrims came to Jerusalem from all over the world, and the scene could only be compared with that of Mecca. Every Jew, even those who lived in distant lands, made a fixed yearly contribution to the maintenance of Temple worship.

The **Temple services and finances** were controlled by the aristocratic priestly caste (Sadducees) under the High Priest. A large number of priests were qualified to conduct services and these were divided into 24 courses. (Note the story of Zechariah in Luke 1:1-25.)

Temple worship ceased, probably for ever, after the **destruction of Jerusalem in 70 AD**. A mosque, sacred to Islam and ranking in importance only after Mecca and Medina, now stands on the platform on which Herod's Temple was built. All that remains of Herod's Temple is a course of huge stones (**the Wailing Wall**) where the Jews regularly meet for prayer.

1.5 The Synagogue

There was only one Temple, but there was a synagogue in every Jewish community. Synagogue is a Greek word meaning **'assembly'**, and strictly speaking it is a congregation of worshippers rather than a building. When the Jews were carried off into captivity in Babylon in 586 BC they found themselves cut off from their only place of worship, the Temple, so they evolved a form of worship which required neither priest, nor altar, nor temple. They gathered together 'by the waters of Babylon' to read the Scriptures and to pray. They found that synagogue worship fulfilled a long-felt need, and on their return from exile in 520 BC synagogue worship continued and became a major influence in religious life.

Worship in the synagogue

Administration was by a council of 'elders' who appointed the 'ruler' of the synagogue whose duty it was to order the services and provide general supervision. There was also an official referred to by Luke as 'the attendant', who took care of the rolls of Scripture and presented the one required to the person appointed to read the lesson. He also signified the beginning and the end of the sabbath by blowing the shofar, a wind instrument made from a ram's horn, a reminder of the ram which Abraham sacrificed in the place of his son, Isaac. The reader could be anyone present. The last lesson was always taken from the Prophets and the reader chose beforehand up to three verses on which he meditated and then interpreted. He stood to read and sat down to speak. (Note Luke 4:16–20 which describes in detail how Jesus took this part of the service in the synagogue at Nazareth.) The prayers could be said by any member of the congregation.

The most important piece of furniture was the **'ark'**, a chest in which the rolls of Scripture were kept. The officiating minister and the reader stood on a raised platform. Seats were provided and the 'chief seats' (Mark 12:39) were those nearest the ark and facing the congregation. They were reserved for the most honourable men.

There were three daily services, 9.00 a.m., 12 noon and 3.00 p.m. Worship was based on three main elements, the reading and expounding of the Scriptures, prayer and praise.

Just as the word **'Church'** stands for both the body of believers and the building, so the synagogue stood for both the congregation and the meeting place. The synagogue was the **centre of Jewish community life**: it governed the daily life of its members, appointed its local magistrates and attended to the education of its children, but it was primarily a house of prayer where men met to hear God speak in the words of His Law. To be expelled from the synagogue was a severe punishment and disgrace (John 9:22).

The Jews of the **Diaspora** (The Dispersion – those who lived outside the Holy Land) also built synagogues for themselves in Jerusalem where they met when on pilgrimage. Thus Acts 6:9 mentions the synagogue of the Cyrenians and Alexandrians.

We cannot overestimate the importance of the synagogue to Judaism. Throughout the Diaspora **it was the synagogue that kept the Jewish faith alive** and prevented believers from being swallowed up in the vast sea of heathenism. The destruction of the Temple could not prevent Jewish communities throughout the empire from meeting on the sabbath day to pray to God and read His Law.

Education

The children of rich and poor were taken to the synagogue school at the age of five. The master was the hazzan, the guardian of the sacred rolls, and the only text-book was the Scripture. The children sat on the ground around the master and repeated by rote the sentences he said aloud. Nothing was taught except religion. Local schooling finished at the age of thirteen but able boys who wished to specialize in religious studies could proceed to Jerusalem to the school where they would be taught by the most famous doctors of the Law. Thus Saul of Tarsus came to Jerusalem to sit at the feet of the Rabbi Gamaliel. The object of such education was to produce future doctors of the Law.

1.6 The Sabbath

The importance of the sabbath in Jewish life is confirmed by the frequency to which it is referred in the Gospels.

The Jews believed that the day was of divine origin, for God Himself had commanded Moses to tell the children of Israel, 'Above all you shall observe my sabbaths. . . that you may know that I am the Lord who hallows you' (Exodus 31:13). Again, in the Ten Commandments, 'You have six days to labour and do all your work. But the seventh day is a sabbath of the Lord your God.' (Exodus 20:8–11).

Keeping the sabbath

During the exile, since there was no Temple in which they might worship, the sabbath became of added importance, taking the place of the Temple as the focal point in the practice of their faith. By their return from exile, the sabbath had been firmly established and regulations for its observance were strengthened. So strict did the observance later become that during the Maccabean revolt, one thousand Jewish soldiers allowed themselves to be killed without resistance rather than break the sabbath by defending themselves. In the time of Jesus, **sabbath breaking was still an extremely serious offence** and to keep it worthily was considered very praiseworthy.

The sabbath began on Friday at twilight, when three stars could be seen in the sky. The hazzan (the synagogue attendant) then blew the shofar (the ram's horn trumpet) from the highest rooftop to signify the beginning of the holy day. At this signal, the sabbath lamp would be lit in every home. Before this, the house would carefully be cleaned and the women would cook the food to be eaten on the day. After the lighting of the lamp a meal was held, following a three-fold blessing, and afterwards no food was eaten until after the synagogue service of the next morning. This would explain the incident of the disciples plucking and eating the ears of corn on the sabbath day (Matthew 12:1–2): they were hungry. After the synagogue service people ate their midday meal, and the final meal was supper at 5.00 p.m. The blowing of the shofar indicated the end of the day.

The spirit of the day

The sabbath was **a day of prayer**, but it was not an unhappy day. People put on their best clothes, the meals were well prepared, and they were invited to share in the joy of God at the **completion of the Creation**. Unfortunately it seems that the truly spiritual significance of the day was at this time being obscured by legalistic prohibitions. It would appear that for some, the essence of sabbath observance was rather negative, that there should be no work done, rather than a sharing in the joy of the Lord. Jesus had no sympathy with this attitude.

The rules governing sabbath observance were laid down with minute attention to detail. The Scriptures decreed that it was forbidden, for example, to light a fire, or walk more than six furlongs (a sabbath day's journey), and from these and similar rulings the rabbis derived further prohibitions, such as carrying a burden or untying a knot, writing more than one letter of the alphabet. The less extreme rabbis were prepared to allow that some prohibitions should be governed by circumstance and thus agreed that it was lawful to fight in self-defence on the sabbath. (In 63 BC during Pompey's siege of Jerusalem, the defenders left their posts as soon as the sabbath began.) They did not consider it unlawful to help a man, or even an animal whose life was in danger, as Jesus himself reminded the Scribes and Pharisees (Luke 14:5). On the other hand, the extremists, notably the Essenes, categorically denied this exception. The attitude of Jesus was that the sabbath should be observed for its spiritual value and that it was not an end in itself. He said, 'The Sabbath was made for the sake of man, and not man for the Sabbath' (Mark 2:27).

Jesus and the sabbath

1 The incident in the cornfield (Mark 2:23–28; Luke 6:1–5; Matthew 12:1–8).The disciples would be fasting since sunset of the previous day. To the Pharisees, plucking the ears of corn would be reaping, and rubbing them out would be threshing. This was work which was forbidden on the sabbath. Jesus pointed out that **David, the national hero**, had, when in need, eaten the holy bread which was reserved for the priests only, and had given it also to his hungry men. In this way, with the connivance of the priest, he broke the letter of the Law. Jesus made it plain that people matter more than things; that circumstances alter cases.

2 See Unit 8.3 for comments on the following miracles of healing on the sabbath:

(a) The Man with the Withered Hand
(b) The Man with Dropsy
(c) The Crippled Woman

3 Jesus seemed to go out of his way to break the sabbath regulations. In fact, he only broke the letter of the Law when it hindered the fulfilment of its spirit.

1.7 The Messiah

The Jews, humiliated by foreign occupation, looked back with deep nostalgia to the days of independence, and especially to the most glorious period of their history, the **reigns of David and Solomon**.

The prophets had taught that their God was the God of history who controlled the destinies of the nations, and it was a crushing blow to their belief when, in **586 BC the Babylonians captured Jerusalem**, destroyed God's Temple and carried off the 'chosen people' into exile. This could only mean either, that the gods of Babylon were stronger and more effective or, that the present disaster was a well deserved punishment for the sins of the nation. God, however, was faithful, would not abandon His People; they remembered the **promise of Moses**, 'The Lord your God will raise up the prophet from among you like myself and you shall listen to him' (Deuteronomy 18:15). God would finally intervene in human affairs to restore Israel to her rightful place of supremacy among the nations of the world. A day would come in which God would establish His Kingdom on earth.

There was a wide diversity of belief about the manner in which God would intervene, and there was no uniform official version, but the main features of popular belief were:

1 God would appoint a vice-king, a Messiah (anointed one) who would be His instrument and whose main function would be to prosecute the **Messianic War** against the enemies of God and of His chosen people.

2 The wicked world would be destroyed and the faithful Jews of the Diaspora (those who lived outside Palestine) and their Proselytes (converts) would return to Palestine.

3 Jerusalem and its Temple would enjoy renewed prosperity and splendour.

4 The reign of the Messiah would inaugurate a period of perfect happiness during which the glory of Israel would be restored and God's justice would rule over all the world.

5 The Messiah would be born of the stem of Jesse i.e. a descendant of David.

This teaching about the final events of the end of the world is given the technical name of **eschatology**. A popular view was that the Messianic Kingdom would last for a thousand years and would come to an end with the direct intervention of God. The present universe would pass away and there would be a new creation of heaven and earth.

The coming of the Messiah

In the time of Jesus, the coming of the Messiah was awaited with intense expectancy, and the Gospels themselves give evidence of this. Thus, 'The people were on tip-toe of expectation, all wondering about John, whether perhaps he was the Messiah' (Luke 3:15).

Again, when John sent messengers to Jesus early in the ministry, their question was, 'Are you the one who is to come, or are we to expect some other?' (Luke 7:19). John 6:15 records that after the miracle of the loaves and fishes Jesus became aware that 'they meant to seize him and make him King'.

To the overwhelming majority of Jews the life and death of Jesus presented a picture of the Messiah which was completely different from that of the traditional Messiah who would fight against the enemies of God and the oppressors of His people, and after a military victory reign in splendour over the whole earth. Jesus chose rather to fulfil the prophecy, 'Tell the daughter of Zion, "Here is your King, who comes to you in gentleness, riding on an ass, riding on the foal of a beast of burden" ' (Matthew 21:5).

Some Jews believed that the Messiah would be more than merely human, and this idea was closely related to that of the 'Son of Man'. Originally, 'Son of Man' was only a poetic expression for 'man', but later Jewish belief held that the Son of Man stands mid-way between God and Man, that he existed before the creation of the world, and his relationship with God is as close as could be held consistent with belief in the one true God.

1.8 Son of Man

Jesus' use of the title Son of Man is generally held to derive from Daniel 7:13 which refers to a mysterious heavenly being, a 'Son of Man' to whom is given 'glory and kingly power', a sovereignty which shall not pass away. Daniel discloses that the Son of Man represents the 'saints of the Most High', that is, the people of God. Jesus saw himself as this Son of Man, the Messiah, the one chosen by God to create His people, and 'the twelve' were its first members. 'Son of Man' was not a popular Messianic title, and because of its mysterious nature Jesus could safely use it without making public claim to Messiahship, which would have brought his ministry to a precipitate end.

In Daniel the Son of Man receives a kingdom from God and Jesus associates the title with his own eventual triumph. 'When the Son of Man comes in his glory . . . he will sit in state on his throne, with all the nations gathered before him' (Matthew 25:31). The people of God are to share in his triumph. Jesus felt that he himself combined in his person the Son of Man of Daniel with the Suffering Servant of Isaiah, for he knew that 'the Son of Man must suffer many things' (Luke 9:22) and that he came 'not to be served, but to serve, and to give his life as ransom for many ' (Mark 10:45). This fusion of the functions of the Son of Man and the Suffering Servant was unique.

1.9 Son of God

In the Old Testament the nation, Israel, which thought of God as Father, is often referred to as the Son of God, 'I brought my Son out of Egypt', Hosea 11:1 quoted in Matthew 2:15, but by the time of Jesus it was regarded as a Messianic title. Thus in Matthew 16:16 we have, in Peter's confession, 'You are the Messiah, the Son of the living God', while the question at the trial in the High Priest's house is, 'Are you the Messiah, the Son of the Blessed One?' (Mark 14:61). That Jesus was conscious of being Son of God is beyond dispute, but did he mean no more by it than that he was the Messiah? Did he claim for himself a unique sonship?

Mark 13:32 throws light on this question. 'About that day or that hour no one knows, not even the angels in heaven, not even the Son; only the Father.' Here the Son is placed above men and angels and has a position subordinate only to God Himself.

The most conclusive evidence is in Matthew 9:27 and Luke 10:22, originating in 'Q' (See Unit 2.4), 'Everything is entrusted to me by my Father, and no one knows the Son but the Father; and no one knows the Father but the Son. . .' Here we see an unmistakeable claim to a unique sonship not shared by anyone else.

2 The Background to the Synoptic Gospels

2.1 Introduction

During this century a new and distinctive view of the Gospels has arisen. The traditional view assumed that they were largely biographies written by well known contemporary Christian authors, recording the story of the life of Jesus with some regard for chronological sequence, while the modern view is that they are in many respects unlike any biographical writing, ancient or modern.

The earliest written Gospel, generally accepted as Mark's, dates from about 66–68 AD, though scholars are not in agreement as to the precise date. Thus for about the first forty years the Gospel (literally the 'Good News') was spread mainly by word of mouth. This **'oral tradition'** of the words and deeds of Jesus was the preaching material of the earliest Christians and was the rich store from which the authors constructed the written Gospels. The tradition had no single author, but was the common property of the early Church, a social possession which was not limited to the recollection of the few, though it did rest ultimately on the memories of those who were 'the original eyewitnesses and servants of the Gospel' (Luke 1:2).

It can be shown that the first twelve chapters of **Mark, the earliest written Gospel**, is composed of completely self-contained units, the position of which could be interchanged without affecting the story. What we have is 'a necklace of pearls of which the string is broken', a miscellaneous collection of self-contained units. Thus, Mark's Gospel cannot be called biography in any real sense, since what he did was to string together already existing units whose background and place in the tradition was unknown, and to arrange them in the way best suited to his purpose in writing. A large number of these incidents and sayings relate to a dozen or so great controversies in which Jesus had been engaged and which still concerned the Church during and after Mark's day. They were selected by Mark out of a large body of oral tradition, to meet the immediate needs of the local church to which he belonged.

The last four chapters of Mark, usually referred to as the **Passion Narrative,** tell the story of the last week in the life of Jesus and is a single unit which was either committed to writing before Mark used it or a 'stereotyped tradition', that is, a fixed tradition faithfully committed to memory. Later, **Luke and then Matthew used Mark's Gospel as a main source** for their own work. Few verses of Mark are not to be found in Luke or Matthew or both. There was no dearth of stories about the ministry of Jesus and these were preserved in the **'oral tradition'** which was based on the evidence of eyewitnesses, particularly that of the Apostles. Many dramatic and unforgettable incidents, especially the details of the last week of his life, the Last Supper, the Crucifixion and the Resurrection were recounted many times, especially when the Church met for the 'breaking of bread', as it regularly did. The Epistles of St Paul bear witness to the fact that they did not forget his teaching, for the Apostle would quote the very words of Jesus to settle problems within the Church. Thus in his first letter to the Corinthians he says, 'To the married I give this ruling, which is not mine, but the Lord's' (I Corinthians 7:10), and in giving us the earliest account of the Last Supper he says, 'The tradition which I handed on to you, came from the Lord himself' (I Corinthians 11:23). Cycles of stories about Jesus must have grown up in all the great centres of early Christianity – Jerusalem, Caesarea, Antioch, Alexandria, and Rome; these stories were used by countless anonymous evangelists in the spreading of the faith.

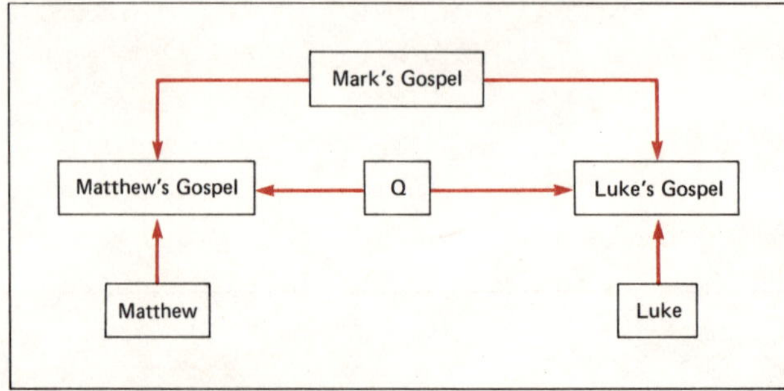

Fig. 2.1 The Synoptic problem

At first Christianity was regarded by the Romans as a sect of Judaism and shared in the tolerance enjoyed by the Jews, but a crisis came in 64 AD with the first great **persecution under the Emperor Nero**. Many prominent Christians were killed, including, according to tradition, Peter and Paul. While eyewitnesses lived, the oral tradition could be preserved and corrected if there was distortion, but they were dying out with the passage of years, and the persecution brought the situation to crisis point: it was now urgent that the oral tradition should be preserved in writing before it was lost or distorted.

2.2 The Synoptic Problem

Matthew, Mark and Luke are called the **Synoptic Gospels** because they have a common outline of the story of Jesus. A comparison of their subject matter discloses not only that they have a great deal in common, but also that in many cases the material is almost word for word the same. In some instances the same material is used by all three, and in others only by two, but what is clear is that there is a strong relationship between all three: the exact nature of their relationship is called the synoptic problem.

It would not be surprising to find the same stories in three different biographies of an important person, but it would be surprising to find such incidents described in practically the same language, unless each writer was using a common source. An analysis of the Gospel material disclosed that only 55 verses of Mark are not used by Matthew, who, in fact, uses 51 per cent of Mark's actual words. Luke has 53 per cent of Mark's actual words, and of the 55 Marcan verses Matthew does not use, Luke has 24. Thus, of Mark's 661 verses, only 31 do not appear in either Matthew or Luke or both. It is now widely accepted that **Mark is an original source**.

2.3 The Priority of Mark

This means that Mark gives us the earliest written account of the ministry of Jesus we now have. Matthew and Luke used Mark as a main source.

1 Matthew and Luke contain large portions of Mark's material, often repeating him word for word.

2 They not only use his material but largely accept his order of events. Where one alters Mark's order, the other always retains it.

3 Both Matthew and Luke 'improve' the style of Mark and refine his language – Mark's Greek is rather colloquial. Thus, in Mark 2:4 the bed of the paralysed man is 'krabattos' – a soldier's word, whereas Matthew uses the more polite 'kline' and Luke, the more refined 'klinidion'. Mark is addicted to the historic present tense, e.g. 'she says to him' and 'he says to her'. Of the 151 examples of this in Mark, Luke retains only one. Mark often quotes the words of Jesus in the original Aramaic, e.g. in Mark 5:41 he raises the daughter of Jairus with the words 'Talitha, cum', and in Mark 7:34 to the deaf stammerer he says, 'Ephphatha.' Both Matthew and Luke omit the Aramaic in their accounts. Matthew uses Mark's version of the cry of desolation from the cross, but corrects Mark's Aramaic.

4 By the time Matthew and Luke were writing, the Apostles had been dead for some time and their memory was held in great reverence by the Church. Matthew and Luke tend to omit or alter anything which would show the Apostles in a poor light, e.g. Mark 10:35–45, the incident of the ambitious request of James and John. According to Mark, it was the Apostles themselves who made this unacceptable request, but Matthew says it was their mother who made it.

5 Matthew and Luke appear to wish to improve Mark's picture of Jesus himself. Thus Mark calls him 'the carpenter', Matthew and Luke, 'the carpenter's son'. In Mark 6:5, Mark says of Jesus' teaching visit to Nazareth, 'he could work no miracle there', while Matthew changes the 'could not' to 'did not', thus suppressing the suggestion that there were limits to the power of Jesus. Both Matthew and Luke omit Mark 3:21, the passage which says the people were saying that Jesus was out of his mind.

There is thus the strongest case for holding that Mark is the earliest of the Gospels and was a main source for the Gospels of Matthew and Luke.

2.4 The Probability of 'Q'

When we examine the parts of Matthew and Luke which do not originate in Mark, we find 250 verses which so closely resemble each other (in many cases they are identical) that a **second**

written source is indicated. Thus we compare e.g. Luke 3:7–9 with Matthew 3:7–10 (the preaching of John the Baptist) and find in the Greek that 60 out of the 63 words are identical. Though many of the sayings of Jesus are quoted by Matthew and Luke in almost identical form, they are often placed in different contexts and we therefore conclude that the document they used, with some exceptions, preserved the sayings of Jesus without reference to the circumstances in which they were spoken; one such exception is the story of the **Centurion's Servant**. It appears that Matthew and Luke were using an early document which preserved many of the sayings of Jesus. In general we may conclude that Matthew and Luke used Mark as their source for the events in the life of Jesus, and this second document for his teaching. Much of 'Q' is preserved in Matthew's account of the **Sermon on the Mount**, almost half of which comes from 'Q'. However, Mark and 'Q' do sometimes overlap, for example both have versions of the Baptism of Jesus and of the Temptation. Both also have material concerning John the Baptist. In such cases Matthew blends one account with the other, while Luke rejects Mark, preferring 'Q'.

This document which scholars call 'Q', arguably an abbreviation from the German 'Quelle', meaning 'a source', is earlier than Mark, being not later than 60 AD and may even be as early as 50 AD. Its place of origin is thought to be Antioch in Syria, which Acts tells us was the first gentile church, and where, therefore, a teaching document would have become necessary very early.

Since 'Q' has not survived in its original form, we have to be careful about what we assume belongs to it.

1 Passages which Matthew and Luke have in common and which do not come from Mark, may not necessarily originate in 'Q'. Proverbial sayings and inspired teaching are easily remembered and many of the sayings of Jesus would have been widely known and would have become the common property of the Church.

2 Both Matthew and Luke are selective in their use of Mark and it can safely be assumed that they used 'Q' in the same way. It is probable that in places one quotes 'Q' and the other does not, so that passages of the teaching of Jesus which occur in one Gospel only may still derive from 'Q'. It is thus not possible to describe the exact limits of the document.

'Q' contains few parables, though among notable inclusions are the parables of the Mustard Seed (Luke 13:18–19) and the Lost Sheep (Luke 15:4–7), but its sayings are full of life and colour. Jesus drew on the common experiences of daily life and of nature to illustrate his message and make it vivid and compelling. He spoke of the harvest, the grass, the figs, the lilies; the foxes, the eagles, the sparrows, the serpents; the weather – lightning, wind and rain; social events – weddings, funerals, feasts; the home – lamps, ovens, loaves, eggs, cups and plates; these and many more, in this way making his message of immediate relevance to his hearers.

2.5 Material Peculiar to Matthew and Luke

When we have accounted for the material originating in Mark and 'Q', a considerable amount remains – almost one third of Matthew and over one half of Luke. The symbols 'M' and 'L' have been given respectively to this material.
Thus:

Matthew = Mark + Q + M
Luke = Mark + Q + L

Material peculiar to Matthew

The material peculiar to Matthew and Luke is of special value in discerning the point of view and emphasis of the Gospel writers. Matthew includes:

1 An account of the birth of Jesus told from the point of view of Joseph which has the exclusive story of the Wise Men and the murder of the children of Bethlehem.

2 Miracles: the Coin in the Fish's Mouth, the Healing of the Blind and Dumb Demoniac.

3 Parables: the Pearl of Great Price, the Hidden Treasure, the Dragnet, the Labourers in the Vineyard, the Marriage Feast, the Talents, the Wheat and Darnel, the Ten Virgins, the Unmerciful Servant.

4 Much information about the teaching of Jesus, especially the Sermon on the Mount.

5 Details of the Trial, Crucifixion and Resurrection; the suicide of Judas, the dream of Pilate's wife, Pilate's hand washing, the appearance of many dead 'saints' after the Resurrection; the guard at the tomb and the bribe they accepted; the earthquake.

The content of the large number of unique sayings of Jesus reveals that Matthew is concerned to stress that the Jewish Law is not abolished by Christianity, but fulfilled. In Chapter 5 he stresses that the Law is to be given a Christian interpretation, and in Chapter 6 he emphasizes the importance of the Jewish devotional practices of prayer, almsgiving and fasting. He believes that the mission of the Church must be first to the Jews, for they have a special place in the plan of God. Lastly, Matthew is interested in the peculiarly Jewish belief in eschatology, the teaching about the final events of the end of the world and the Last Judgment.

Material peculiar to Luke

Most of this material is generally referred to as 'L'. Its existence as a document is thought to be doubtful, and it was probably an oral source which Luke preserved. It has been described as the contents of his notebook, something like a modern author's 'commonplace book'. This material was of particular value because Luke, as the companion of Paul, was in intimate contact with most of the great personalities of the early Church and therefore had personal access to their recollections. Probably during Paul's two-year imprisonment at Caesarea, circa 60 AD, his companion, Luke, wrote down these memories which included some of the most beautiful of the Gospel stories. What we do know is that 'L' is a distinct source of the tradition and includes:

1 Certain incidents, including the meeting with Zacchaeus, Mary and Martha, additional information about the Last Supper, Trials, Crucifixion (words from the cross – 'Father, forgive them. . .' and the word to the Penitent Thief), Resurrection stories (the journey to Emmaus and the appearance at Jerusalem), the Ascension.

2 Miracles: the Widow of Nain, the Man with Dropsy, the Ten Lepers, the Crippled Woman.

3 Parables:

(a) The Good Samaritan, The Prodigal Son, The Lost Coin, The Two Debtors, The Rich Fool, The Rich Man and Lazarus, The Pharisee and the Publican, The Fig-Tree.

(b) Three unique parables – The Friend at Midnight, The Unjust Judge, The Unjust Steward. In each of these, the principal character does not, as is usual, represent God, instead his character is contrasted with that of God.

Many scholars hold the opinion that the **Birth Stories** are a separate source, probably Hebrew or Aramaic. They include the Promise and Birth of a Son to Zacharias and Elizabeth, the Annunciation (Visit of the Angel Gabriel to Mary), the Birth of Jesus and the Visit of the Shepherds, the Circumcision and Presentation in the Temple, the Visit of the Boy Jesus to the Temple.

There are distinct parallels in the stories of the births of Samson (Judges 13:1–14) and of Samuel (1 Samuel 1 and 2). This is especially true of The Magnificat (The Song of the Virgin Mary) and Hannah's Hymn of Gratitude (1 Samuel 2: 1–10).

Such a long continuous narrative is unusual in the Gospels, apart from the **Passion Narrative**, which is itself a separate source in Mark, and it is probable that the **Nativity Stories** are also a separate source which is called 'I'. The atmosphere of Jewish piety and the style of the stories are markedly different from the Preface to the Gospel which precedes them, and from the 'historical' emphasis of Chapter 3 which follows. It is likely that this material was not available to Luke when he wrote the first draft of his work, and that he added it sometime after his stay at Caesarea.

3 The Characteristics of the Gospels

3.1 St Mark

Authorship, date and purpose

Eusebius, Bishop of Caesarea, circa 320 AD in his history of the Church, quotes Papias, Bishop of Hierapolis, circa 130 AD, who recorded a tradition which he claimed was handed down by an elder, that Mark, who had been the interpreter of Peter, wrote down accurately Peter's account of the sayings and doings of Jesus, though 'not in order'.

Irenaeus, Bishop of Lyon from circa 178 AD and who had been at Rome, states that after the deaths of Peter and Paul, Mark, the disciple and interpreter of Peter, handed down in writing the preaching of Peter. It is probable that Irenaeus is bearing witness to a local Roman tradition. These two pieces of evidence suggest that the author of the earliest written Gospel was Mark and that the place of origin was Rome.

The occasion of writing was the first great persecution of the Christians after the **Great Fire of Rome in 64 AD**. The blame for starting the fire is attributed by the Roman historian, Tacitus, to the Emperor Nero himself, who, to avoid the consequences of his crime, made the Christians the scapegoats. They were arrested in hundreds and put to death with every refinement of cruelty. Many of the leaders of the Church in Rome were thus killed, including eyewitnesses of the ministry of Jesus, and it now became essential to commit the tradition to writing before it was lost or distorted.

We can thus date Mark's Gospel with some accuracy to circa 66–68 AD. There is a strong tradition that Mark is the John Mark of Jerusalem whose mother's house became the meeting place of the leaders of the Church after Pentecost (Acts 12:12); it may well have been the house where the Last Supper was held. It is possible, or even probable, that Mark may have been the 'young man' wearing nothing but a linen cloth in the garden of Gethsemane who, leaving his garment in the clutches of the soldiers, ran away naked to escape arrest (Mark 14:51). It is difficult to account for this quite irrelevant detail in the middle of a highly dramatic story, unless it is a personal reminiscence – a way of saying 'I was there.'

Acts tells us later that Mark accompanied Paul and Barnabas on the first missionary journey. Paul refused to take him on the second journey because he had turned back at Perga, but the quarrel must have been healed because, years later, we find Mark serving Paul during the Apostle's imprisonment in Rome.

The main purpose of Mark's Gospel was to encourage and strengthen a Church faced with destruction at the hand of tyranny.

The characteristics of Mark

1 Mark's Greek is **colloquial**, that is, in the language of common speech, the kind spoken by the lower classes in Rome, or possibly the imperfect Greek spoken by a Jew whose first language was Aramaic: hence the use of Aramaic words and phrases. His use of the historic present is typical of his colloquial speech.

2 Such language can, however, because of its informality, convey acute perception, and may in part account for the vividness of Mark's record. The detail of his descriptions, such as that of the Feeding of the Five Thousand – they sat 'in ranks' on the 'green grass' (Mark 6:48), or his account of the blind man whose recovery of sight was gradual and who first declared that he saw men who looked 'like trees' walking about (Mark 8:24), suggest that **Mark was drawing on the first hand accounts** of eyewitnesses.

3 The **orderly arrangement** of his material. Contrary to the opinion of the Elder of Papias, Mark, in fact, has a definite order, but it is according to subject and only chronological in a broad way. He was writing, not for the Church of a distant future, because the **Second Coming** was imminently expected, but to meet the current needs of a martyr Church. Many of the problems of the Church in Rome in Mark's day were those faced by Jesus during his ministry, and Mark selected out of the large body of oral tradition those incidents in the life of Jesus which gave answers to these problems, such as the attitude to the Jewish Law, the food regulations, social contact with pagans, civil obedience, the payment of taxes and sabbath observance.

4 The story of the Passion of Jesus takes up almost one third of the Gospel. There is evidence that by the time Mark wrote this was an established tradition, an orderly sequence of the events of the last week in the life of Jesus which we now call the **Passion Narrative**. Paul himself refers to the tradition which he had received about the death and resurrection of Jesus, and which he had handed on (I Corinthians 15: 1–18). It is noticeable that Matthew and Luke, though they do not hesitate to alter and correct the earlier chapters of Mark, show great respect for his order of the Passion. The events were too firmly established for any serious changes to be made by later authors.

5 Mark's Gospel is traditionally associated with Peter. It is noticeable that **his picture of Peter is very frank**, and he is hardly mentioned as an individual except as an occasion for a rebuke, for example after his inspired confession at Caesarea he earns first approval and then rebuke, 'Away with you, Satan' (Mark 8:33), and on the night of the betrayal he is singled out as the

disciple who boasted of his loyalty, but who would, in the event, deny his Lord in the hour of need.
Two explanations may be offered:

(a) Who but Peter himself would report events which were so little to his own credit?

(b) The Gospel was written after Peter's martyrdom, at a time when, by his death he had redeemed his past failures. Peter had learned to take up the Cross, and the effect of the stories would be to encourage those Roman Christians who, in perilous times, might well be called upon to witness to their faith in the same way, and to accept martyrdom. Those conscious of their own weakness should be strengthened by the example of Peter.

6 Mark has a special **interest in exorcism** (casting out evil spirits), but in the healing miracles, faith is either required or can safely be assumed. The reply of Jesus to the father of the epileptic boy is typical, 'Everything is possible to one who has faith' (Mark 9:23).

7 Mark lays **emphasis on the Messiahship** of Jesus. At his Baptism he is called the beloved Son of God; at Caesarea Philippi he approves Peter's insight that he is the Messiah, the Son of God (Mark 8:29); at his trial his reply to the High Priest's question 'Are you the Messiah?' is the firm 'I am.' (Mark 14:62).

8 The Gospel is addressed to **a martyr Church** threatened with extinction by tyranny. The **shadow of the Cross** lies even over the early pages. The way of discipleship is the way of the Cross. 'Anyone who wishes to be a follower of mine, must leave self behind: he must take up his cross and come with me' (Mark 8:34).

3.2 St Luke

Authorship, date and purpose

Written sources dating from the second half of the second century claim Luke as the author, (e.g. the Muratorian Canon, circa 180 AD, a document giving a list of New Testament books accepted by the Church in Rome at that time). Some of the language used by the author is consistent with the tradition that he was a doctor, though this is not conclusive. The tradition also mentions his close association with Paul.

Autobiographical details gleaned from the New Testament are in accord with the tradition. In Colossians 4:14 Paul sends greetings 'from our dear friend Luke, the doctor', and in 2 Timothy 4:1, names him as his companion, while in Philemon 24, Luke is listed among Paul's fellow workers. If the writer of the passages of Acts, which appear to have been recorded by an eyewitness and participant in the events described, was indeed Luke, he first appears on the scene in Troas, the port in Asia Minor from which Paul sailed to Macedonia during the second missionary journey (Acts 16:11).

It is not possible to date the Gospel with any real precision. Since Luke used Mark, his Gospel cannot be earlier than 65–70 AD and most scholars say that it was **probably written some time between 75–85 AD**.

If Mark writes for a Church faced with martyrdom, Luke writes for a no less serious situation at a later date. Before the Great Fire of Rome and the subsequent persecution of Christians, Christianity had been treated as a sect of Judaism and had shared the exceptional tolerance which Rome accorded to Jews. The Roman government had ended its attitude to tolerance however, and was becoming more and more repressive. **Luke's purpose** is to prove that the Christian movement was in no way dangerous to the state. Neither Jesus nor his disciples were guilty of insurrection. The charges against him were trumped up by his enemies, and Pilate, the Roman governor, who had made a positive effort to release him, had been pressurized by a mob stirred up by the Jerusalem High Priests. Against the better judgment of Pilate, who found no fault in him, Jesus had been condemned to death, while Barabbas, the leader of a revolt, had been released. The purpose of Luke is to show, by reference to the facts, how unfounded were the charges against Christ and his Church. As the Preface to the Gospel declares, Luke's purpose is not only to provide a continuous narrative, but also to provide 'authentic knowledge'.

The characteristics of Luke

1 Luke the historian. It is Luke's intention to fix the life and death of Jesus firmly in history. In the Preface he claims to be giving authentic knowledge of the events he describes, and in

Chapter 3 he uses the device common to historians of his time, to fix dates by giving the year in the reign of the Roman Emperor or ruler. Augustus died in 14 AD, so the fifteenth year of his successor would be 28–29 AD. **Pontius Pilate was a governor of Judaea from 26–36 AD.**

2 Luke was a Greek and his Gospel is the most Greek of the synoptics. From very early times Luke's Gospel was regarded as having been written for gentile converts. Mark is quite content to keep a borrowed Latin word, but Luke gives its Greek equivalent. He addresses his work to 'Theophilus', who appears to have been a high government official since Luke refers to him as 'Your Excellency', and this is the most popular theory, but since this Greek name can mean literally either 'lover of God' or 'one whom God loves', it is a title which could describe any Christian: it could therefore, be a general title, though this is unlikely.

3 Its note of **universalism**, i.e. the Gospel is good news for all peoples, not only for Jews. The descent of Jesus is traced back to Adam, rather than as in Matthew to Abraham, the Father of the Jewish people. The message of the angels is of goodwill to men; Simeon sees in the infant Jesus a light for the gentiles; John the Baptist restates the prophecy that all mankind shall see God's deliverance; Samaritans are twice singled out for high praise, and a Roman centurion is found to have a faith stronger than any Jesus had yet discovered in Israel. The Risen Lord commands that his Gospel shall be preached to all nations, beginning at Jerusalem.

4 Luke stresses the deep sympathy of Jesus for the outcast and poor. Jesus is the friend of publicans and sinners and has come to save that which was lost. Luke tells us about the woman living an immoral life (Luke 7:36ff); of Zacchaeus, the rich superintendent of taxes, and of the Penitent Thief at Calvary (Luke 23:39ff). Jesus brings to them all the forgiveness of God. Luke stresses the **deep sympathy of Jesus for the poor**, while he warns the rich about the dangers of covetousness, e.g. the parable of the Rich Man and Lazarus. In the Parable of the Great Feast, it is the rich who reject the king's invitation and those in need, the poor, the lame and the blind who become his guests (Luke 14:15ff).

5 Of special importance is the place Lukes gives to women. In New Testament times women were uneducated, had no legal rights and no place in public life, but in Luke's Gospel they play a large and important part. Luke gives us the story of Elizabeth, the mother of John the Baptist, and the loveliest of all pictures of Mary, the mother of Jesus. He recounts the incident of Anna, the prophetess, who greeted the infant Jesus in the Temple, the stories of the notorious woman in the house of Simon the Pharisee (Luke 7:36ff), the Widow of Nain (luke 7:11ff), the women who provided for the needs of Jesus out of their own purses (Luke 8:2–3), Mary and Martha (Luke 10:38–42), the poor widow and her offering of two small coins (Luke 21;1–4), the weeping daughters of Jerusalem at Calvary (Luke 21:1–4), and the very important part played by the women in the story of the Resurrection.

6 It is the **Gospel of Prayer**. Each of the Gospels shows Jesus at prayer, but Luke describes seven occasions not mentioned by the other evangelists. In Luke only do we find the parables about prayer, The Friend at Midnight, The Unjust Judge and The Pharisee and the Publican.

7 It is the **Gospel of the Holy Spirit**. The first two chapters are a record of the action of God, the Holy Spirit. The father of John the Baptist was told that his son would be filled with the Holy Spirit from birth (Luke 1:15). The Virgin Mary was told that the Holy Spirit would come upon her (Luke 1:35); Elizabeth was filled with the Spirit when Mary, her kinswoman, came to visit her (Luke 1:41); the Holy Spirit disclosed to Simeon that he would not die before he had seen the Messiah, and guided him to the Temple when the time came (Luke 2:25–27). After the Baptism, Jesus was full of the Holy Spirit (Luke 3:22) and was led by that Spirit into the wilderness (Luke 4:1–2); armed with the power of the Spirit he began his ministry in Galilee (Luke 4:14).

3.3 St Matthew

Authorship, date and purpose

It is certain that the author of the Gospel as it stands could not have been Matthew, one of 'the twelve'. No one with personal knowledge of the ministry of Jesus would base a Gospel so firmly on the secondhand account of Mark. We know nothing about him, except that from the content of his work and his detailed knowledge of the Jewish religion, we deduce that he was a **Jewish convert to Christianity**. The Gospel was written by a Jew, mainly for Jews. An accurate date is not possible, but it cannot be earlier than Mark, and its references to the Church suggest a period of development in its growth and understanding of itself. **The prominence of the apocalyptic**, i.e.

teaching about the end of the world, also suggests a later date. In the early days, baptism was in the name of the Lord Jesus, but Matthew 29:19 quotes the Trinitarian formula which was not used until later. Most scholars date Matthew in the period 85–90 AD.

The purpose of Matthew was:

1 to defend the Christian faith against Jewish opponents;

2 to instruct converts from paganism who, unlike Jewish converts, had no tradition of a highly moral religion;

3 to encourage those inside the Church to live a disciplined life consistent with the life and teaching of Jesus;

4 to provide an orderly record of the words and deeds of Jesus for reading at the worship of the Church.

Characteristics of Matthew

1 Matthew's Gospel is appropriately placed first in the New Testament because more than any other Gospel it is a **bridge between the Old and New Testaments**. The prophetic expectations about the coming of the Messiah are realised in the coming of Jesus.

2 Matthew sees the Gospel as the fulfilment of the Old Testament religion.

(a) He begins by tracing the descent of Jesus back to Abraham, the father of the nation, rather than to Adam as Luke, the gentile, had done. Each of the three stages in the genealogy have fourteen names; since the name David has fourteen letters in Hebrew, Matthew finds evidence for the Messiahship of Jesus in each of the three stages. This kind of evidence makes little appeal to us, but it would have impressed Matthew's first readers. Eight times Jesus is referred to as the Son of David.

(b) The **Nativity Stories** – Matthew alone records the story of the visit of the Magi and the star of Bethlehem. He tells it from the point of view of Joseph, not Mary, typical of a male dominated Jewish society.

(c) Parallels with the Old Testament dominate the Nativity Stories.

(*i*) The slaughter of the children of Bethlehem by Herod has a parallel with the story of Moses – the killing of the Hebrew children by the tyrannical Pharaoh.

(*ii*) The flight of the Holy Family to Egypt. God called 'His Son' out of Egypt as He had called His son Israel out of Egypt at the Exodus.

(d) Further parallels are seen in the Sermon on the Mount.

(*i*) Moses received the Old Law on a mountain (Sinai) and Jesus promulgates the New Law for men of the New Age which has dawned, upon a mountain. (Luke's setting is 'on level ground'.)

(*ii*) Matthew sets out the teaching of Jesus in five blocks, suggesting a parallel with the Pentateuch, the first five books of the Bible which Jews attributed to Moses.

(*iii*) Jesus talks with Moses on the Mount of Transfiguration and with Elijah, thus identifying with both the Law and the prophets.

(e) Dreams play an important part in the story. Joseph, the Magi and Pilate's wife.

(f) Matthew believed, in agreement with the scholars of his time, that everything relating to Christ had been, either explicitly or in a hidden manner, foretold in the words of the Old Testament.

3 While the Gospel is the **fulfilment of the Old Testament**, it is not merely a continuation of improvement, but a new and decisive word of God. Thus though, while he claims that 'not a letter, not a stroke, will disappear from the Law' (Matthew 5:18), and that 'the Pharisees sit in the chair of Moses' (Matthew 23:2), the Gospel is not an exclusively Jewish possession. While it is firstly for the Jews, it is to be preached to the whole world. The Jews are warned in the Parable of the Vineyard (Matthew 23:43) that the Kingdom would be taken from them and the Jewish leaders are often severely condemned.

4 Matthew has a prominent interest in the apocalyptic, the teaching about the return of Christ to earth at the Last Judgment. This is especially noticeable in such parables as Matthew alone records – the Sheep and the Goats (Matthew 25:31ff); the Ten Virgins (Matthew 25:13) and the Talents (Matthew 25:30). He adds considerably to the **Apocalyptic Discourse** of Mark 13.

5 Matthew displays great interest in the Church. He alone uses the word 'ecclesia'.

6 Matthew tends to heighten the miraculous or supernatural (e.g. Mark's young man in white becomes the angel of the Lord whose appearance was like lightning).

7 Matthew displays a **tendency to duplicate**, thus Mark's blind man and demoniac become in Matthew, two blind men and two demoniacs; two animals take part in the Palm Sunday procession.

4 The Nativity

4.1 St Matthew

Chapter 1 illustrates Matthew's purpose in emphasizing that Jesus is the fulfilment of the Old Testament prophecies. Matthew is writing to Jews to convince them that Jesus is the Messiah, and quotes texts from the Old Testament which support this belief. Modern scholars would not accept that the use of the Old Testament in this way is legitimate, but Matthew's readers would find it convincing.

Mary was only betrothed to Joseph, but a betrothal was an essential part of the marriage process and far more binding than an engagement. It would not be ended without a written declaration in the presence of witnesses. Joseph was thinking of a private annulment to prevent painful publicity for Mary. Joseph's dilemma is solved by a dream.

Old Testament prophecies fulfilled in Matthew's account

1 'A virgin will conceive and bear a son and he shall be called Immanuel' (Matthew 1:23). Matthew is quoting Isaiah 7:14 but from the Septuagint, the Greek translation of the Old Testament, not the original Hebrew. The Hebrew word for a 'young woman' was inaccurately translated 'parthenos', i.e. a virgin, and the original means no more than that a young woman of marriageable age would bear a son whom she would call Immanuel (meaning 'God with us'). This child would be a sign from God to King Ahaz who felt himself threatened by his powerful neighbours, the Syrians. The land would be made desolate for a time, but while the boy was still an infant the enemies of Ahaz would be defeated.

Matthew, writing for Jewish Christians, felt that this prophecy applied to Jesus who was to be a sign from God. The name Immanuel (God with us) is especially significant for it is the heart of the Christian Gospel that **God became man in the person of Jesus**. 'The word became flesh; he came to dwell among us' (John 1:14).

2 'Bethlehem in the land of Judah, you are far from least in the eyes of the rulers of Judah; for out of you shall come a leader to be the shepherd of my people Israel' (Matthew 2:6; Micah 5:2). Bethlehem was the city of David and Joseph was 'of David's line', so it was the obvious place for the birth of the Messiah.

3 'I called my son out of Egypt' (Matthew 2:15; Hosea 11:1). The flight of the Holy Family to Egypt was in order that this prophecy might be fulfilled. God called His Son (Jesus) to return from Egypt as in the Old Testament He had called His Son (Israel) out of Egypt at the Exodus.

4 'A voice was heard in Rama, wailing and loud laments; it was Rachel weeping for her children, and refusing all consolation, because they were no more' (Matthew 2:18; Jeremiah 31:15). The prophet was, in fact, describing an event of his own day, the sad sight of hundreds of Jewish prisoners being taken into captivity in Babylon. As they pass the tomb of their ancestress Rachel, Jeremiah feels that she must be weeping in her grave. The words are very appropriate to the massacre of the children of Bethlehem.

Very much in the mind of Matthew was that Moses, the greatest of the Jewish prophets and the first great deliverer of God's people, was himself saved from death in a massacre ordered by a tyrant.

5 'He shall be called a Nazarene' (Matthew 2:23). This prophecy is not to be found in the Old Testament and its source is unknown.

The Gospel writers relied on two sources of evidence:
(a) eyewitnesses of the ministry and especially the death and resurrection of Jesus;

(b) prophecies of the Old Testament which Jesus in his life fulfilled.
These 'proof texts' were regarded as of equal value to the evidence of eye-witnesses.

The Visit of the Magi (Astrologers)

Their origin is obscure, but they were probably priests of the Persian religion of Zoroaster, of which astrology was an important feature. Isaiah 60:3 and Psalm 72:10–11 have inspired the tradition that they were gentile kings to whom the name Caspar, Melchior and Balthazzar have been given. Matthew does not say how many there were, and there is no New Testament evidence for any of this.

Their **gifts of gold, frankincense and myrrh** have been given symbolic meanings in Christian tradition: gold symbolizes kingship, frankincense worship, and myrrh, suffering and death.

The **star of Bethlehem** may have been a conjunction of two planets (i.e. when from earth they appear to be almost in line and overlap) resulting in what appeared to be a single star of extraordinary brilliance. Such a conjunction appeared three times in 7 BC and this phenomenon would suggest to the peoples of the east, who were eagerly expecting a deliverer, that a special event was to take place.

Fig. 4.1 'The Adoration of the Kings' *Gossaert* called *Mabuse*
Reproduced by courtesy of the Trustees, The National Gallery, London

God reveals Himself to men in the manner they best understand. To the Jewish shepherds the angels were appropriate, to the gentile Magi, a star.

If the time sequence of Luke is accepted, the visit of the Magi took place some time after the birth. The Holy Family were 'in the house' (Matthew 2) having apparently moved from the stable when better accommodation became available. The fact that Herod killed the children of Bethlehem of two years old and under, in accordance with the information given to him by the Magi, also suggests a passage of time.

The account says that the star went ahead of them: this need mean no more than that it continued to be ahead of them as they journeyed. Herod's religious advisers had directed them to Bethlehem.

4.2 St Luke

The Annunciation (Visit of the Angel Gabriel to Mary)

Those who accept the story of the Virgin Birth must hold that Mary herself preserved it. There are difficulties, and among them the fact that Paul, whose writings are the earliest in the New Testament never mentions it in his recital of the main elements of the first Christian preaching (I Corinthians 15:1–11) though he undoubtedly believed Jesus to be the Son of God. Some scholars feel that the story is from Luke's later sources. Luke appears to have made use of the **prophecy of Isaiah 7:14** quoted explicitly by Matthew, 'A virgin will conceive' etc, though he does so indirectly. The early Christians, most of whom knew Greek, found the Greek version of the Old Testament ready to hand and would accept its version of the prophecy.

Another mistranslation, this time in the fourth century by St Jerome who translated the Bible into Latin (the Vulgate), has had far-reaching effects. Gabriel's salutation to Mary is given as, 'Hail Mary, full of grace', which can be taken to mean that, like Jesus, she is a source of grace, whereas the New English Bible's 'Greetings, most favoured one' means that she has received favour above all others.

The doctrine of the Virgin Birth has had far-reaching effects in raising the status of women. The account implies that Mary was free to refuse her high vocation but she accepted it in glad obedience.

The **promise of Gabriel to Mary** is in terms of the popular Jewish view of the Messiah. Her son is to inherit the throne of his ancestor David, and reign for ever as King of Israel. However, at the birth there was no room at the inn, just as later the Son of Man had nowhere to lay his head; the crown of the King of the Jews was of thorns and his throne a cross. Thus the note of rejection is struck early in the story.

The Magnificat (The Song of the Virgin Mary) (Luke 1:39–56)
There are parallels in the Song of Hannah in I Samuel 2:1–10.

The Circumcision and Purification (Luke 2:21 and 2:22–24)

To a Jew, circumcision is a symbol of the covenant (solemn agreement) between the people and God through Abraham. The Jewish child becomes a member of Israel by the rite of circumcision as a Christian becomes a member of the Church through baptism.

Eight days after birth came the circumcision, and forty days after birth, the purification. The mother had to offer a sacrifice of a lamb and a pigeon, but the poor could offer a second pigeon instead of a lamb. The account draws attention to the lowly birth of Jesus.

The Song of Simeon (Luke 2:25)

Simeon was a devout servant of God. It had been revealed to him that he would see the Messiah before he died, and he was patiently waiting at the Temple for the fulfilment of the promise. When he saw the infant Jesus he knew he could die in peace. Simeon was inspired to understand that Jesus would be the Saviour not only of the Jews but the gentiles as well. His prophecy about the Virgin Mary, 'You too, shall be pierced to the heart', was fulfilled at Calvary when she stood at the foot of the cross.

4.3 Differences Between the Two Accounts of the Nativity

1 In Matthew the story is told from the point of view of Joseph, in Luke, from that of Mary.

2 Mary was visited by an angel; Joseph was told in a dream.

3 **Matthew's emphasis is on Old Testament prophecy**, and underlying the account are parallels from the story of Moses. (Both Moses and Jesus are saved from death at the hands of a tyrannical king; the return of the Holy Family from Egypt is a parallel to the Exodus of the Jews under Moses.)

In Luke the emphasis is on Messiahship in terms of popular Jewish belief, i.e. He will be a military conqueror who will restore the Jewish nation to its rightful place of pre-eminence.

4 In Matthew, Jesus is given the name **Immanuel** i.e. **'God with us'**. This is the heart of the Christian faith, that Jesus is not merely a very good man, but God Himself in human form.

5 Matthew alone mentions the massacre of the children of Bethlehem and the flight to Egypt.

6 Matthew alone records the visit of the Magi and the star of Bethlehem. Luke alone tells the story of the Shepherds and the Angelic Host, the visit to Elizabeth and to the Temple for the rite of purification.

7 The visit of the Magi took place later than that of the Shepherds, for the Holy Family by this time had moved from the stable to a house. Herod killed the children of two years old and under, according to the information given him by the Magi.

8 In Matthew, the visitors are gentiles (Magi), in Luke, Jews (shepherds).

5 John the Baptist and Jesus

5.1 The Vision of Zechariah

All male descendants of Aaron, the brother of Moses, were priests and entitled to officiate at services in the Temple. So many thousands were qualified that the number was divided into twenty-four groups, each of which served twice in a year for a week at a time. Selection of individual priests for the morning and evening sacrifices was decided by drawing lots and the greatest honour, which could not fall more than once in a lifetime, was to enter the sanctuary to offer incense. The worshippers thought of the smoke as carrying their prayers through the curtain into the Holy of Holies, the innermost shrine, and heavenward to God. The priest was expected to emerge from the sanctuary after the ceremony and to pronounce God's blessing on the people.

As he performed his duty the **angel Gabriel appeared to Zechariah** and told him that his prayers for a son had been answered. Gabriel then foretold the purpose of the life of the child:

1 he would be called John – 'God's gracious gift';

2 he would be great in the eyes of the Lord;

3 he would be filled with the Holy Spirit from birth and be called to a life of self denial;

4 he would have the spirit and power of the prophet Elijah and would be the forerunner of the Messiah (Many Jews believed that Elijah would return to announce the coming of the Messiah – Malachi 4:5);

5 he would be the means by which many lapsed Israelites would return to God;

6 he would fulfil the prophecy of Malachi 4:6 by reconciling fathers and sons;

7 he would prepare the people so that they would be fit to receive the Messiah.

Because Zechariah could not believe the vision, he was struck dumb and did not recover his speech until the naming of the child.

5.2 The Visit of the Boy Jesus to the Temple (Luke 2:41–52)

(This incident is not to be confused with Jesus' visit to the synagogue at the beginning of his ministry.)

At the age of twelve, a Jewish boy becomes Bar-mitzvah, a son of the Law, accepting the responsibilities laid on him at his circumcision. After a course of intensive training from a rabbi, he reads the Law at the synagogue service. As a fully adult member of Israel he is entitled to wear the tallith (prayer shawl) and the tephillin (phylacteries), small leather boxes containing the Shema, texts from Deuteronomy and Numbers (see Unit 32). One is worn bound to the forehead and the other to the left arm close to the heart, as a reminder that God's Law should rule heart and mind.

The law laid down that Jews should visit the Temple each year at the Feasts of Passover, Pentecost and Tabernacles, but for those who lived at great distances from Jerusalem this was not possible, and Joseph's custom was to attend only at Passover.

The visit to the Temple is the only reference in the Gospel to the childhood of Jesus.

He did not instruct his teachers, but was rather listening to them and asking questions. What astonished his teachers was the intelligence of his questions as well as his answers.

The journey from Nazareth would have been in the company of relatives and neighbours, so a boy would not be missed for some time. The family had travelled a whole day before they realized that Jesus was not with them. The reference to 'my Father's House' suggests that from an early age Jesus was aware, however dimly, of a special relationship with God. This awareness was fully realized at the Baptism.

5.3 The Ministry of John the Baptist (Mark 1:1–8; Luke 3:1–18; Matthew 3:1–22)

Many Jews believed that a prophet would herald the coming of the Messiah. The prophet Malachi had said (Malachi 4:5) that it would be Elijah. John was, in fact, very similar to Elijah in that he lived a life of self denial and bravely spoke out against wrong. His outspoken condemnation of Herod and his Queen put his life at risk as had Elijah's condemnation of Ahaz and Jezebel. He was **in the line of the great Old Testament prophets**.

John was **a man of the desert**, eating desert food and wearing desert clothes of rough camel hair. He preached and baptized at the edge of the desert so that people had to go out to meet him; Jesus worked among the people in their towns and villages.

Luke typically insists that what he is about to relate is historical fact. It took place in the fifteenth year of the reign of the Emperor Tiberius, i.e. 29–28 BC and when Pontius Pilate was governor of Judaea. He refers also to local rulers and to the High Priests of the day.

Mark and Luke stress that John's message was not his own but was **'the word of God'**, which would therefore be fulfilled.

5.4 The Message of John the Baptist

1 'Repent, for the Kingdom of heaven is upon you.'

2 John is 'A voice crying aloud in the wilderness, "Prepare a way for the Lord" '. The analogy is of a king making a tour of his domain. He sends out a herald to give warning of his approach and to warn his subjects to prepare for his visit by repairing the roads, filling in the pot-holes, levelling the rough places, re-aligning where necessary.

3 The coming of the Messiah is imminent so they must repent of their evil ways and be baptized. There is no time to lose; if they are to qualify they are to submit to baptism as a public admission that by their sin they have failed God.

4 God is coming in judgment, so you must prove your repentance by your deeds. It will not be enough to say that God's promise to your forefather Abraham will save you. Neither his personal faith and obedience nor God's promises to him will be sufficient for you.

5 The message that the judgment is imminent is repeated. The analogy is of the woodman who decides to cut down an unfruitful tree and has thrown down his axe at its foot preparatory to starting work.

6 John had a special message for:

(a) the common people – share what you have with those in need;
(b) the tax gatherers – take no more than the fair assessment;
(c) the soldiers – no intimidation, no blackmail and be content with your wages.

7 John denied that he was the Messiah but declared that the Messiah was one who already stood among them. He is 'mightier than I, and I am not fit to take off his shoes.' (Removing a guest's shoes was the task of a humble servant.)

8 'His shovel is ready in his hand.' The analogy is of the threshing floor. The grain is beaten out, the straw removed and the remaining mixture of grain and husk is shovelled up and thrown into the air for the unwanted husk to be blown away by the wind. **The Messiah will come and gather to himself a new and purified Israel**.

5.5 John the Baptist's Question (Luke 7:18–35; Matthew 11:1–19)

John was in prison because of his outspoken condemnation of Herod Antipas for marrying Herodias, the divorced wife of his brother. This was in defiance of the Jewish Law (Leviticus 18:16).

From his prison John sent messengers to ask, 'Are you the one who is to come, or are we to expect some other?' The question could indicate that John had begun to doubt that Jesus was the Messiah, or that he had become impatient at Jesus' delay in publicly declaring himself: the baptism of fire John had spoken of had not arrived. John's question is indirect (he does not mention the word 'Messiah') and the answer of Jesus is also indirect. Jesus reminds him of the marks of Messiahship by 'there and then' working miracles of healing, and implying that what the messengers had seen and heard was a fulfilment of what the prophets had foretold about the Messiah.

Luke identifies John with the Elijah prophesied by Malachi, who was to be the herald of the Messiah.

Jesus likened the Jews to children who, whether the game was about sad or happy things, did not want to play. Neither the ways of the ascetic John nor the social Jesus met with approval.

Matthew's account comes from the same source as Luke's, but he does not say that Jesus performed miracles in the presence of the messengers.

5.6 The Baptism of Jesus (Mark 1:9–11; Luke 3:31–33; Matthew 3:13–17)

The stories of the Baptism and of the Temptation must have been told by Jesus himself.

Why was Jesus baptized if he was without sin? Matthew records that John tried to dissuade him, but that Jesus insisted. It would appear that since **John had called upon the whole of Israel to repent**, Jesus would not exclude himself. He chose rather to identify with them as later he was to be 'numbered among the transgressors'.

Mark's account suggests that the experiences of Jesus at the Baptism were private (subjective) but Luke's account can bear the interpretation that the opening of the heavens, the descent of the dove and the voice from heaven were witnessed also by others, though not necessarily so.

The Baptism marks the end of a long period of spiritual development. Jesus' mental and physical growth were those normal to any boy, and while we cannot say when he first realized his high calling, it was at the Baptism that he fully realized that God was his father in a unique sense, 'Thou art my Son, my Beloved.'

5.7 The Temptation (Mark 1:12–13; Luke 4:1–15; Matthew 4:1–11)

The experiences at the Baptism confirmed the growing awareness of Jesus that he was the Messiah. Full of the power of the Holy Spirit he went alone into the wilderness to reflect on the kind of Messiah he was called upon by God to be. (Spiritual experience is often difficult to describe without recourse to picture language. That Jesus was 'tempted by the devil' could be a way of describing his inner spiritual conflict.) Forty days would be a very long time to sustain a total fast, and the figure is probably a conventional Biblical way of saying 'for a long time'. It is best to follow Luke's version.

1 'If you are the Son of God, tell this stone to become bread.' We can interpret this temptation on two levels:

(a) It was a temptation to use his miraculous power to satisfy his own personal needs. If he had done so, he could not have claimed fully to share the limitations of our earthly life, for we cannot perform such miracles to supply our needs.

(b) Jesus later taught his disciples to pray, 'Give us this day our daily bread.' Did not compassion demand that he should provide food for the hungry millions? The Messiah was expected to give bread from heaven as Moses had done in the wilderness. Was this a means of winning popular approval for his ministry by supplying the physical needs of the people? Would not this, however, be to give priority to physical things, to treat man as an animal and rob him of his spiritual dignity? Would it not be open to the charge of bribery?

Jesus overcame the temptation with a quotation from the Scriptures, 'Man cannot live on bread alone' (Deuteronomy 8:3). A Chinese proverb says, 'If I had two pence, I would spend one to buy bread that I might live; I would spend the other to buy a flower that I might have a reason for living.'

2 Next, the devil takes him in imagination high in the air (Matthew says 'on a mountain') and shows him all the kingdoms of the world, claiming that it is all in the power of evil. Jesus may have it all, if only he will hold it in the devil's name. Rome ruled the world by right of conquest, and if he were Caesar, what might he not achieve? The Zealots, the fanatical nationalists expected a Messiah who would lead them in a war of liberation against the Roman empire. After the conquest, he could become a benevolent dictator. Jesus saw, however, that the **Kingdom of love** could not be brought into being as a result of a bloody war. He defeated the temptation with words from Deuteronomy 6:13 to the effect that an alliance with the devil was excluded if the will of God was to be done.

3 Thirdly, Jesus imagines himself to be standing on a parapet of the Temple high above the crowds in the courts below. As he later told the father of an epileptic boy, 'All things are possible to those who believe'; should he not leap down and prove to the people by working a spectacular miracle, that he was indeed, the Son of God? The angels would save him from disaster. Jesus defeated this last temptation by another quotation from the Old Testament, Deuteronomy 6:16, 'You are not to put the Lord your God to the test.' To test God is the opposite of trusting Him: faith does not consist of expecting God to do what we want, it means putting ourselves in God's hands and trusting Him to give us the power to do what He wants.

Luke tells us that in the end 'the devil departed, biding his time'. This implies that temptation was to recur throughout the ministry of Jesus and a notable example is found in the Agony in the Garden of Gethsemane. Note that Matthew reverses the order of the last two temptations.

We may ask the question 'How could Jesus be sinless if he was tempted?' **Temptation is not sin**: sin consists in giving way to the temptation or enjoying it in the mind. Jesus gave an example of this latter in the Sermon on the Mount (Matthew 5:28) 'If a man looks on a woman with a lustful eye, he has already committed adultery with her in his heart.' The person who wrestles against temptation knows more of its power than one who soon gives in: Jesus knew the full power of temptation because he did not give way.

6 The Ministry in Galilee

6.1 The Visit to the Synagogue in Nazareth

Mark places the rejection of Jesus by the people of Nazareth later in the ministry (Mark 6:1–6) but Luke places it at the beginning of the Galilean ministry because it announces the form which the ministry is to take (See Unit 1.5).

At the Temptation Jesus rejected the contemporary popular ideas about the Messiah and announced that the Messiah was one who would fulfil the prophecy of Isaiah 61:1–2. Jesus claimed to be the chosen messenger of God sent to announce to Israel that the **Messianic Age had dawned**. God was offering to His people forgiveness, freedom and healing. It is noticeable that in his reading he stopped short of the severe words which follow, namely, 'the day of vengeance of our God'. All this was accepted by his hearers with enthusiasm until they realized that he was claiming a central place for himself in the setting up of God's reign. The stir of admiration gave way to hostility. The question, 'Is not this Joseph's son?' suggests that they could not accept that someone who was just one of themselves should be the fulfilment of God's promises to Israel. The words of Jesus in verses 23 and 24, 'We have heard of all your doings in Capernaum; do the same here in your own home town', suggest that Jesus is again aware of the temptation to prove his claim by working a spectacular miracle.

Verse 24, 'No prophet is recognized in his own country', suggests that 'while his own would not receive him' (John 1:12) those outside Israel would be able to recognize him. The stories of Elijah and Elisha (1 Kings 17:9–16 and 2 Kings 5) are relevant here. It was to meet the need of gentiles not Jews that these prophets exercised God's power. The synagogue could not permit the admission of gentiles to the Kingdom and the attitude of the Nazarenes was typical of the narrow religious self-centredness of Israel at this time. The outraged sensitivities of the congregation could have ended in his death: they led him out to the brow of the hill intending to kill him, a foreshadowing of the time when they did take him out of a town and kill him.

6.2 The Call of the Disciples (Mark 1:14–20; Matthew 4:18–22; Luke 5:1–11)

Note that there are considerable differences in the accounts of this given in the first three Gospels.

Matthew follows Mark. The call of the disciples was of great interest to early Christians, firstly because the Apostles were held in great reverence throughout the Church, but it was also highly relevant in a missionary Church to stress that acceptance of the Gospel could involve a call to leave home and friends. The call is first to be disciples, with a promise that later they will become fishers of men. The emphasis of the account is on the sacrifices they were ready to make.

They left their nets 'at once'; the **call of Jesus** was one of authority and required unconditional obedience. It is likely, however, that these men had encountered Jesus previously. According to John 1:35–42 Andrew, who was a disciple of John the Baptist, spent several hours in the company of Jesus, presumably before the call.

Luke's version contains the miracle of the great catch of fish. The command to 'Let down your nets', is highly symbolic and the miracle is also an acted parable. 'Put out into deep water', probably signifies the wider mission of the Church. The message to missionaries dispirited by lack of success is to have faith in the promises of their Lord. Miracles will follow if only they trust in Him.

Peter's response, 'Leave me, sinner that I am', suggests that he recognized the supernatural power of Jesus.

The emphasis is again that the disciples 'left everything' and followed him.

6.3 The Call of Matthew (Levi) (Mark 2:13–14; Matthew 9:9–13; Luke 5:27–32)

Levi was a collector of tolls in the employ of Herod Antipas rather than of the Romans. Capernaum was an important centre of communications on the route from the sea to the east, and custom was levied on all goods carried by caravan.

Matthew's Gospel alone identifies Levi as Matthew the Apostle. He probably had two names. The call is in the same terms as for the other disciples and the response is identical.

Both Mark and Matthew describe the feast which follows, but only Luke specifies that the house in which it was held was Matthew's. The Pharisees draw attention to the fact that many 'tax gatherers and sinners' were present. To mix with such people was bad enough, but to eat with them was considered to be particularly improper. Jesus said, 'It is not the healthy that need a doctor, but the sick; I did not come to invite virtuous people, but sinners.' Luke adds, 'to repentance'.

6.4 The Choosing of the Twelve (Mark 3:13–19; Matthew 10:2–24; Luke 6:12–16)

Luke says that Jesus went up to a mountain and spent the night in prayer. In the morning he called his disciples to him and from among them chose twelve Apostles who, Mark says, 'he appointed twelve as his companions, whom he would send out to proclaim the Gospel, with a commission to drive out devils'. Matthew adds that they were given authority 'to cure every kind of ailment and disease'.

Note

1 There were many disciples (followers) but only twelve Apostles (messengers or envoys).

2 They were to live in intimate companionship with him throughout his ministry, and be witnesses to his life, teaching, death and resurrection.

6.5 The Sermon on the Mount (Matthew 5, 6 and 7)

The Old Testament records how the Law had been delivered to Moses on Mount Sinai and Matthew sets the scene of the giving of the New Law to the New Israel, the Church, on a mountain, by one greater than Moses. This device would not be lost on Matthew's Jewish Christians. Luke, on the other hand, sets the scene on the level ground. Both place the Sermon immediately after Jesus had chosen 'the twelve'.

The material for both versions is taken largely from 'Q', but whereas Luke attaches some of the teaching to suitable incidents throughout the Gospel story, Matthew uses it en bloc, as though Jesus gave the teaching in a single complete sermon.

When listening to a sermon or long speech we expect a logical structure, point smoothly following point, but we do not find this in the Sermon on the Mount. Sometimes passages do follow on from one another, but often sudden changes take place in which the subject is unrelated to what has gone before. It is therefore not likely that Jesus gave all this teaching on a single occasion.

The heart of the Sermon

The main concern of the Sermon is the behaviour appropriate to those who believe the Gospel. Though it does tell us of the response called for in certain circumstances, it does not provide a code of moral rules for all occasions. The main concern is with a **Christian's attitude**, the motives which lie behind the outward act. The popular idea that if only people would follow the moral teachings of the Sermon on the Mount, the world would be a better place, may be true, but it misses the point, for the heart of the Sermon on the Mount is not morals, but religion. Jesus said, 'A good tree always yields good fruit, and a poor tree bad fruit' (Matthew 7:17); Christian behaviour is the fruit of religion, and could not exist without it. What God requires of the Christian is not obedience to a code of conduct, but a response to His love which expresses itself in love towards men.

6.6 The Beatitudes

Eight in number, they describe those who are happy at the deepest level of their nature. They are fortunate not in the way the world assesses this, but from the point of view of the Kingdom of heaven. 'How blest' bears the meaning 'How happy'.

1 *'How blest are those who know their need of God; the Kingdom of Heaven is theirs.'*
The Gospel is for those who know their need, not for the self-satisfied and self-righteous. Note the parable of the Pharisee and the Publican (Luke 18:9–16) where the prayer of the Pharisee is not a prayer at all, but a self-congratulatory speech to God. The publican knows his need of forgiveness and his prayer is accepted.

2 *'How blest are the sorrowful; they shall find consolation.'*
The sorrowful here are those who take a caring attitude towards the sins and sorrows of the world. An example is found in Luke 19:41–44 where Jesus is recorded as weeping over Jerusalem because he foresaw the destruction which would come upon her because she rejected the Gospel of God.

3 *'How blest are those of a gentle spirit; they shall have the earth for their possession.'*
These are the humble and unassuming; those who do not stand on their rights. They are not weak nor necessarily submissive. Thus the picture of Jesus as 'meek and mild' is very misleading, for though meek, he could never be described as mild, as the demands he made upon himself and his followers show. Those of a'gentle spirit' are those whose strength and confidence are not in themselves, but in God.

4 *'How blest are those who hunger and thirst to see right prevail; they shall be satisfied.'*
This describes the deepest longing that God's will shall be done in oneself as well as in society. This prayer has been described as 'Taking part with God in the struggle for the kind of world God wants.'

5 *'How blest are those who show mercy: mercy shall be shown to them.'*
The Gospels give us many examples of how, by his attitude towards sinful and anxiety-ridden people, Jesus restored broken lives. The conduct of the Good Samaritan is described as an act of mercy towards a man in dire distress. The parable of the Unforgiving Debtor (Matthew 18:21–35) teaches that our claim on the mercy of God depends upon our willingness to show mercy to others.

6 *'How blest are those whose hearts are pure: they shall see God.'*
In the Old Testament the heart was the seat of thought rather than of emotion. To be pure in heart, therefore, means to be pure-minded. When we ascribe to others the poor standards of thought and motive which are to be condemned, we may well be reflecting our own low standards. The pure in heart do not readily attribute evil motives to others. More than this, they are single-minded in their service of God.

7 *'How blest are the peacemakers; God shall call them His sons.'*
The peacemakers are not only those who strive to end the strife between individuals and nations, but those who have peace within themselves, those whose anxieties are relieved because of their absolute trust in God. Because God is a God of peace, the peacemakers possess an attribute of God and to that extent are sons of God.

8 *'How blest are those who have suffered persecution for the cause of right; the Kingdom of Heaven is theirs.'*
Jesus warned his disciples that they would suffer persecution for their loyalty to him. Their

suffering was to be seen as a cause for rejoicing, for they were in the company of the prophets of God, who had suffered in the past.

6.7 Jesus and the Law

Jesus, by his conduct and teaching, greatly angered those who upheld the Law of Moses because they believed that he wished to destroy it. For the devout Jew the **ritual law**, which was concerned with the ceremonies of their religion, was just as binding as the **moral law**, which was concerned with their duties towards their fellow men, so that when for example, Jesus did things on the sabbath day which broke the ritual law they were scandalized. He not only failed to observe such ritual laws as ceremonial washings (Matthew 15:2; Mark 7:1–4), but also claimed the right to reinterpret the laws of right behaviour; yet he could say that he had not come to abolish the Law, and even said that the Scribes and Pharisees sat in the chair of Moses and that attention should be paid to their words (Matthew 23:2).

He observed the Law about attendance at synagogue worship on the sabbath (Luke 4:16) and also commanded a leper he had healed to appear before the priest and to offer the sacrifice required by the Law after the cure had been certified (Matthew 8:48; Mark 1:44; Luke 5:14). He also upheld the laws of right conduct and criticized the Pharisees (Matthew 23) because they made no distinction between the ritual and moral laws. Thus he accused them of allowing a ritual law to take precedence over the moral law which set out the duty of children towards their elderly parents. He reminds them that Moses said, 'Honour your father and your mother', but the ritual law allowed them to set apart for the work of God 'Corban', money or goods which ought to be used for the support of needy parents (Mark 7:9ff). In this way the ritual law contradicted the Law of Moses.

Jesus, then, upheld the Law, saying that 'not a letter or stroke' would disappear from it, and yet he felt free to criticize it and even to amend it.

The key to this apparent contradiction is that **Jesus made a careful distinction between the ceremonial and moral or ethical law**. He was not against ritual in principle, but he saw that it could be self-defeating. Thus he criticized those who made an outward show of their religion by standing to pray at street corners so that their piety could be observed by as many people as possible (Matthew 6:5) and also those who, when fasting, tried to make their self-discipline as obvious as possible by looking gloomy and who made 'their faces unsightly so that other people may see that they are fasting' (Matthew 6:16). The only reward such people will have is on earth. He called such people hypocrites, a Greek word which means 'someone acting a part'. Above all, however, he criticized the ritual law when it stood in the way of human need. Thus he did not hesitate to heal the sick on the sabbath day, even in the synagogue (Mark 3:1–6). His attitude to sabbath observance was that 'The sabbath was made for the sake of man and not man for the sabbath' (Mark 2:27).

Thus Jesus taught that the Law was to be observed in accordance with its original purpose, duty towards God and also towards men. If the work of God, or human need was being hindered because of the Law, then the work of God or human need claimed priority.

6.8 Re-interpreting the Rules

In the Sermon on the Mount Jesus gives five examples of how his teaching fulfilled the Law of Moses by revealing its deeper meaning. Each example begins with the words, 'You have learned that our forefathers were told. . .', and the particular law is quoted; Jesus then continues ,'But what I tell you is this'. He then proceeds not to deny the Law but to give it a deeper interpretation. He was, in fact, claiming the right to revise the sacred Law of Moses, and it is little wonder that he aroused such hostility.

1 **Murder** (Matthew 5:21–26)

The Jewish Law condemned murder, but Jesus condemned also the hatred which gave rise to it. The phrase, 'Anyone who nurses anger', must refer to the longstanding animosity of a bad relationship rather than a flash of indignation. It means nurturing a grudge. Such a person is liable to punishment by the legal court if he abuses his brother, but if he expresses his contempt in a sneer he will answer to God. Jesus is saying that anger is the root from which violence grows, or, as Shakespeare puts it, 'Hates any man the thing he would not kill?' We may never take a murderous weapon to anyone, yet we can assassinate a character with the tongue and so destroy a life. Jesus says that it is from the heart that evil arises (Mark 7:20–23), 'Evil thoughts, acts of fornication, of theft, murder, adultery' etc. Since St Paul records that among the gifts of the

Holy Spirit are patience and self-control (Galatians 5:22) we can assume that the early Christians had learned this lesson. In Matthew 5:23 and 26 we are given two examples of the harm anger can give rise to:

(a) Those who quarrel must seek to be reconciled before they approach God in worship.

(b) Those who are being pressed for settlement of a debt must pay up if they are to escape the consequences. This situation is used as an illustration of the point that it is important to be reconciled with an opponent.
A relevant parable is that of the Unforgiving Debtor (Matthew 18:23–35).

2 Adultery (Matthew 5:27–28)

The same emphasis on attitude as the spring of action is found here. Under the Law, adultery was punishable by death, but Jesus condemns also the lustful thoughts which underlie the sin. Thus the man who harbours lustful thoughts which would involve him in adultery if only he had the opportunity or the resolution, is guilty of the sin. (This is a very searching precept. We often have little control over our temptations, but when they do come to us we can refuse to dally with them, or to enjoy them vicariously by, for example, reading pornography etc.) What Jesus says raises the whole subject of sexual morality for our own generation.

3 Oaths (Matthew 5:33–37)

The Law forced a man to fulfil a promise made in a sacred name, but what Jesus says, in effect, is that any promise is binding, and that it is not acceptable to be tied to some promises and not to others. In spite of the saying of Jesus that his followers were not to take oaths, Christians (except for Quakers) do not refuse to take the oath in courts of law, for it is a legal requirement. The point of the saying is that the character of a Christian should be such that others know that his word is his bond.

4 The Law of 'An eye for an eye and a tooth for a tooth' (Matthew 5:38–42)

This was a great advance on the law of revenge, in that it limited retaliation; for revenge in practice knows no limit. Jesus, however, sets a higher standard than the Law of Moses on the principle that our treatment of others should be based on love; 'How blest are those who show mercy', he had said, and this is a principle which overrides justice. To retaliate is to sink to the level of the aggressor, whereas restraint may result in the aggressor convicting himself, recognizing that his victim is more generous than he. In Roman-occupied Palestine a legionary could compel a Jew to carry his pack for a mile (note that Simon of Cyrene was even compelled by a Roman soldier to carry the cross of Jesus); Jesus says that his followers should be ready to do much more than they were required to do. It is important to realize that Jesus taught much more than non-resistance, for refusal to retaliate may still leave resentment and even hatred for the oppressor. What Jesus requires is **reconciliation**, 'Love your enemies and pray for your persecutors' (Mark 5:43). God bestows His gifts on good and bad alike without discrimination, so Jesus says, 'You must therefore be all goodness, just as your heavenly Father is all good.'

6.9 Religious Duties (Matthew 6)

Acts of charity, prayer, fasting

Matthew Chapter 5 is concerned mainly with the Christian's attitude towards the Jewish Law, while Matthew Chapter 6 is concerned mainly with religious duties. Each subject is treated from the point of view of the same principle, namely, that the inner motive is more important than the external observances which carry with them the spiritual danger inherent in advertising one's own goodness. The principle stated in the first verse is, 'Be careful not to make a show of your religion before men.'

1 Acts of charity (Almsgiving A.V.)

This must be done as secretly as possible. 'Blowing your own trumpet' has been a metaphor for drawing attention to oneself from ancient times. The word 'hypocrite' is the actual Greek word used in the text; it originally meant an actor, and eventually came to mean someone who pretends to be better than he really is.

Acts of charity should not be self-conscious, and if the motive is merely to gain the good opinion of men, there is no other reward. To be good to others in order to feed one's own self-esteem is to be good to oneself rather than to others.

2 Prayer

Here too, the right motive is to give glory to God and not to try to impress others with a showy outward act. The attitude of prayer was standing with arms outstretched, and to pray at a street corner was to indulge in a spectacular public act of piety. It must be stressed that Jesus is not here denying the value or duty of public prayer, for he regularly took part in synagogue worship (Luke 4:16) and also, when at Jerusalem, in the worship of the Temple. Our duty to pray in private is not a substitute for the duty to identify with our brothers and sisters in God, in public prayer. Jesus also warned against the mechanical repetition of prayers where no conscious thought is required. He drew a distinction between saying prayers and actually praying.

3 Fasting

All the great religions of the world commend the discipline of fasting. When the Jews fasted they put ashes on their heads and looked gloomy so that all might see. Jesus taught that there was no virtue in keeping a fast in an ostentatious manner; the fast must be known only to God so that it might be to the glory of God and not to self. The Pharisees criticized the disciples because they did not fast as did the followers of John the Baptist, but Jesus defended them by saying that the friends of the bridegroom do not fast at a wedding, but that the day would come when he, the bridegroom, would be taken away from them, and then they would fast.

Jesus himself fulfilled each of these three duties, but he draws attention to the fact that even such commendable practices can be corrupted if men's hearts are not right. Those who perform these duties as in God's sight only, store up for themselves treasures in heaven 'which are incorruptible', while those who perform them for the sake of their own pride have their reward (the good opinion of their neighbours) on earth only.

6.10 The Lord's Prayer

Matthew places the Lord's prayer in the Sermon on the Mount while Luke says that it came as a response to the request of the disciples that he should teach them to pray as John the Baptist had taught his disciples. The prayer can be divided into two sections:

1 that God's name should be held in reverence (hallowed) and that His will be done on earth as it is perfectly done in heaven;

2 that He will supply our daily needs both physical and spiritual.

'Our Father' Luke's version omits 'our', which suggests that by the time Matthew was composed, the prayer was established as that of the community, the Church.
'In heaven', The prayer is addressed to One who is infinitely above us in holiness and yet infinitely near to all His children.
'Thy name be hallowed'; that is, held in reverence.
'Thy kingdom come', The Kingdom is central to the prayer as it is to the Gospel as a whole. Jesus uses the phrase 'the Kingdom of God' in two ways.

1 It was an event which would take place in the future when the Son of Man would come with his angels and gather his chosen 'from the four winds'. Then would come the final judgment, the end of all things, and the Kingdom would be completely established. This teaching was consistent with popular belief.

2 Jesus made an entirely new contribution by teaching that the Kingdom had come already, but spiritually, in the hearts of all who accepted God as King. While the world was still in the power of evil, Jesus was weakening its hold and reclaiming the world for God. His miracles of healing were a demonstration that the power of God was stronger than that of evil – 'If it is by the finger of God that I drive out the devils, then be sure the Kingdom of God has already come upon you' (Luke 11:20). Jesus taught his followers to pray that God's sovereignty might be as complete on earth as in heaven.

'Thy will be done, On earth as in heaven'. This phrase is omitted by Luke, perhaps because he regarded it as similar in meaning to the previous petition. It is best explained by the use Jesus made of it in his prayer during the Agony in the Garden of Gethsemane (Matthew 26:42). It means the complete surrender of the human will to God. 'Thy will be done' is a favourite inscription on tombstones, where it implies the acceptance of suffering as the will of God. This, however, represents only a part of its meaning for, perhaps more importantly, it means the

desire to co-operate with God in His great plan for the world. In the prayer Christians offer themselves as 'fellow-workers' with Him. Prayer has been defined as 'taking part in the struggle for the kind of world God wants'.

Consistent with this is the modern prayer, 'O God, give us grace to accept the things that cannot be changed: give us strength to change the things that can and should be changed: and give us wisdom to distinguish the one from the other.'

'Give us today our daily bread'. At his temptation Jesus said that 'Man cannot live on bread alone', but he did not say that man can live without it. While in the Sermon on the Mount (Matthew 6:31 – 34) he warned his followers against anxiety about bodily needs, the basis for this is his confidence that God will provide. God bestows His gifts upon His children without distinction, but the followers of Jesus are not to presume on God's generosity. They are to ask for the gifts that their heavenly Father is glad to supply.

'Forgive us the wrong we have done, As we have forgiven those who have wronged us.' Elsewhere, as for example in the Parable of the Unmerciful Servant (Matthew 18:21 – 35) reconciliation with men is made a condition of the forgiveness of God. Matthew's version implies that this condition has already been fulfilled.

'And do not bring us to the test, But save us from the evil one'. The wording appears to imply that God is a tempter rather than a deliverer, but the Epistle of James makes it clear that this cannot be so. 'No one under trial or temptation should say "I am tempted by God" for God is untouched by evil, and does not Himself tempt anyone.' (James 1:13). The words of Jesus to his disciples in the garden of Gethsemane shed light upon the difficulty – 'Stay awake and pray that you may be spared the test.' The tests of faith originate within ourselves, they are self-generated because of our temptation to live our lives in independence from God. The life of faith is bound to appear to us as a series of tests, but when we turn to Him, God enables us to face these trials; He can 'deliver us from evil'.

Note that Luke's version of the Lord's Prayer (Luke 11:1–4) is placed in a different setting and is in a briefer form. The omissions may be in order to delete words and phrases which, though meaningful to Jewish Christians, might be misunderstood by gentiles. If this is so, Matthew rather than Luke, has preserved the prayer in its original and more Jewish form.

6.11 Teaching about Anxiety (Matthew 6:25–34)

This passage is summed up in the final sentence, 'Do not be anxious about tomorrow; tomorrow will look after itself.'

Jesus is certainly not warning his followers against prudent provision for the future, but against the anxiety which becomes a paralysing obsession. Those who put God first, rather than material possessions, will find little cause for worry. Jesus points out that to worry about matters which are outside our control is useless. God provides even for the flowers and the birds: much more then, will He provide for His children.

6.12 Judging Others (Matthew 7:1–6)

'Pass no judgments, and you will not be judged.' On the face of it, this seems very impractical, for it is impossible for us not to make judgments of many kinds, including judgments of other people. For a Christian, however, such judgments must be in the nature of a diagnosis. Thus, when a doctor tells us we have an illness, he is not condemning us, but simply stating the facts of the case; his diagnosis makes it possible for him successfully to treat the patient. Similarly, a Christian's judgment is not a condemnation, it is a generous and merciful assessment of what appears to be wrong, with the intention of offering help. When making judgments, a Christian must bear in mind that he is himself, by no means without fault.

6.13 The Golden Rule

The emphasis of the Ten Commandments is on 'Thou shalt not', a negative, while Jesus sums up the teaching of the Law and the prophets in a positive command, 'Always treat others as you would like them to treat you.' (It is not enough to do no evil, we must also do positive good.) We can only claim rights for ourselves if we are ready to accord the same rights to others.

6.14 The Principles of Christian Moral Teaching

Jesus used the teaching methods of his time, including the use of short stories (parables) and also striking proverbial sayings which were easily remembered. Such sayings are not of universal application for they are often contradictory. Thus well known proverbial sayings of our own day, such as 'Look before you leap' and 'He who hesitates is lost', while they both express a truth over a limited field, cannot each be true on every occasion for they contradict each other. Similarly, the proverbial sayings of Jesus cannot be held to be of universal application to Christian conduct. There are, however, two principles which apply to the behaviour of Christians at all times:

1 'Love the Lord your God with all your heart, with all your soul, with all your mind and with all your strength.'

2 'Love your neighbour as yourself.'

7 The Parables

7.1 Introduction

Teaching is largely a matter of explaining the unknown by reference to the known and familiar; even in science much is made of models, for science teachers often explain a concept by saying it is 'like' something else. Jesus used this method of communicating spiritual truth in vivid stories which often began with the words, 'The Kingdom of Heaven is like. . .' We call these stories 'parables'.

It is important to distinguish between a parable and an allegory.

1 A parable has one central point and the details of the story serve only to make that point more vivid and clear. By contrast, in an allegory, each detail of the story has a meaning which needs to be understood if the complete message is to be conveyed. An allegory is a story whose message is hidden in a kind of code. Bunyan's *The Pilgrim's Progress* is a good example.

2 A true parable must relate to real life, the world in which things are what they appear to be. Thus in the parables of Jesus, fish are real fish, sheep are sheep and the lost coin is local currency, whereas in an allegory this need not be so. It may conform to real life, but it may depart into the world of fantasy for example, in Genesis 41:14–36 Pharaoh's dreams of the fat and lean cows and the ripe and shrivelled ears of corn are given a point by point interpretation by Joseph. By contrast, a Gospel parable would certainly not feature lean cows eating fat cows or shrivelled ears of corn devouring fat ones. The parables of Jesus are grounded in the lives of ordinary men and women, and draw on the everyday experience of his hearers.

The parables are more than a means of illustrating an important point in a sermon and more than mere aids to memory: they are designed to challenge and to provoke decision. Jesus sometimes begins, 'What do you think?' and the hearers are invited to see the relevance of the story to his teaching about the Kingdom.

It is very important to note that the **theme of all the parables is the central theme of the Gospel, namely, the meaning of the Kingdom of God**. The Jews thought that God would come to reign on earth sometime in the future: they called it the Messianic Age, and the heart of the message of Jesus was that this intervention had now taken place. **The Kingdom of God had arrived**, God was visiting and redeeming His people in the person of Jesus himself. This is the original setting, not only of the parables, but also the miracles, and it is important to recognize this, for the present settings of some of the parables in the Gospels are not original, but those given them by the early Church in the new context of its missionary endeavour.

The parables may conveniently be divided into groups according to the four main themes of Jesus' teaching about the Kingdom.

7.2 Group 1 The Kingdom of God has Arrived

1 'The New Cloth and the New Wine' (Mark 2:21ff; Luke 5:36–39; Matthew 9:16–17)
The Gospel is not a patch that can be placed on the worn garment of Judaism, nor can the new wine of the Gospel be contained within the old wineskins of the Jewish religion. A new patch

would shrink and tear the old cloth; the new wine would, as it fermented, put a strain on an old wineskin which had lost its elasticity, and the wine would be lost. There was no way that the strict legalism of the old Law could contain the Gospel of the Kingdom.
What then, is the nature of the Kingdom and what are the laws of its growth?

2 The Mustard Seed (Mark 4:30–32; Luke 13:18–19; Matthew 13:31–32)

To many, the proclamation of the Kingdom must have been very unconvincing – there had been no earth-shattering events, only a wandering teacher with a handful of unremarkable followers, and this parable is the answer of Jesus to the doubtful. The mustard seed is very small, but grows into a large shrub, and this miracle is about to be repeated in the spiritual world.

3 The Leaven (Luke 13:20–21; Matthew 13:33)

When the yeast is placed in the warm, moist dough, it grows silently and irresistably, the bubbles permeating the whole dull lump, making it more palatable. Similarly, the transforming power of God was now at work, and nothing could stop it.

4 The Seed Growing Secretly (Mark 4:26–29)

There were those who were impatient with the slow rate of growth. Jesus says that a growing crop has to develop stage by stage: once the ground is prepared and the seed sown it has to be left to nature. In a similar way the growth of the Kingdom has to be left to God.

5 The Darnel (Tares A.V.) (Matthew 13:24–30, 36–43)

The Pharisees, who separated themselves from all who did not observe the Law, would require to know why, if the Kingdom had arrived, there had been no separation of sinners from saints in Israel. In the early stages of its growth the darnel is difficult to distinguish from the corn and more harm than good can be done by weeding. Whenever the Church has indulged in 'witch-hunting' it has led to the gravest injustice and to scandal of the worst kind, for God only knows the secrets of each heart and therefore only He can judge. We cannot judge each other by appearances.

6 The Dragnet (Mark 13:47ff)

Jesus had called some of his disciples to be 'fishers of men'. Should they try to be selective in their evangelism? Jesus tells them that the Kingdom gathers in all kinds just as a dragnet does not discriminate between the kinds of fish it enmeshes. Only when the shore is reached can the sorting be done. So with the Kingdom: the separation will come in God's good time, in the final judgment.

7 The Sower (Mark 4:1–9; Matthew 13:1–9, 18–23; Luke 8:4–8, 11–15)

Like the parables of the Darnel, the Mustard Seed and the Seed Growing Secretly, this is a parable of the growth of the Kingdom. It may be interpreted in two ways:

(a) There will be difficulties and losses, but in spite of all, there will be a splendid harvest. The parable sounds a strong note of encouragement to the discouraged. Mark, writing soon after the catastrophic disaster of Nero's persecution at Rome when many Christians, including probably Peter and Paul, had died, applies the parable to the situation of his own day and uses it as an exhortation to steadfastness.

(b) The seed is good, but depends for its fruitfulness on the soil in which it falls. The parable presents a personal challenge to the hearers: each must ask himself, 'What kind of soil am I?'

Many scholars now believe that the 'explanations' of the parables of the Darnel, the Dragnet and the Sower in the text are not original. The whole point of a parable is the clear simplicity of its message. A joke which has to be explained is a very poor one and a parable which needs an explanation is likewise a poor parable. The 'explanations' miss the central point of the parables and reflect a much later period in the development of the Church.

The subject of each of these parables is taken from country life and a quite wrong impression is given if we take this to mean that the ministry in Galilee was rather idyllic, with no conflict to disturb the calm of the scene. **The ministry was, in fact, a campaign against the powers of darkness** which was to lead, in the end, to the dereliction of the Cross and the victory of the Resurrection.

8 The Parable of the Strong Man Bound (Mark 3:23–27; Luke 11:14–22; Matthew 12:25–30)

The miracles of Jesus were demonstrations that the power of God was stronger than that of evil. His success resulted in the accusation that he was in league with the powers of darkness. 'The finger of God' means the power of God, and in the parable Jesus makes the point that if his accusers speak true, the devil's forces are divided and therefore doomed to defeat. He himself has invaded the stronghold of the devil, made him captive and released his prisoners. By exorcizing evil spirits he has proved himself master of the devil.

9 The Parable of the Empty House (Luke 11:24–26; Matthew 12:43–45)

This parable is a sequel to the previous one. The expulsion of the evil spirit is only the first stage, for it will return in greater strength if the 'house' is left empty. It is not enough merely to accept forgiveness; the Christian must also grow in holiness, by dedicating his life to the Kingdom.

These parables each contribute to the message that the Kingdom of God has arrived, that its growth is slow but irresistible, that the powers of evil are reeling in defeat before the power of God.

7.3 Group 2 Parables of Reconciliation

The Pharisees drew attention to the fact that Jesus was eating with tax-gatherers and sinners, a practice which they thought highly improper. He replied, 'It is not the healthy that need a doctor but the sick; I did not come to invite virtuous people, but sinners.' Some of the best known parables are comments on this central truth of the Gospel.

1 The Parable of the Labourers in the Vineyard (Matthew 20:1–16)

On first reading, this parable presents difficulties, for by the strict rules of justice it is unfair, and the complaint of the men who had worked all day is justified. The key, however, is to be seen in the social and economic conditions of the time. There was a considerable slave population and also a class of free labourers who depended for their livelihood largely on casual labour. In this respect at least, the slave was better off, because he had security. For the free labourer, no work meant no food. The meaning of the parable is God's generosity. It was not the fault of the labourers that they had worked for only a short time, and the Lord gave them more than they had earned because of their need. We cannot earn the gifts of God; they are given to us not because we deserve them, but because He loves us and is merciful.

From the time of the prophet Isaiah, the vineyard had been a recognized symbol for Israel, and the parable would originally refer to the 'publicans and sinners' who were answering the call of Jesus to God's Kingdom. The early Church would tend to identify the late-comers as the most recent gentile converts and use the parable to teach that the gifts of God are not earned by merit or long service.

2 The Two Sons (Matthew 21:28–31)

This was an answer to the Pharisee who complained that the Kingdom was being opened to tax gatherers and prostitutes. Jesus in effect replies that it is those who do the will of God rather than merely talk about it who are approved. The Pharisees 'say one thing and do another', whereas he found that the notorious sinners, on a practical level, often lived more in accord with the spirit of true religion.

3 The Two Debtors (Luke 7:41–43)

The circumstance of this parable is the story of the sinful woman who anointed the feet of Jesus in the house of Simon the Pharisee. We must assume that Jesus had previously assured her of the forgiveness of her sins. Jesus tells Simon that the person who has the greatest debt remitted has the greatest reason for gratitude. The woman's overwhelming display of gratitude indicated that her many sins had been forgiven. The parable has much in common with that of the **Two Sons**.

4 The Great Supper (Luke 14:16–24) and **The Marriage Feast** (Matthew 22:1–14)

These parables can be regarded as variations of the same story, differing only in detail. The banquet which stands prepared and ready means the Kingdom of God. (The great banquet was a popular symbol of the Messianic Kingdom.) The following points need to be noted:

(a) The guests had been invited in good time. After accepting the invitation they had allowed personal interests to take priority.

(b) The first two excuses are very weak. The field would not run away and the purchaser merely wanted to gloat over it. Similarly, a man would hardly buy expensive draught animals without first trying them out. The third excuse is better, because a Jew was excused even military service for a year after marrying. This, however, could not excuse the breaking of a long-standing social engagement.

(c) The parable is introduced by the remark of a man who said, in effect, how happy it would be to be invited to God's banquet, and to accept it. Jesus is telling the Pharisees that they have actually received this invitation and have rejected it. The invitation is now transferred to those who have no claim on God except their need, the sinners and the gentiles.

(d) 'Invite them to come in.' The authorized version says 'compel them' and this phrase has, in the past, been made the excuse for enforced 'conversions'. People cannot be persecuted into the Kingdom, for this is to do violence to their freedom. Indoctrination is a subtle form of this, because it induces people to believe in such a way that no argument, however conclusive, can shake that belief.

(e) The parable has allegorical elements. The 'servant' in Luke who was sent to the highways and byways to seek new guests, would appear to represent Jesus, and the guests, sinners and gentiles. Matthew refers to 'servants' rather than 'servant', and the point is lost in his version.

(f) Matthew's account presents the difficulty of the new wedding guest (Luke's 'supper' is, in Matthew, a wedding feast) who had no wedding garment. He could hardly be expected to be wearing one if he had been called in from the field. It has been suggested that the king provided the proper garment and the guest's neglect to wear it was therefore a studied insult. This interpretation seems unlikely. It may be another allegorical inset. Matthew may have felt that the parable as it stood, made entry into the Kingdom too easy and that a minimum requirement would be to wear the garment of penitence.

5 The Wedding Guests (Luke 14:7–11)

Jesus turns this rule of social etiquette into a parable. He sees that the Pharisees love to sit at the places of honour and to claim preferential treatment. On the spiritual level this represents the presumption that one's own high opinion of oneself is shared by God. Jesus says, in effect, that God honours those who make no claim on His favour. Those who judged themselves entitled to the best places would be disappointed, for the self-righteous are far from the Kingdom of God.

6 The Pharisees and the Publican (Luke 18:10–14)

The Pharisee does not actually pray at all, for what he utters is a boastful catalogue of his own virtues which he apparently expects God to appreciate. The words, 'even as this tax collector', perfectly express the contempt of the self-righteous for the sinner. The Pharisee stands in self-confidence before God while the tax collector stands 'afar off', overwhelmed by his own unworthiness. The word 'justified' recalls Paul's teaching of 'justification by faith'. The sinner's prayer, not that of the Pharisee, is acceptable to God. None is as far from the Kingdom as the self-righteous.

7 The Lost Sheep (Luke 15:4–7; Matthew 18:12–14)
The Lost Coin (Luke 15:8–10)
The Prodigal Son (Luke 15:11–32)

Three parables of the lost, each carries the message of God's forgiveness of the repentant sinner.

The sheep is lost through its wayward foolishness while the coin is lost by accident; unlike the sheep, the coin has no initiative. The message is that if the shepherd shows such joy over the recovery of a lost sheep, and a woman over her lost coin, how much more does God rejoice over the recovery of one of the lost sheep of the house of Israel.

The Parable of **the Prodigal Son**, with that of **the Good Samaritan**, is the best known of all the parables, and is the third about God's concern for the lost. Here, the son is lost through his own deliberate choice, unlike the unintelligent sheep or the inanimate coin. Note however, that it is a parable about **two sons**, one lost in a foreign land and one lost at home behind a fence of self-righteousness.

The parable has certain elements of allegory; for example, the father obviously stands for God, the prodigal for sinners and the elder brother for the Scribes and Pharisees. It is, however, too close to real life to be an allegory.

Note

1 Pigs were and are 'unclean' to the Jew, and this underlines the point that the prodigal was driven to the depths of degradation.
2 'When he came to his senses' he had reached his own personal gutter. He was now truly himself. He knew that he deserved nothing and determined to throw himself on his father's mercy after confessing his fault.
3 'Father, I have sinned against heaven' etc. This is the classic statement of repentance. He had offended against God and man.
4 The father did not wait for him to arrive, but seeing him a long way off went out to meet him. ('His heart went out to him')
5 The destitute son is reclothed, an eloquent symbol of forgiveness and restoration to the family.

The elder son represents the Scribes and Pharisees. He was working to support the family while the younger son was wasting the family inheritance. On the face of it, his bitterness is not hard to understand, nor is his rejection of his brother with the words 'this son of yours'. The father's approach is persuasive, calling him 'my boy'. 'This your brother', he says, reminding him of the blood relationship, 'who was dead and has come back to life, was lost and is found.'

God loves the sinner even while he is still in his sin, and when he repents God forgives and restores him to His family. The message for the Scribes and Pharisees is that their way of treating sinners is not God's way. Note however, that the rebuke of the elder brother, though firm, is a speaking of the truth in love.

7.4 Group 3 The Kingdom's Claims

1 The Hidden Treasure (Matthew 13:44)
The Pearl of Great Price (Matthew 13:45–46)

These twin parables are a challenge to decision. Their general meaning is that no price is too high to pay to gain the Kingdom. A subtle difference is worth noting, however, that the discovery of the treasure was unexpected, while the pearl was found after a long, deliberate search. Conversion can appear to come suddenly or after a long, conscious search and a rejection of alternatives.

2 The Tower Builder (Luke 14:28–31)
The King going to War (Luke 14:31–33)

Jesus warns those who wish to enter the Kingdom that it involves complete commitment. They must calmly count the cost of discipleship in terms of loss of family ties, and a hard and uncertain future, before reaching a firm decision.

3 The Unworthy Servants (Luke 17:7–10)

Jesus knew himself to be a Messiah who came not to be served but to serve, and his disciples must be, above all, servants of God. The parable was aimed at the Pharisees who were preoccupied with trying to earn God's favour by good deeds. His own disciples were to be servants of God without thought of reward. To serve God with a reward in mind is to serve ourselves.

4 The Unjust Bailiff (Steward A.V.) (Luke 16:1–8)

Jesus once told his disciples that they must be 'wise as serpents and harmless as doves'. In this parable Jesus stresses that those who belong to the Kingdom will need to be no less shrewd and astute than the worldly. In other parables the king or master represents God, but since here the master approves of a servant who behaves dishonestly, though shrewdly, we must look for another interpretation.

(a) The rich man is himself dishonest for he finds in the bailiff a man after his own heart.

(b) The bailiff has cheated his master, and consistent with this, goes on cheating him to protect himself from the full consequences of his dishonesty.

(c) The debtors are dishonest because they collaborate with the dishonest steward, and, consistent with this, are by implication prepared to harbour a thief.

Thus, all the characters in the parable are dishonest and, moreover, consistent in their dishonesty. The meaning of the parable is revealed in the verse, 'The worldly are more astute than the unworldly in dealing with their own kind.' The children of light have, in this respect, something to learn from the children of darkness.

5 The Friend at Midnight (Luke 11:5–8)

On first reading this would seem to mean that the sons of the Kingdom must be persistent in prayer. In view of the fact that Jesus warned that his followers were not to imitate the heathen who believe that 'the more they say, the more likely they are to be heard' (Matthew 6:7), the meaning must go deeper. It is a parable of faith. If a reluctant friend can be persuaded by persistence to do what is required, how much more will a loving heavenly Father hear the prayers of His children.

6 The Unjust Judge (Luke 18:2–8)

If an unjust judge who needs to be pestered before he gives the protection of the law to a widow, can be prevailed upon by sheer persistence, how much more shall a loving heavenly Father hear the prayers of His children.

Both the Parable of the Friend at Midnight and that of the Unjust Judge teach that those who claim a place in the Kingdom must have strong faith.

7 The Unmerciful Servant (Matthew 18:23–35)

Those who belong to the Kingdom must show to others the same forgiveness that they themselves have received. This forgiveness must be 'from the heart', that is, utterly sincere. The unmerciful servant owed a vast sum which he would be quite incapable of repaying, whereas his 'fellow servant' owed a comparatively trivial sum. Our indebtedness to God is infinitely greater than the debt others owe to us.

8 The Speck (Mote A.V.) **and the Plank** (Matthew 7:5)

This makes a similar point to the previous parable. Before judging others, those of the Kingdom must look to their own faults and remember how much God has forgiven them.

9 The Good Samaritan (Luke 10:30–37)

Jesus taught that the two great principles of the Kingdom were love of God and love of neighbour. The parable of the Good Samaritan is his answer to a lawyer's question. 'Who is my neighbour?' A neighbour is anyone in need, irrespective of religious, racial, social or political divisions.

Few parables have suffered more from allegorical interpretation, but the parable has a clear, single meaning. (For background see Unit 1.3 Samaritans.) It is essential to the point of the story to realize that Samaritans were the most hated of the non-Jews.

The traveller was left 'half-dead'; the priest and the Levite seeing him may have been too concerned for their own safety to offer help. If he were dead, to touch him would involve ceremonial defilement and make them temporarily unfit for their ritual duties. If this was why they passed by, it means that they put the claims of the Law before their duty to their fellow men. In the parable Jesus is saying that if obedience to the Law takes precedence over human need, a heretic half-breed could fulfil the will of God better than the Scribes and Pharisees.

10 The Two Foundations (Matthew 7:24–27; Luke 6:47–49)

Those who belong to the Kingdom must be doers, not merely hearers. Their lives must be built on the firm foundation of obedience to the teaching of Jesus.

7.5 Group 4 Gathering Clouds

Many of the parables are to be understood in the context of the crisis that was eventually to fall upon the nation. Jesus saw that the failure of the Jewish nation to accept that the Kingdom had come, and their rejection of God's Messiah, would accelerate the collision course with Rome and lead to the destruction of Jerusalem and the Temple. The death of the Servant Messiah was now inevitable, but it would lead to resurrection and the rise of the New Israel.

1 The Rich Fool (Luke 12:13–15)

It can be taken as a warning against greed, but in origin it may well have been a warning about the catastrophe which was to come. Why squabble about money when the end might come any day?

2 The Faithful and Unfaithful Stewards (Luke 12:42–46; Matthew 24:45–51)

The early Church saw this parable as a warning to its leaders to be faithful in the interval before the Second Coming, but the original message was probably addressed to the Jewish leaders who had burdened the people of God with so many rules and regulations that they were hindering rather than advancing the Kingdom. The day of God's judgment was near when an account would be made of their stewardship.

3 The Talents (Matthew 25:14–30)
The Pounds (Luke 19:12–27)

Matthew and Luke have different versions of the same parable and both interpret it as a warning to the Church to be faithful to its calling during the interval before the Second Coming of Christ. The message in the original setting is directed to the Scribes and Pharisees, identified as the wicked and slothful servant, for they kept the word of God to themselves with their policy of narrow exclusiveness, and so hid it from the world at large. This policy not only embittered other nations against Israel, but also hindered the spread of the knowledge of God to all mankind.

4 The Barren Fig-Tree (Luke 13:6–9)

The fig-tree is a symbol of Israel. Jesus is the man who came in search of fruit, but found none. Israel's time for repentance is short, but the people are being given one last chance.

5 The Rich Man (Dives) and Lazarus (Luke 16:19–31)

The parable is addressed not to the Pharisees but to the Sadducees, because the rich man in his lifetime, like them, did not believe in an after-life, though he did believe in Moses and in the Law.

It is not intended that we should take literally the picture of heaven and hell. The point is being made that there is an after-life, that it is continuous with this one, and that our behaviour here on earth has eternal consequences. Believing that this life was the only one, the rich man felt that it could not be better spent than in luxurious living. He had no qualms about spending his wealth irresponsibly because he never believed he would have to answer for his conduct before God. He was also too callous, too lacking in humanity, to care for the sufferer at his very gate.

After death the roles are reversed, but the Rich Man, even then, does not entirely accept this. He says, 'Send Lazarus' first to bring him water, and afterwards to take a message for him. Having ignored Lazarus during his lifetime, he still feels entitled to treat him as a servant. At this point the story takes a new twist. The Rich Man claims to have acted in ignorance of the consequences and cares enough for his brothers to request that a message be sent to them, an unmistakable sign, so that they might be saved from the terrible consequences of their selfish lives. The reply is that they already have the testimony of Moses and the prophets and if they will not hear them, neither will they be persuaded by an apparition.

6 The Watchful Servants (Luke 12:35–38)

The early Church quite naturally took this parable to mean that Christians must be in a continual state of preparedness for the Parousia, the Return of Christ in Glory. The disciples of Jesus may well have given it a more immediate interpretation. A great crisis of the Kingdom might come at any moment, and they must be alert for any eventuality.

7 The Thief at Night (Luke 12:39)

The meaning is very similar to that of the parable of the Watchful Servants.

8 The Ten Virgins (Matthew 25:1–13)

As it stands, the parable is about the Second Coming and the need for vigilance. The expected Second Coming had not yet taken place and the early Church saw in this story of a village wedding at which the bridegroom was delayed, a parable stressing the need for watchfulness. The bridegroom would come unexpectedly and the Church should not be found unprepared.

However, as in the parables of the Watchful Servant and the Thief at Night, the original message was probably that a great crisis of the Kingdom was close at hand, and Israel was being given its last opportunity. The door would be shut: the decision to reject the Kingdom would be irrevocable.

9 The Wicked Husbandmen (Mark 12:1–9; Luke 20:9–19; Matthew 21:33–46)

From the days of the prophet Isaiah, the vine had been a symbol of Israel. The Exodus is referred to as the transplanting of a vine: 'You took a vine out of Egypt and when it had taken root it filled the land.' The faithlessness of Israel is expressed in terms of a neglected vineyard which produces no fruit (Isaiah 5:1–7).

Jesus uses the symbol of the vineyard to tell the story of Israel's failure to respond to the call of God throughout its history. The servants sent by God are the prophets who had been rejected by Israel. There is an echo of this in the judgment of Jesus on Jerusalem: 'O Jerusalem, Jerusalem, the city that murders the prophets and stones the messengers sent to her.' (Luke 13:34).

Jesus himself is the central figure of the story, the beloved Son of the Father who in a few days time is to fulfil the prophecy in the parable. He is to be rejected by the leaders of Israel and condemned to death.

10 The Sheep and the Goats (Matthew 25:31–46)

The Last Judgment is to involve all nations, not only the Jews, so by implication the parable is the answer to the question of what principle will those who have not known Christ be judged by at the Great Assize. The answer is that they will be judged by the compassion they have shown to those in need. To serve the needy is to serve Jesus himself. 'Anything you did for one of my brothers here, however humble, you did for me' (Matthew 25:40). Many modern scholars take the view that in this case, 'my brothers here' refers not only to disciples but to all men everywhere. The heathen who show compassion have a share in the Kingdom.

8 The Miracles

8.1 Introduction

In stark contrast to modern Science, the Bible knows nothing of a creation which is bound by a closed system of natural laws of cause and effect. The universe is **God's Creation** and everything that happens within it is the handiwork of God Himself. Nature is, however, only the backcloth against which the major work of God, His 'Grand Design', takes place, and for the men of faith pictured in the Bible, God is not a proposition discoverable by reasoned argument, but a personal reality known in the experience of their daily lives. They were as conscious of living in His presence as they were of living in the natural world, and they would certainly say, with Paul, 'In Him we live and move and have our being.' God is, in the Bible, not the God of the philosophers, but 'the God of Abraham and Isaac and Jacob', a living, present reality.

The writers of the Bible did not believe that God had left the Creation to run itself as a totally independent scheme of things: on the contrary, they believed that He was in control of events, intervening directly from time to time, saving men in times of crisis and also judging them for their disobedience. His interventions into human affairs were discernible to the eye of faith, and are the 'miracles' of the Bible, events which were sufficiently dramatic for faith to recognize as acts of God. Faith did not make them acts of God, but recognized them for what they were.

The writers of the Bible accepted miracles because they believed in a living, personal and all-powerful God who expressed His will in the course of nature and history. Since He was the creator and sustainer of the universe, all things were possible to Him, though He did not exercise His power in an arbitrary or capricious way, like an oriental despot, for He was just and holy.

Is it reasonable to believe in the miracles of the New Testament today?

Much depends on our definition of a miracle. If it is defined as an event which is contrary to the laws of nature, miracles cannot happen. (To the believer, the laws of nature are also the laws of God, and therefore unchangeable.) If, however, with St Augustine, miracles are defined as events which appear to be contrary to what we know of the laws of nature, we are in a different case, for the clear implication is that science does not claim to know all the secrets of the universe. This does not entirely dispose of our difficulty, for it is plain that the Church in its early history rejected some of the reported miracles of Jesus recorded in the so-called apocryphal Gospels.

The early Christians found that there were apparently believable and unbelievable stories of the miraculous. Most critics nowadays would accept the New Testament miracles of healing in view of the discovery by psychologists of psychosomatic illnesses, that is, illnesses which are caused by worry, anxiety, distress or guilt, rather than physical causes. Such illnesses can be cured by psychiatrists and even by faith healers: remove the anxiety and the illness is cured. It must be stressed, however, that Jesus was not a faith healer. Nowhere in the Gospels is it suggested that Jesus could not have worked a miracle if the subject of the miracle had no faith in his power to heal.

The most important thing about the miracles is their meaning, and this can only be understood in the context of the whole ministry of Jesus. The New Testament miracles have their roots in the prophecies of the Old Testament and the hope of Israel that God would finally intervene in human affairs by sending His Messiah, the Anointed One. This is made clear in the first account we have of the preaching of Jesus when in the synagogue of Nazareth he expounded the opening passage of Isaiah Chapter 61,

> The spirit of the Lord is upon me because He has anointed me;
> He has sent me to announce good news to the poor,
> To proclaim release for prisoners, and recovery of sight for the blind (Luke 4:18ff).

The miracles of Jesus were fulfilments of such prophecies as Isaiah 35: 4–6. When God comes:

> Then shall blind men's eyes be opened
> and the ears of the deaf unstopped.
> Then shall the lame man leap like a deer
> and the tongue of the dumb shout aloud.

They are evidence of the fact that the long-awaited Kingdom had come, and point to the fact that Jesus himself is the Messiah, the promised one. They make the same point as the parables and, like them, are not a secondary element of the Gospel story which can be omitted without damage, but are an essential part of the whole. **Over thirty per cent of Mark's Gospel is devoted directly or indirectly to the subject of miracle.**

1 Jesus made it clear that he regarded his miracles as evidence of the arrival of the Kingdom. 'If it is by the finger of God that I drive out the devils, then be sure the Kingdom of God has already come upon you.' (Luke 11:20). The New Testament writers see the power of Jesus as displayed in his ministry, not as the power of an ordinary man, but the power of God in action. Peter, in his speech at Pentecost (Acts 2:22) speaks of the miracles and signs done by Jesus as actions 'which God worked among you through him'. Again, in the account of the conversion of Cornelius in Acts 10:38 Peter declares of Jesus, that 'God anointed him with the Holy Spirit and with power.' while Paul in 1 Corinthians 1:24 describes Jesus as 'the power of God and the wisdom of God'. The miracles of Jesus are seen not as acts of a human wonder-worker, but as a revelation of the power and purpose of God.

2 The miracle stories were used by the earliest Christian preachers to illustrate the essential features of the Gospel.

The miracles of healing provide good examples: the restoring of sight to the blind, hearing to the deaf, the cleansing of the lepers, all had deep religious significance. In spite of the message of the Book of Job, the general assumption of the Bible is that sickness, handicap or even accident are the result of, or even punishment for, sin (though Luke 13:1–5 makes it clear that Jesus himself repudiated this belief). Thus, to take Mark's story of the healing of the leper (Mark 1:40–45):

1 The leper was unclean not only because he was contagious, but also in a religious sense: he was ceremonially defiled, but also as a sinner he was excluded and his life was a living death.

2 God's attitude to the sinner is made clear by the fact that Jesus accepted ceremonial defilement by touching the leper. He became a bearer of sin.

3 The sinner in his defilement has no right to approach a Holy God, just as the leper has no right to approach Jesus. Jesus touched the leper; God also puts forth his hand to save.

4 The leper is told to present himself to the priest and to make the offering laid down by Moses. What he formerly could not offer by his own merit he can now offer to God through Christ. **The miracles of healing were thus parables of God's forgiveness in action.**

It is clear from the charge given by Jesus to his disciples that they were given power to heal as well as to preach. Mark says that Jesus gave the twelve disciples power over unclean spirits (Mark 6:7–13) while Matthew adds that he gave them power to heal all manner of sickness and disease (Matthew 10:1). Luke says that they were sent 'to proclaim the Kingdom of God and to heal' (Luke 9:2). Thus the ministry of healing was placed side by side with the ministry of preaching. When John the Baptist sent some of his followers to ask Jesus if he were indeed 'the one who is to come', Jesus replied in words which reflect the language of Isaiah 35:5ff, 'Go and tell John what you have seen and heard: how the blind recover their sight, the lame walk, the lepers are cleansed, the deaf hear, the dead are raised to life' (Luke 7:22ff). The message to John was clear – there could be no doubt that the Messianic Age prophesied by Isaiah had arrived.

8.2 Miracles of Healing

The Paralysed Man (Mark 2:1–12; Luke 5:17–26; Matthew 9:1–8)

This is one of an important group of miracles in which the faith of others rather than that of the sick person is accepted.

The words of Jesus to the paralysed man, 'Your sins are forgiven', imply that Jesus accepted the popular belief that all illness is caused by sin, though his comment on the fate of the Galileans described in Luke 13:1–5 suggests that he did not. However, everyone present, especially the Pharisees, would have accepted the popular belief.

The miracle lends itself to a 'modern' interpretation, namely, that the paralytic is so overwhelmed by his sense of guilt that it has completely incapacitated him. It is his way of opting out of his situation, albeit unconsciously, of course. Jesus, who 'knew what was in man' correctly diagnoses the trouble, convinces the man that his load of guilt is removed, and restores him to wholeness of life, physical and spiritual. Our knowledge of psychosomatic illnesses (those caused by such things as anxiety, a sense of inferiority or guilt, rather than physical causes) makes this interpretation seem plausible. Neat as it may be, however, this interpretation misses the essential point, for 'faith' in the New Testament does not have the modern sense of faith-healing. Jesus does not heal by the power of suggestion like some rather civilized witch-doctor. **'Faith' in the Gospel means faith in the power of God which is active in Jesus.** The appeal of Jesus is to faith in God, not to faith in himself.

Jesus does not offer the miracles of healing as proofs that he is the Messiah, for this would be contrary to his refusal at the Temptation to perform a miracle to prove his own authority as Messiah. The miracles are, rather, consequences of the dawning of the Kingdom of God on earth; they are an essential part of the breaking of the powers of evil and the inauguration of the Kingdom of God. The Jews were well known throughout the Roman world for their skill in exorcism, and 'miracles' of this kind were not uncommon. Those who witness the miracles of Jesus are free to decide that they are done by the power of Beelzebub, the prince of devils, as the Pharisees accused, or by the power of God, as he himself claimed.

The Woman of Canaan (Syro Phoenician Woman A.V.) (Mark 7:24–30; Matthew 15:21–31)

Here again the faith of someone other than the sick person is accepted. The incident presents great difficulties because it gives us a picture of Jesus which seems thoroughly inconsistent with what we learn about him in the rest of the Gospel. It is difficult to understand how one who could be so compassionate towards the outcasts of society could be so harsh to a suffering gentile woman.

Jesus is saying that his work was, at the time, directed to the Jews, and mainly to the twelve disciples whose work it would be to spread the Gospel after his death. Priority was to be given to the proclaiming of the Kingdom to the Jews, where such a proclamation would be understood, rather than to gentiles.

The woman's first request is met by silence: she is completely ignored, but since she continues to make a scene the disciples beg him to send her away. The woman now kneels and implores his help. 'Help me, Sir' she says, to which Jesus replies, 'It is not right to take the children's bread and throw it to the dogs.' The picture is softened somewhat by noting that the word used for 'dog' is that for the domestic animal not the scavenging wild dog of Palestine.

Possibly Jesus is here expressing not his own sentiments, but those of his Jewish disciples who had no time for a gentile. Jesus knows the depth of the woman's faith and she is equal to the challenge. Her reply is not only a superb statement of faith, but also a rebuke to the disciples for it is a reply to their attitude rather than to that of Jesus. Jesus, in fact, commends the gentile woman for her faith.
It is to be noted that:

1 through faith, the barrier between Jew and gentile is broken down, and gentiles become, with Jews, the people of God;

2 the miracle is an example of the power of Jesus to heal at a distance;

3 a parallel miracle is that of the Centurion's Servant in Luke 7:1–10;

4 the faith of someone other than the sick person is accepted.

The Centurion's Servant (Luke 7:1–10; Matthew 8:5–13)

A centurion was the highest ranking non-commissioned officer in the Roman army, a position earned by long service and bravery in action. Centurions are always mentioned favourably in the New Testament, e.g. the centurion at the Cross. The first recorded gentile convert was Cornelius, a centurion.

In this incident the centurion was a God-fearer, i.e. one of the many gentiles attracted to the Jewish religion by its high moral standards, but not a convert. To become a proselyte, or full convert, involved circumcision, (a painful and even dangerous process), the acceptance of Jewish food laws, and social separation from gentiles. Understandably, many preferred not to take this final step. This unnamed centurion must have been wealthy for he had built a synagogue. The special lessons of this miracle are:

1 as an officer in the Roman army he knew that his commands would be obeyed because he held the delegated authority of the Emperor. His faith did not consist of trust in the ability of Jesus to heal, but in discerning that Jesus held authority from an even higher source, that of God. Jesus brought to bear the power of God and this showed that God's Kingdom had come

2 the healing of the servant was made possible by the faith of someone other than the sufferer

3 the miracle was performed at a distance without any contact with the patient

4 the centurion, a gentile, received the highest praise ever uttered by Jesus, 'Nowhere, even in Israel, have I found faith like this.'

The Epileptic Boy (Mark 9:14–29; Luke 9:37–43; Matthew 17:22–23)

Both Luke and Matthew abbreviate Mark. Though the disciples had been succesful in exorcizing evil spirits on their mission tour, they were powerless in this case. The symptoms described clearly indicate some form of epilepsy.

The heart of the incident is the dialogue between Jesus and the father. The appeal for help is met with the reply, 'All things are possible to one who has faith.' It is an invitation to accept the healing power offered by God. The father, however, is forced to fling himself still further on the mercy of God for he feels his faith to be unequal to the occasion, saying, 'Help me where faith falls short.' His acknowledgment of his own weakness and his need of help, his confidence that Jesus, by the power of God, is able to strengthen him, makes the miracle possible.

The testing of the faith of the disciples to the uttermost was, at this point, not far ahead. The suffering and death of the innocent has always been a severe test of faith and the disciples were to remember this event when their own time of testing came.

In each of these three miracles, the Woman of Canaan, the Centurion's Servant and the Epileptic Boy, it is the faith of someone other than the sick person which makes the healing possible.

8.3 Healing on the Sabbath

The observance of the sabbath was of vital importance to the survival of the Jewish religion in a hostile gentile world. Many of the requirements of the Jewish Law could be fulfilled or neglected in the privacy of the home, but the command to keep the sabbath made it obligatory to make a public witness by abstaining from work. Any threat to the sabbath was a threat to Judaism as a whole and was bound to meet with hostility.

It appeared that Jesus went out of his way to break the sabbath regulations. In fact, the enemies of Jesus were wrong to accuse him of breaking the Law: he broke only the letter of the Law when it obstructed the fulfilment of its spirit.

The Man with the Withered Hand (Mark 3:1–6; Luke 6:6–11; Matthew 12:9–14)

The rabbis permitted medical treatment on the sabbath if life was endangered, but for no other reason. Jesus argued that to do good could not be wrong at any time. The miracle is important for the fact that Mark records that the Pharisees and the Herodians, two most unlikely allies, made common cause in plotting to kill Jesus. In Mark 12:13 the Pharisees and Herodians once more combine forces to trap him.

The Man with Dropsy (Luke 14:1–6)

This miracle is similar, but the grounds of its justification are different. It was permitted by the Pharisees to rescue an animal which, through accident, was in danger of death on the sabbath; how much more justified was it, therefore, to heal a man on the holy day.

The Crippled Woman (Luke 13:10–17)

The president of the synagogue did not rebuke Jesus directly, but addressed himself to the congregation, forbidding them to present themselves for healing on the sabbath. Jesus replied that if it is permitted to untie the farm animals to take them to water on the sabbath, how much more fitting it is to release this woman 'who has been kept prisoner by Satan for eighteen long years'. His enemies were covered in confusion, but the congregation as a whole was delighted. ('A daughter of Abraham' means a Jewess, a descendant of Abraham; the overtone is 'one of us', a member of the family.)

8.4 Raising the Dead

The Daughter of Jairus (Mark 5:21–43; Luke 8:40–56; Matthew 9:18–26)

The noise in the house was probably being made by hired mourners whose wailing was designed to drive away evil spirits. Mark records the actual Aramaic words Jesus used, 'Talitha cum' meaning 'Get up, my child', and this provides a note of authenticity.

Some scholars suggest that the girl was only in a coma and was not dead. They point out that it was the messengers only who said she was dead, and that Jesus himself said that she was not dead but asleep.

The request of Jesus that she should be given something to eat displays thoughtfulness and concern for her full recovery.

Whatever the explanation, it is plain that the early Christians saw this incident as an act of God. It is evidence that God has intervened in human history in that 'the dead are raised up'. Death is no longer final. It is significant that the father was an official of the synagogue.

The Widow's Son (Luke 7:11–17)

The miracle shows the compassion as well as the power of God. Not only had the widow lost her only means of support, but as a childless widow would be regarded as one who had lost God's favour.

To touch even something which had been in contact with the dead involved ceremonial defilement. Jesus touched the bier and again became a sin bearer. His concern throughout is for the woman rather than the son. The last line of the story, 'He gave him back to his mother' is a direct quotation from the story of Elijah and the raising of the son of the widow of Zarephath in 1 Kings 17:23. This explains the comment of the crowd that 'A great prophet has risen among us.'

The bringing to life of the dead is as well authenticated as any of the miracles of Jesus, because evidence for it is drawn from the earliest sources, Mark and 'Q' as well as from Luke's special source. Mark contains the story of the raising of the daughter of Jairus, 'L' contains the story of the Widow of Nain, and 'Q' records that among the achievements of the ministry of Jesus was that 'the dead are raised to life' (Luke 7:22). While it is not possible to prove to the doubter that the people concerned were not dead but in a trance or coma, there can be no doubt that the early Church was convinced that Jesus had brought back to life people whom others believed to be dead. Early Christians saw the spiritual counterpart in the raising to life, through the power of God, of those spiritually dead.

The Deaf Man at Decapolis (Ten Towns) (Mark 7:31–37)

The man's speech impediment would be a natural result of his deafness, since he would be unable to hear the sounds he was making.

Note

1 Jesus took the man aside from the crowd so that the healing was in private. He asked the crowd not to publicize the cure, but to no avail. It seems that his success as a healer was in danger of making impossible the main object of his ministry. Mark says earlier that he was unable to show himself in any town because of this kind of publicity.

2 He touched the man, putting his fingers into his ears and touching his tongue with saliva. It was a common belief that saliva had healing properties. Jesus did not need to use outward actions of this kind, but a stone deaf man would need help to understand what was to happen. Mark alone records the actual word of Jesus, the Aramaic 'Ephphatha' – 'be opened'.

The Blind Man at Bethsaida (Mark 8:22–26)

This miracle, like that of the healing at Decapolis, is found only in Mark, and there are similarities. In this case Jesus laid his hands on him and spat on his eyes, outward actions which indicated or perhaps even aided, the healing.

Note that the cure was not instantaneous: as his sight began to return, the man said, 'I see men; they look like trees, but they are walking about.' Jesus laid his hands on him again and his sight was fully restored. These details again bear the mark of authenticity.

Once again Jesus made the request that the miracle should not be made known.

Blind Bartimaeus (Mark 10:46–52; Luke 18:35–45; Matthew 20:29–34)

At this late stage in the ministry Jesus had resolutely set his face to go up to Jerusalem where all the prophets had said of him would be fulfilled. The disciples did not understand the implications of this. This miracle records the first time anyone other than one of 'the twelve' had given Jesus a Messianic title, in this case, 'Son of David'. After healing Bartimaeus Jesus did not forbid him to tell others: **the time for secrecy was past.**

Matthew records that there were two blind beggars; this may be due to his tendency to exaggerate, or that he combined the versions of Mark and Luke. Note the contrasting features of the miracles of the blind man at Bethsaida and that of Bartimaeus.

1 At Bethsaida the blind man was brought to Jesus, whereas Bartimaeus was discouraged by the crowd.

2 Bartimaeus was healed in public, whereas at Bethsaida the blind man was taken aside.

3 At Bethsaida Jesus used outward signs as an aid to healing, whereas Bartimaeus was healed with a word.

4 Bartimaeus was healed instantaneously; at Bethsaida the cure took some time and a second laying on of the hands was necessary.

5 Bartimaeus was not forbidden to tell others and he became a follower; at Bethsaida the blind man was told to return home and not to advertise his cure.

The Woman with a Haemorrhage (Mark 5:25–34; Luke 8:43–48; Matthew 9:20–22)

The account of this miracle is embedded in the story of the raising of the daughter of Jairus.

The woman felt it necessary to come to Jesus secretly because her ailment made her 'unclean' according to the Law. Jesus was not aware of her need, but though many were touching him, hers was the touch of faith which released the power of God to heal and Jesus felt the power leaving him.

The Ten Lepers (Luke 17:11–19)

This is one of several instances in the Gospel of a Samaritan being singled out for high praise.

It is noticeable that in this miracle there is no actual word of healing, nor did Jesus touch anyone; he simply said, 'Go and show yourselves to the priests.'

While they were all cured, the Samaritan is made whole, both physically and spiritually.

The Gergesene Demoniac (Mark 5:1–20; Luke 8:26–39; Matthew 8:28–34)

The symptoms of his madness are described in detail and have a ring of truth – morbid interest in death, abnormal physical strength, insensitivity to pain, refusal to wear clothes. All forms of madness were believed to be caused by demonic possession.

Jesus first asks, 'What is your name?' – a friendly question indicating personal interest and compassion. It was also a common belief that to know the name of a demon was to have power over it. 'Legion' suggests that the man had many demons, a very severe case of madness. As in other miracles the demon recognized the identity of Jesus, calling him 'Son of the Most High God'.

The miracle probably took place in a mainly gentile part of the country, for pigs were unclean to the Jews. The local inhabitants begged Jesus to leave the district after the destruction of the pigs. There is a telling contrast between the frenzy of the madman and the man restored, 'sitting there clothed and in his right mind'. The story ends on a note of mission: the restored man is told to return to his family and to tell them what God had done for him. He spread the news throughout the Ten Towns.

8.5 The Nature Miracles

There are four nature miracles:

1 **Walking on the Water** (Mark 6:45–56; Matthew 14:22–36)

2 **The Stilling of the Storm** (Mark 4:35–41; Luke 8:22–25; Matthew 8:23–27)

3 **The Feeding of the Five Thousand** (Mark 6:30–44; Luke 9:10–17; Matthew 14:13–21)

4 **The Withering of the Fig-Tree** (Mark 11:12–14; 20–24; Matthew 21:18–22)

These miracles present great difficulties, since:

(a) they appear to be inconsistent with what we know of the laws of nature;

(b) they seem inconsistent also with the decisions Jesus took at the Temptation (Luke 4:1–13), not to use his miraculous powers in his own self-interest, not to obtain a hearing for his message by providing for people's physical needs, and not to compel belief by working a spectacular miracle.

The following explanations have been suggested:

1 Walking on the Water (The word 'on' is equally well translated 'by')

The boat was 'immediately' at the land which suggests that the incident took place very near the shore. It was also a dark and windy night (John 6:16–21). Peter's failure could easily be explained, given the not uncommon feature of an irregular sloping beach.

2 The Stilling of the Storm

(a) Sudden storms are common on the Sea of Galilee because of its geographical position. They tend to subside as quickly as they arise.

(b) Many scholars regard this as an allegory, probably an illustration from an early Christian sermon which has, in the telling, been taken literally and recorded as a miracle. (The Gospel was preached for almost forty years before the first complete Gospel appeared.) During and after the first great persecution under Nero in 64 AD, the infant Church was in peril of being engulfed by a sea of heathen reaction. (Christians were falsely blamed for the Great Fire of Rome.) In the 'miracle' the ship is the Church and the storm is the violence of the persecution. Those who may lose heart are being reminded that Christ, their Captain, is in command, and he will steer the ship safely into calm waters. Meanwhile they must keep their faith.

3 The Feeding of the Five Thousand

(a) This is the only miracle recorded in all four Gospels, so the evangelists must have felt it to be of great importance.

(b) It seems mainly to have been a symbolic meal. It was believed that the reign of the Messiah would be inaugurated by a banquet, and the purpose of the 'miracle' may have been to admit those present to the Messianic fellowship by a symbolic sharing in the Messianic Banquet. Each received only a fragment.

(c) To early Christians the story was an anticipation of the Last Supper and the Eucharist, since Jesus blessed and broke the bread.

(d) The accounts describe a large meal and we can only speculate. Possibly the young lad (John 6:9) who offered to share his food inspired the others to share the food they had brought and had intended to keep for themselves.

(e) The miracle of the feeding of the Four thousand is generally regarded as a duplicate.

4 The Withering of the Fig-Tree

The details of this story are quite inconsistent with the character of Jesus described in the rest of the Gospel.

(a) It was not harvest time and it was unreasonable to expect the tree to be in fruit.

(b) The 'miracle' did not help anyone, least of all any man who may have owned the tree.

(c) It was as public and spectacular as it was unnecessary and unfair. The action of Jesus would appear to be a piece of mindless spite, completely uncharacteristic of him.

(d) The miracle probably began life as a parable, perhaps the parable of the Unfruitful Fig-Tree quoted by Luke (13:6–9) who, perhaps significantly, omits the miracle.

9 Important Events in the Life of Jesus

9.1 The Confession of Peter at Caesarea Philippi
(Mark 8:27–30; Luke 9:18–22; Matthew 16:13–20)

After the Temptation Jesus resolved to be a very different kind of Messiah from that expected by the Jews; to have declared himself Messiah at the beginning of the ministry would therefore have involved great risk, for the people might well have made him a rallying point for an insurrection against the Romans.

After their missionary tour, the disciples would be in close touch with public opinion and know the people's estimate of Jesus; he was also anxious to know how far their own understanding had progressed. The event at Caesarea is a turning point in the Gospel story: the **ministry in Galilee was almost over** and from now on the shadow of the Cross loomed over events. When Peter's declaration, 'You are the Messiah, the Son of the Living God' was accepted by Jesus, the secret was out. The disciples were warned to keep the knowledge to themselves, for the time was not yet appropriate for the challenging of the Jewish authorities. There also remained the danger of an uprising.

Matthew alone records the promise to Peter, 'You are Peter, the Rock; and on this Rock I will build my Church.' (The name Peter means 'a rock'.) The use of this word has been the cause of great controversy and there are several interpretations, the main ones being:

1 Peter is the rock on which the Church is built;

2 the truth about Jesus which had been revealed to Peter (i.e. 'You are the Messiah . . .') is the rock on which the Church is built;

3 with St Augustine of Hippo in the fourth century, a model of orthodoxy, the text will bear either interpretation equally well.

'The keys of the Kingdom' (Matthew 16:19)

The crossed keys have become a popular symbol of Peter. As keeper of the keys, Peter's position in the Church was to be one of rule and authority as we see in the early chapters of Acts. He was also given authority to interpret Christian law, to forbid or allow, which meant, in practice, to reconcile or excommunicate members of the early Church. This same authority was later given to all 'the twelve' (Matthew 18:18).

9.2 The First Prediction of the Passion (Mark 8:31 – 38; Luke 9:22 – 27; Matthew 16:21 – 28)

Immediately after accepting the title of Messiah and swearing them to secrecy Jesus began to tell them what his claim to Messiahship would mean, namely, suffering and death, to be followed by the Resurrection. The reaction of the disciples is expressed by Peter, 'No, Lord, this shall never happen to you' and the rebuke to Peter is meant for them all. The disciples still cherished popular ideas of what the Messiah would be. They would have to reject military and political ambitions, accept the principles of the Kingdom of God and be ready to pay the high price of discipleship.

The protest of Peter is an echo of the Temptation in the wilderness, so that the words of the rebuke, 'Away with you, Satan', are most appropriate.

After predicting his own suffering and death, Jesus warns his disciples that anyone who wishes to follow him must 'take up his Cross'. He interprets his messiahship in terms of the Suffering Servant of Isaiah 53.

9.3 The Transfiguration (Mark 9:2–8; Luke 9:28–36; Matthew 17:1–8)

Matthew specifies that this event took place six days after the event at Caesarea Phillipi, and may be seen as its sequel. The disciples were privileged to see the Messiah for a short time in his heavenly glory. The details of the story are full of meaning:

1 The appearance of Moses and Elijah indicate that his coming death will fulfil the Law and the prophets. Elijah was expected to return before the arrival of the Messiah (Malachi 4:5–6).

2 When Moses received the Ten Commandments on Mount Sinai his face shone so that the people could not look at him when he came down from the presence of God: on the Mount of Transfiguration the face of Jesus 'shone like the sun' (Matthew 17:2).

3 The cloud that overshadowed them was the cloud that led the Hebrews out of Egypt, and covered Mount Sinai when Moses met with God. It was the 'Shekinah', the cloud of glory which surrounds and indicates the presence of God: it was to appear again at the Ascension.

4 Luke says that Moses and Elijah spoke to Jesus of his coming 'departure', a translation of the word 'exodus' in the original Greek. The word clearly suggests that what is to happen in Jerusalem will be a New Exodus, that is, the Cross and the Resurrection are to be the spiritual counterparts of the deliverance of Israel from slavery in Egypt: mankind is to be saved from the greater slavery of sin. This illuminates the remark of John the Baptist (John 1:29) in which he refers to Jesus as 'the lamb of God. . . who takes away the sin of the world', and also Paul's statement, 'our Passover has begun; the sacrifice is offered – Christ himself' (1 Corinthians 5:7). The Passover lamb was the means by which the Hebrews were saved from bondage in Egypt, and Jesus is the 'lamb of God' by whose death the world is to be saved from the bondage of sin.

5 The implication is that Moses and Elijah, great as they were, only pointed the way to Jesus the Messiah.

6 The words spoken from the cloud, 'This is my Son, my Beloved, on whom my favour rests' are similar to those spoken by God at the Baptism.

Peter did not wish this ecstatic religious experience to come to an end, and wanted to make shelters or tabernacles to provide for a long stay, but Jesus led the disciples down the mountain. He had to return to serve suffering humanity. If religion is only a rarified spiritual experience it runs away from life.

9.4 The Mission of the Twelve (Mark 6:7–15; Luke 9:1–10; Matthew 10:5–42)

Jesus believed that his disciples should share in his mission as he himself could not hope to fulfil the demands of all the crowds that followed him. The news of the death of John the Baptist probably emphasized his sense of urgency, reminding him of the fate he knew he too, would face. Because of this, **the disciples were told to conduct their campaign in haste**, carrying no unnecessary equipment, relying on people's hospitality, and wasting no time on those who would not listen to their message.

They were to go in pairs, and whereas Mark says they might wear sandals and take a stick, Matthew's account denies the use of both. They were to take no money or food. The message they were to proclaim was, **'The Kingdom of Heaven is upon you'**, and they were to heal the sick and raise the dead. They were not to linger where they were not wanted,but to 'shake off the dust' from their feet, just as devout Jews would shake off the dust after travelling through gentile lands.

The mission was limited to Galilee and they were expressly forbidden to enter Samaritan or gentile towns.

9.5 The Mission of the Seventy (Luke 10:1–20)

The account of this mission, found in Luke only, is similar to that of 'the twelve', and some scholars believe that it is another version of the same event. However, Luke includes both accounts. Seventy was regarded as the number of nations of the world, so it has been suggested that the mission of 'the twelve' represents the preaching of the Gospel to the twelve tribes of Israel, and the mission of 'the seventy' to the rest of the world.

The account is notable for the response of Jesus to the news of success brought by the returning missionaries, 'I watched how Satan fell, like lightning out of the sky' (Luke 10:18). The Kingdom had come and had been proclaimed by the disciples both by word of mouth and the working of miracles.

9.6 The Request of James and John (Mark 10:35–45; Matthew 20:20–28)

The incident shows that 'the twelve', despite all warnings to the contrary, still believed in a conquering Messiah in whose earthly glory they hoped to share.

The word 'cup' is a symbol of suffering, as in the story of the Garden of Gethsemane (Mark 14:36). The word 'Baptism' is a symbol of death, for example in Romans 6:3–4 we find, 'by Baptism we were buried with him, and lay dead'. James was the first of the Apostles to be martyred, being put to death by Herod Agrippa (Acts 12:1).

The incident of the request of James and John is important because it spells out the true meaning of greatness in the Kingdom. **The great must be the willing slaves of all**, their greatness lying not in status, but in humility and service, for Jesus came not to be served but to serve, and 'to surrender his life as a ransom for many'.

Matthew follows Mark, but it is noticeable that the request to Jesus is made not by the brothers but their mother. Matthew's account was written at a time when all the Apostles were revered figures, and Matthew declines to place them in an unfavourable light: the blame for their ambitious request is placed on their mother.

10 The Passion Narrative

The Passion Narrative was the earliest consecutive narrative of the events in the life of Jesus to take a fixed form, containing, as it did, the events which lie at the heart of the Gospel, the death, burial and resurrection of Jesus. Mark preserved it, and Luke and Matthew show great respect for Mark's order of events, which is also confirmed independently by John.

10.1 The Triumphal Entry (Mark 11:1–11; Luke 19:29–40; Matthew 21:1–11)

Jesus' claim to be the Messiah was now made public in Jerusalem, the capital. He chose to make this claim in action rather than in words, in terms of the prophecy of Zechariah 9:9, 'riding upon an ass' rather than a horse, thus emphasizing the fact that **he was a peaceful Messiah rather than a military leader**. Mark does not quote this prophecy, and the blessings are called down not upon the 'King' as in Matthew and Luke, but on 'the Lord' and upon the 'coming Kingdom'. He does, however, record that Jesus began his ride from the Mount of Olives, which fulfils the prophecy of Zechariah 14:4 that when 'the Lord will come. . . his feet will stand on the Mount of Olives'. Arrangements had obviously been made previously with the owner of the ass and the phrase 'Our master needs it' may be a kind of password.

It is a matter of dispute as to how far the ordinary people understood the significance of Jesus' action. Some undoubtedly did. At about this time the Feast of Dedication was being celebrated in Jerusalem and the carrying of palm branches was part of the ritual. According to Matthew, some of the people were merely giving a welcome to the prophet of Nazareth. The

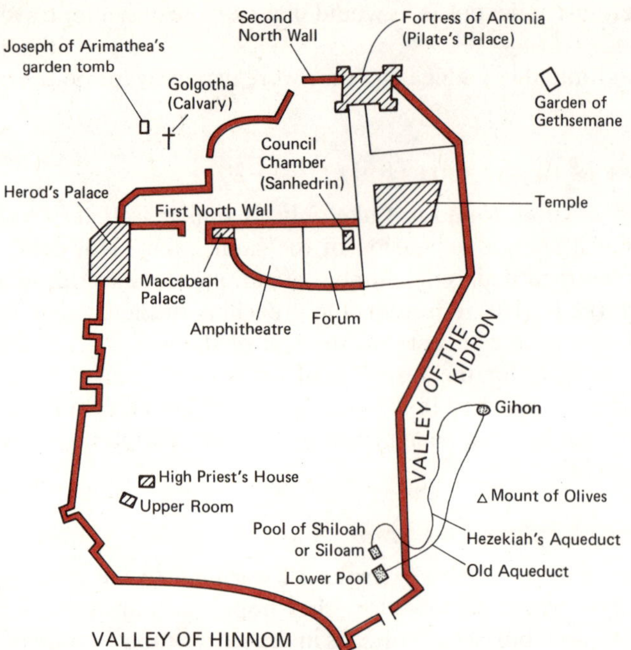

Fig. 10.1 Jerusalem in the time of Jesus

words 'Blessing on he who comes in the name of the Lord', were used generally as a welcome to pilgrims, but when directed to Jesus they took on a deeper meaning.

Luke follows Mark, but the crowd cries, 'Blessings on he who comes as King'; instead of 'Hosanna' we have 'Peace in heaven, glory in highest heaven'. Luke records the protest of the Pharisees and the refusal of Jesus to restrain his disciples.

Matthew alone quotes Zechariah 9:9. Writing long after the event, his account is probably influenced by the interpretation the early Church placed upon the event, in the light of the Resurrection, and he gives Jesus the Messianic title 'Son of David'. Matthew strangely fails to take note of the poetic parallelism in Zechariah's words, 'riding on an ass, riding on the foal of a beast of burden', and puts two animals into his account.

10.2 The Cleansing of the Temple
(Mark 11:15 – 19; Luke 19:45 – 48; Matthew 21:12 – 17)

The prophecy of Malachi 3:1–3, though not quoted in the accounts, probably lies behind this event: 'Suddenly the Lord whom you seek will come to his Temple.' The prophet goes on to tell of the manner of his coming, that 'He will take his seat refining and purifying .' The event is seen as the throwing down of a major challenge to the religious authorities at the very centre of their power.

Jesus was not in principle objecting to the use of the Court of Gentiles for the essential facilities needed by pilgrims. People often travelled great distances and it was convenient that they should be able to purchase on the spot sacrificial animals which conformed to the strict rules laid down in the Law of Moses. Temple dues had to be paid in Jewish coinage because foreign coins were stamped with the images of rulers or heathen gods, thus offending against the commandment. Many years later we find Paul (Acts 21:26), then a Christian, observing the Law and purchasing the birds necessary for the sacrifice to fulfil his Nazarite vow.

The profits of this lucrative trading went to the chief priests and the impression of commercialism given to foreigners was unattractive. Jesus quoted the words of Isaiah 56:7 and Jeremiah 7:1 which were inscribed on the wall of the Court of Gentiles. While it is true that the chief priests gained much profit from the commerce of the Court of Gentiles and therefore strongly objected to the action of Jesus, their main objection may well have been the claim to authority which his action implied. Their profits and their authority were both being challenged.

10.3 The Question of Authority (Mark 11:27–33; Luke 20:1–8; Matthew 21:23–27)

1 'By what authority?'

The cleansing of the Temple court caused great offence to the authorities and the public nature of the demonstration made it impossible to ignore. Challenged to disclose the ground of his

Fig. 10.2 'Christ driving the traders from the Temple' from the studio of *El Greco*
Reproduced by courtesy of the Trustees, The National Gallery, London

authority, Jesus turned the tables on his questioners by asking a question of them, 'The Baptism of John: was it from God, or from men?' To reply to a question by a counter-question was a device commonly used among the rabbis, so Jesus was not being evasive but astute. If they said they believed it came from God, they would be accusing Herod of a crime against God. On the other hand, if they denied John's authority they feared popular disapproval. John claimed to be the herald of the Messiah, so any admission of the validity of his authority supported the claim of Jesus to be exercising the authority of God. Luke and Matthew follow Mark.

2 The Wicked Husbandmen (See Unit 7.4)

This is the only parable which says that when the authorities heard it, they saw that it was 'aimed at them'.

3 The Tax paid to Caesar (Mark 12:13–17; Luke 20:20–26; Matthew 22:15–22)

The Pharisees and Herodians, who hated each other, made an unholy alliance. They had also combined against Jesus earlier (Mark 3:6).

The story owes its preservation to its continuing importance for Christians living in the Roman Empire, since all were required to pay taxes to Rome. The practical value of the story was that it made it plain that loyalty to Jesus did not involve disloyalty to the emperor. The Jews, on the other hand, bitterly resented the tax, which when first imposed in 6 AD led to a serious revolt. The coin was the silver denarius which was stamped with the image of Caesar – in itself, an offence to the Jew. Since they also paid a Temple tax the Jews felt not only that they were paying double, but that the Roman tax was a reminder that the nation was under foreign rule.

The question was calculated to force Jesus to give an answer which would either make him unpopular among the people and forfeit their support, or put him in serious trouble with the Roman authorities. His answer, again, was astute rather than evasive, and was in fact a challenge to both parties. The Jews, though they would be loath to admit it, derived considerable advantage from Roman rule, and it was not unjust that they should pay something towards it. The Herodians, on the other hand, were casual about their duty to God, for a condition of the rule of Herod Antipas was that he should allow the erection of temples devoted to emperor worship, and make no attempt to win converts to Judaism. The supporters of Herod could well be reminded of their duty to God.

10.4 A Dispute with the Sadducees A Question of Life after Death (Mark 12:18–27; Luke 20:27–40; Matthew 22:23–33)

This is the only recorded conflict with the Sadducees. The story they imagine is technically possible under a marriage law which, by the time of Jesus, had become obsolete. The practice originated at a time when it was believed that man's only immortality was in his children. If a man died childless it was his brother's duty to marry his wife, and the first-born of the union would then be the heir of the dead man.

The question of life after death was often hotly debated by Sadducees and Pharisees and we find an excellent example of this at Paul's examination by the Sanhedrin in Acts 23:6–8. Uproar broke out between the rival parties when Paul claimed that 'the true issue of this trial is our hope of the resurrection of the dead'.

In this incident Jesus supported the Pharisees and shrewdly met the argument of the Sadducees with a text from Exodus 3:2–6, the story of the Burning Bush. God does not say, 'I was the God of your forefathers' but 'I am the God', as though the patriarchs, who lived many years before Moses, were still alive. The Sadducees held the Five Books of Moses, so called, in great veneration and accepted the authority of no others; an appeal to any of them would carry great weight. This kind of argument would be very persuasive at this time, if not for us, but the important point is that Jesus is defending belief in life after death. He points out that marriage is for this world only.

10.5 The Great Commandment (Mark 12:28–34; Luke 10:25–37; Matthew 22:34–40)

The lawyer was in this case a sympathetic listener, for he acknowledged how well Jesus answered his questioners, and in Mark's version his question is sincere. Jesus in his answer linked together the words of the Shema (Deuteronomy 6:4–5) and Leviticus 19:18, 'Love your neighbour as yourself.' While it is possible that Jesus was the first to link these commands, we cannot be certain, but for the Jew the term 'neighbour' was restricted to other Jews. The lawyer's approval of the answer of Jesus gains for him a commendation, 'You are not far from the Kingdom of God.' This level of agreement between the two is evidence that Jesus was as good and loyal a Jew as his enemies.

Luke places the incident after the return of 'the seventy' and it is the lawyer who answers his own question, linking the texts from Deuteronomy and Leviticus. In Luke, the lawyer is 'testing' Jesus, and the answer to his supplementary question, 'Who is my neighbour?' is the parable of the Good Samaritan which is peculiar to Luke. This in effect defines our neighbour as anyone in need. (For an examination of this parable see Unit 7.4.) Matthew is a shorter version of Mark.

It is important to note that Jesus did not lay down detailed rules to govern the behaviour of Christians in every possible circumstance. He gave two unchangeable principles which apply to every situation and by which all actions are to be judged. Even his own proverbial sayings are to be applied by the rule of love of God and love of neighbour. In contrast the Pharisees laid down detailed rules to cover every conceivable circumstance: the Jewish Law contained 248 'do's and 365 'dont's.

10.6 The Apocalyptic Discourse (Mark 13; Luke 21:5–38; Matthew 24:1–51)

The prophets of Israel who lived before the **Fall of Jerusalem in 586 BC** and the subsequent exile in Babylon, taught that the disasters which had befallen the nation were God's punishment for its sins, especially the sin of disloyalty in following other gods. They believed that when the nation returned to its allegiance, the Day of the Lord would arrive in which He would intervene in history and vindicate the nation's faith in Him: He would right the wrongs they had suffered, punish those who oppressed them, and restore Israel, His chosen people, to their proper status as supreme among the nations of the world. **The Day of the Lord** would inaugurate a golden age of righteousness, justice and peace, and the eyes of all men would look to Jerusalem where true knowledge of God was to be found.

The prophets saw God as the Lord of history, who controlled all events upon earth, and whose purpose would be fulfilled in history. Their teaching about the climax of history is called **prophetic eschatology** (Greek 'eschatos' – last, at the last, finally).

After the exile there came a change in outlook. The national disaster had dealt a crushing blow to Israel's self-confidence, but reflection taught the nation how true the prophet's warnings had been. Suffering engendered repentance and on their return in 537 BC there was a resolve to reform the pratice of their religion: idolatry was rooted out and the moral and ceremonial Law of Israel was established in the land. Yet the promised day did not arrive and the situation was in many ways worse, for the nation was impoverished and there was no end to oppression and persecution. Far from attaining supremacy, the nation was conquered by Alexander the Great and worse was to come, for in the second century BC there was the bitter confrontation with Antiochus Epiphanes, a ruler aggressively committed to Greek religion and culture, who used appalling brutality in an attempt to root out the religion of Yahweh from the land.

Belief that God would finally intervene in human affairs was not, however, abandoned, but took a new direction: there came a new emphasis on an already existing form of eschatology called 'apocalyptic' (from the Greek word to unveil, to reveal a secret). Whereas the prophets taught that God would vindicate His people in the course of historical events, the apocalyptists believed that the intervention would be in the form of **a universal catastrophe**. This wicked world would come to an end in a supernatural cataclysm of violence and devastation.

Whereas the prophets tend to speak of the Day of the Lord in general terms, the apocalyptists go into great detail. They believed that they had access to God's secret designs, and revealed them in highly symbolic language. Theirs is an eschatology of supernatural wonders, of angels, demons, monsters and fantastic visions. They generally forecast the end of all things to be very near.

The Old Testament contains several examples of apocalyptic writing, notably **the Book of Daniel**, and in the New Testament there is one complete book, **Revelations**. The Apocalyptic Discourse found in the Synoptic Gospels is probably a piece of Christian apocalyptic writing.

Mark 13:1–2 The Temple was that of Herod the Great. It was not complete until 64 AD and was destroyed by the Romans in 70 AD.

13:13 'All will hate you for your allegiance to me.' The first great persecution of Christians had taken place in 64 AD during the reign of the Emperor Nero. Many Christians had been killed, including, according to tradition, Peter and Paul.

13:14 '. . . the abomination of desolation'. This is an echo of Daniel 11:31, 'the abominable thing which causes desolation', which refers to the statue of Zeus, the chief of the Greek gods, set up by Antiochus Epiphanes in the Temple in Jerusalem in 168 BC. Mark may have in mind the coming destruction of the Temple by the Romans, but it might also refer to the attempt made by the mad Emperor Caligula to have a statue of himself placed in the Temple in 38 AD.

After signs in heaven and earth the Son of Man will come 'in the clouds with great power and glory' (13:26) and the angels will gather in his 'Chosen' from the 'farthest bounds of earth to the farthest bounds of heaven'.

13:32 This verse says explicitly that no one knows the hour or the day when this shall be, 'not even the Son', which, strange to say, does not discourage some people from confidently predicting when it will be.

Luke, writing after the Fall of Jerusalem, uses Mark as a basis but omits reference to the 'abomination of desolation'. He describes the city encircled by armies and refers to the Jews who were killed or taken captive (Luke 21:19–24). Luke refers obliquely to a delay in the Parousia (Second Coming of Christ). He did not come as Mark seems to predict, immediately after the Fall of Jerusalem, but Luke says that 'the present generation will live to see it' (Luke 21:32).

Matthew follows Mark but adds that the Parousia will be seen from one end of the earth to the other (Mark 24:27).

10.7 The Anointing at Bethany (Mark 14:1–9; Matthew 26:1–13)

Simon the leper had apparently been cured, possibly by Jesus himself. The name of the woman is given only in John 12:3 as Mary, the sister of Martha and Lazarus. The action of the woman was highly symbolic:

1 Kings were (and in England still are) anointed with oil at their crowning.

2 The word Messiah (Greek 'Christ') means 'anointed one'.

3 Bodies were anointed before burial. Mark would have known that there had been no time to anoint the body of Jesus before burial. The woman's action was therefore a confession that Jesus was the Messiah.

Luke has a parallel story (Luke 7:36 – 8:3) which differs considerably from Mark. The host is Simon the Pharisee, and the woman a prostitute who not only anointed Jesus, but wet his feet with her tears and wiped them with her hair.

Matthew follows Mark, but says that the disciples complained about the waste of the ointment, whereas Mark says that it was 'some of those present'.

10.8 The Betrayal (Mark 14:10–11; Luke 22:1–6; Matthew 26:14–16)

Matthew alone specifies the amount of money paid to Judas. Thirty pieces of silver was the amount laid down by the Law as the value of a slave. The motive of Judas has been the subject of much speculation.

1 Matthew says it was greed (so does John 12:6), but the amount of money was relatively small.

2 A deeper motive has been suggested, that Judas, impatient that opportunities had, in the past, been missed (e.g. John 6:15 records that after the Feeding of the Five Thousand, the people wanted to proclaim him king) wished to force Jesus publicly to proclaim himself the Messiah. According to Judas' mistaken ideas about what Messiahship involved, this would mean an open challenge to Rome, a violent conflict, and the result would be to 'establish once again the sovereignty of Israel' (Acts 1:6). It may be that Judas saw the advantage to himself in this eventuality, envisaging himself as one of those who would 'sit as judges of the twelve tribes of Israel' (Matthew 19:28).

What did Judas actually betray?

It seems unlikely that he merely betrayed the whereabouts of Jesus and identified him to the Temple police: they could easily have discovered where he was because his movements were well known, and his identity would have been no secret, so it is thought that what Judas betrayed was the fact that Jesus had disclosed to the disciples that he was the Messiah and had asked them to keep it secret. This knowledge in the hands of his enemies was fatal to him, for it gave the Jewish authorities a capital charge to place before Pilate.

10.9 The Last Supper (Mark 14:12–26; Luke 22:7–38; Matthew 26:17–30)

The Preparation

The room probably belonged to John Mark's mother, for Acts 12 informs us that it was a regular meeting place of the disciples: its many associations would probably cause it to be used by them. (Carrying water was not normally a man's job, so this was probably a secret sign.)

There is considerable speculation as to whether the Last Supper could have been at a Passover meal. It is argued that Jesus could not have been crucified on the actual day of the Passover, if only because the priests would have been too busy to take part in the trials, and according to Mark 14:2 they had in any case decided to avoid a confrontation on that day. There was, however, a dispute between the Pharisees and the Sadducees as to when the festival began, and Pharisees were allowed to hold the meal a day before the official date. Jesus would be following the custom of the Pharisees by observing the festival on Thursday evening. John 13:1 records that the meal took place 'before the Passover'.

There is evidence from the Dead Sea Scrolls that at Qumrân they followed the solar rather than the lunar calendar which regulated the Temple services. Other non-conformist groups among the Jews did likewise and Jesus would be celebrating the Passover three days before most Jews if he followed the custom of Qumrân. Many scholars believe that it was from the 'non-conformist' section of Judaism that Christianity sprang.

If the Last Supper was not a **Passover meal** it could have been a **Chaburah meal**. This took place when private groups of men met together as friends early on Friday evenings to prepare for the sabbath, but the meal would take place on Thursday if the sabbath were also a special feast day, so as not to clash with the observance of the eve of the feast.

The meal proper began with a blessing; the leader then broke bread and consumed a fragment before giving a piece to everyone at the table. After the meal came the grace, a long prayer said by the leader on behalf of all present, a **thanksgiving** for all the mercies God had bestowed upon His people. It ranged from the Creation to the Exodus from Egypt, the Covenant and the inheritance of the promised land.

After the thanksgiving a blessing was recited over a cup of wine called the **cup of blessing**, and after being sipped by the leader it was handed round to all present. The meeting ended with the singing of a **psalm**. Jesus presided at this meal as head of the family.

If the Last Supper was a Chaburah meal and not a Passover, it is still evident that the **Eucharist** (one of the earliest names the Church gave to the service) originated in the atmosphere of thanksgiving for all God's blessings to the nation, and especially for the Exodus from Egypt and the Covenant. It is not difficult to see why the early Church therefore understood the meaning of the Eucharist in terms of the Passover and the Covenant.

The meaning of the Last Supper is best seen from the account of Paul in 1 Corinthians 11:23–26, written about ten years earlier than Mark and which Paul claims to be the tradition handed down from Jesus himself. Body and blood together represent life so when Jesus said, 'This is my body', and later, separately, 'This is my blood', he was telling his disciples that he was to give his life for them. He was, in fact, using the **language of sacrifice**: the breaking of the bread and the pouring out of the wine symbolized his death. He said that by this means a **New Covenant** between God and man, a new relationship, would be established. (Note that the NEB, in each of the Synoptists' accounts, omits the word 'new' when referring to the 'Covenant', but we may fairly accept it as implied.)

The Old Covenant, the solemn agreement between God and Israel, had been made at Mount Sinai (Exodus 24:1–8). The conditions of that agreement, the Ten Commandments, were read aloud by Moses and accepted by the people; he then flung half of the blood of a sacrificed animal on the altar and sprinkled the remainder upon the people. The agreement was thus sealed by the blood of the sacrifice.

The Old Covenant failed because the people were unable to keep their side of the agreement and the Old Testament in the Bible gives us a record of their failure. The prophet Jeremiah said that God had revealed that one day there would be a New Covenant (Jeremiah 31:31–34), 'I will set my law within them and write it on their hearts.' Jesus said that this New Covenant was being brought into being at the Last Supper and would be sealed by his blood on Calvary.

The Passover was and is a remembrance of **God's deliverance** of His people from bondage in Egypt: the Eucharist is for Christians a remembrance of God's deliverance of His people from the greater bondage of sin, through the death and resurrection of Jesus. The word **'remembrance'**, however, has a much deeper meaning in the Bible than in present day usage. When we say we remember, we mean to bring into mind a past event which is dead and gone, whereas to the Hebrew at the Passover, it means bringing an event out of the past and into the present as a living experience. Thus a Jewish family celebrating the Passover is not merely remembering the Exodus, it is there, with Moses, reliving it, taking part in it, experiencing it as a present event. So for the Christian, the Eucharist is not so much a remembering but a reliving, a re-presentation of the Last Supper, the Cross and Resurrection.

In the NEB the words 'Do this in remembrance of me' appear in a footnote only in Luke's account, on the grounds that the best manuscripts omit them. Neither Mark nor Matthew records the words. Matthew's account is based on Mark.

10.10 The Garden of Gethsemane
(Mark 14:32–42; Luke 22:39–46; Matthew 26:36–46)

As a man Jesus knew the experience of acute apprehension at the thought of the ordeal of the next day. If at the Transfiguration we see his glory as Son of God, at Gethsemane we see his true humanity: he shrank from the Cross as any normal man would. He prayed that the 'cup' of suffering would be taken away, but he conquered his fear and prayed 'yet not what I will, but what Thou wilt', in complete submission to the will of God.

There are points of comparison with the Temptation: there were three in number, they were private experiences, and according to Luke only, there came an angel to bring him strength. At the Temptation the devil left him 'biding his time' and here is another instance of acute temptation. Luke does not mention the garden, or that the three disciples were singled out, or that the prayer of Jesus was repeated three times. He alone mentions the sweat of blood and the ministering angel. Matthew closely follows Mark.

Luke's account of the arrest differs from that of Mark in some respects. Judas does not actually kiss Jesus though he tries to. Luke alone records that the wounded man had his ear restored.

Matthew is based on Mark, but after the kiss of Judas, Jesus says, 'Friend, do what you are here to do.' Jesus rebukes the disciple who attacked the High Priest's servant with the words, 'All who take the sword die by the sword.'

Fig. 10.3 'The Agony in the Garden of Gethsemane' from the studio of *El Greco*
Reproduced by courtesy of the Trustees, The National Gallery, London

10.11 The Jewish Trials

The appearances of Jesus before the Sanhedrin are called trials only for convenience. They were, in fact, interrogations intended to see if a case could be made. The second of these was only a short consultation at which it was agreed to present the case to Pilate.

The reasons for seeking the death of Jesus were probably:

1 The challenging of their authority, the latest incident of which was the cleansing of the Temple.

2 The attitude of Jesus to the Jewish Law, and especially sabbath observance.

3 The danger of a popular rising against the Romans as a result of his claim to be the Messiah. The privileged position of the chief priests, granted by the Romans, would be endangered if they were seen not to have acted decisively in a case of treason.

If the Jewish authorities had, in fact, been given permission to hold a formal trial, we can point to serious irregularities in the proceedings:

1 The Sanhedrin could, in principle, sit at night, but if it did so, could not pass a sentence of death.

2 A false witness always suffered the penalty which would have been awarded to the innocent victim.

3 In the case of a unanimous verdict of guilty, it was compulsory to defer the sentence for twenty-four hours.

4 The proper procedure was to make a charge and to call witnesses, rather than questioning to induce a prisoner to condemn himself.

5 It was highly irregular to allow the beating of a prisoner in a court of law.

6 If witnesses disagreed, a trial ended.

10.12 The Trial (Examination) before the High Priest and Sanhedrin (Mark 14:53–65; Matthew 26:69–75)

A threat to the Temple was blasphemy, and punishable by death, but the evidence of the witnesses should have been inadmissible because it was contradictory. They claimed that Jesus said, 'I will destroy this Temple', but in John's Gospel his words are 'Destroy this Temple', meaning 'if you destroy this Temple'. In the second case, Jesus was perhaps foretelling the destruction of the Temple in 70 AD during the Jewish revolt, or of course it might be a product of hindsight. Jesus remained silent during the proceedings, but to the question 'Are you the Messiah, the Son of the Blessed One?' he could not remain silent. His reply, 'I am', was enough, for it provided ample reason for the case to be referred to the Roman governor with a recommendation for a sentence of death. The question may well have been prompted by information given by Judas Iscariot.

Matthew follows Mark but says that the answer of Jesus to the question of the High Priest was not 'I am' as in Mark, but 'The words are yours'. Vague as this answer appears to be, the Sanhedrin clearly took it to be an admission of guilt.

10.13 The Denial of Peter (Mark 14:66–72; Luke 22:54–62; Matthew 26:69–75)

Peter, with the other disciples, had boasted that he would face death rather than betray his Lord. In the event, he was unequal to the task. According to a strong tradition Mark was the companion of Peter in Rome and it is likely that some, at least, of the teaching of Peter is contained in Mark's Gospel. Mark does not spare Peter and this may indicate that in his preaching Peter did not spare himself.

Peter had a distinctive northern accent and it was this which caused him to be recognized.

Luke's story is based on Mark, but whereas in Mark Peter is twice questioned by the same serving maid, Luke records three questionings, the first by the serving maid, the second by another person and the third by a man. Luke omits Peter's cursing. Matthew follows Mark, but Jesus is questioned by two different maids.

10.14 The Questioning at Daybreak (Mark 15:1; Luke 22:63–71; Matthew 27:1–2)

A claim to Messiahship could be presented to Pilate as a case of treason which he could not ignore.

Luke records this as the only trial, but the subject matter is very close to the accounts in Mark and Matthew of the questioning at night. Matthew follows Mark.

10.15 The Death of Judas (Matthew 27:3–10)

One theory is that when the plan of Judas to force Jesus to overcome the opposition by violent action had failed, he was genuinely appalled at what he had done. Pride and selfishness lay at the root of his betrayal, and consistent with this, he committed the ultimate act of pride, by treating his life as though it belonged to himself rather than to God.

10.16 The Roman Trial (Mark 15:1–20; Luke 23:1–25; Matthew 27:11–31)

The accounts substantially follow Mark, even in John's Gospel, and this is taken as evidence that Mark has recorded a very early tradition revered throughout the Church.

All the Gospels portray a Pilate who is most reluctant to condemn Jesus and would have preferred to release him. At a time when the Church was being persecuted both by Rome and by the Jews, it would have been inadvisable to lay the blame directly at the door of Pilate, and more politic to blacken the Jewish authorities.

Pilate began with the direct question, 'Are you the King of the Jews?' to which Jesus replied, 'The words are yours', and made no further reply. Barabbas was a patriot who had led an uprising against Rome and this explains his popularity with the crowd. Pilate was willing 'to satisfy the mob'; he released Barabbas and handed Jesus over to be crucified. He was probably influenced by the fact that serious complaints about him had already been sent to Rome, and he could not afford to be tolerant over a charge of treason.

The soldiers in mockery dressed Jesus in the 'royal purple' and a crown of thorns as a substitute for the laurel wreath worn by emperors. The cane with which they beat him became a mock sceptre.

Luke 23:2 specifies the charges made against Jesus:

1 subverting the nation;

2 opposing the payment of taxes to Rome;

3 'claiming to be Messiah, a king'.

Luke alone records the trial before Herod Antipas and says that Herod's soldiers, not the Romans, mocked Jesus.

Matthew follows Mark but adds the story of Pilate's wife's dream. The story of Pilate's dramatic action of literally washing his hands of responsibility for the case is also found only in Matthew.

Matthew alone records that the mob cried, 'His blood be on us and on our children'; this has been the pretext for the infliction of untold cruelty on the Jewish people ever since.

10.17 The Crucifixion (Mark 15:20–41; Luke 23:26–49; Matthew 27:31–56)

The cup of drugged wine was provided by some women of Jerusalem to alleviate the pain of crucifixion, and the Romans permitted its use. Jesus would not drink it.

It was customary for the executioners to divide the prisoner's effects among themselves.

Fig. 10.4 'Christ on the Cross' *Delacroix*
Reproduced by courtesy of the Trustees, The National Gallery, London

The nailing up of a titulus or notice stating the crime of the prisoner was common practice: Jesus was described as King of the Jews, possibly a deliberate insult to the Jews on the part of Pilate.

Those who passed by, and the chief priests and Pharisees, mocked him and challenged him to come down from the cross. The temptation to do this was of the same kind as that in the wilderness, namely, to compel belief by working a spectacular miracle. It was because he was saving others that he could not save himself.

The darkness over the land from noon to 3.00 p.m. is not explained, but it was seen by the disciples as highly significant.

Mark records only one saying of Jesus from the cross, the Aramaic 'Eli, Eli, lema sabachthani', meaning, 'My God, My God, why hast thou forsaken me?' This is one of the most difficult 'Words from the Cross', for it suggests that Jesus felt that God had deserted him and left him to his fate; his trust in God had therefore been misplaced. In view of the Gospel, the 'Good News', this interpretation is hard to accept. One explanation is that Jesus was on the cross bearing the sin of the whole world and could not therefore escape the full consequences of that sin – an overwhelming sense of separation from God. Jesus came through that experience with his trust in God unbroken.

The cry is a quotation of the first verse of Psalm 22, a psalm which, though it begins in despair, ends on a strong note of triumph. It is significant that this psalm also refers to the casting of lots and abuse by passersby; it would seem that Mark is implying that Jesus, while quoting the first verse, had the whole psalm in mind, including the final note of triumph.

The curtain of the Temple referred to is probably that which hung over the opening to the Holy of Holies where God dwelt; only the High Priest had access and that only once in a year. The rending of the veil was a powerful symbol – that which was hidden is now revealed and God is seen face to face in Jesus. Through the death of Jesus, all men, not only the High Priest, had access to the presence of God.

In Mark the centurion calls Jesus a 'Son of God'.

Luke expands the single verse in Mark about the two criminals into a dialogue in which one taunts Jesus while the other reproves his companion. To the request, 'Remember me when you come to your throne' Jesus replies, 'Today you shall be with me in Paradise.'

Luke alone records the prayer of Jesus for his executioners, 'Father, forgive them; they do not not know what they are doing.' He also alone records the dying words of Jesus, 'Father, into thy hands I commit my spirit.' The centurion in Luke's account said, 'Beyond all doubt, this man was innocent.'

Matthew follows Mark including the cry from the cross in Aramaic, but adds that after the death of Jesus there was an earthquake. Note that the appearances of the dead took place after the Resurrection. Matthew adds the name of Jesus to the title on the cross, and the wine offered to him was mingled with gall not myrrh as in Mark.

10.18 The Seven Words from the Cross

Three are found in John's Gospel only, and the traditional order of the seven is as follows:

1 Luke 23:34 'Father, forgive them; they do not know what they are doing.'

2 Luke 23:43 'I tell you this: today you shall be with me in Paradise.'

3 John 19:27 'Mother, there is your son. . . there is your mother.'

4 Mark 15:34 'My God, my God, why hast thou forsaken me?'

5 John 19:28 'I thirst.'

6 John 19:30 'It is accomplished.'

7 Luke 23:46 'Father, into thy hands I commit my spirit.'

10.19 The Burial (Mark 15:42–47; Luke 23:50–56; Matthew 27:57–66)

In Palestine, out of consideration for the Jewish Law on the matter, the Romans did not leave dead bodies to rot on the cross, but buried them in a common grave. Sometimes permission to bury their dead was given to relatives or friends. Joseph of Arimathaea, who was probably a member of the Sanhedrin, took down the body of Jesus and laid it in his own tomb. The women watched and saw where he was laid, intending to complete the burial arrangements when the sabbath was over.

Luke states that Joseph was a rich man.

Matthew says that Joseph was a disciple.

Matthew alone tells the story of the guarding of the tomb, and it is not easy to account for. The disciples themselves seem not to have understood that Jesus would rise from the dead, even though he had repeatedly told them, so it seems unlikely that the chief priests would anticipate that it might take place.

10.20 The Empty Tomb (Mark 16:1–8; Luke 24:1–11; Matthew 28:1–15)

The majority of scholars agree that Mark's Gospel finishes abruptly at verse 8 and that the remaining verses are a later addition, not the work of Mark.

The sabbath ended at sunset on Saturday so the women could buy the spices and perfumes they required for the Sunday morning.

In **Luke,** two men in dazzling garments appeared to the women, not one, as in Mark. According to Luke, the disciples refused to believe the women's story.

Matthew characteristically heightens the account considerably. There was an earthquake and an angel of the Lord rolled away the stone and sat upon it.

Matthew alone says that the women met Jesus himself when on their way to give the disciples the message of the angel. To explain the ineffectualness of the guard Matthew says that they were stunned by the earthquake and were later bribed by the authorities to say that the body of Jesus had been stolen.

10.21 The Ending of Mark's Gospel

Since most of Mark's Gospel is contained in Luke and Matthew it was once thought that Mark was a later summary of the other two. This may account for a certain neglect of Mark and that consequently the original ending was lost.

Some manuscripts add verse 8b (the Shorter Ending) while others add verses 9–20 instead (the Longer Ending). Some manuscripts insert both. Most of the ancient manuscripts bring the book to an end, abruptly, at verse 8a. The author's style in both of these endings is different from that of Mark.

10.22 The Road to Emmaus (Luke 24:13–35)

For many, this is the most beautiful of all the Resurrection stories. They possibly did not recognize him because of their preoccupation with their grief. From their remark, 'We had been hoping that he was the man to liberate Israel' we may gather that they still held the traditional view that the Messiah was to be a conqueror.

Two important points stand out:

1 Jesus told them that all that had happened was in fulfilment of the Law and the prophets. Thus this basic ingredient in the preaching of the early Church can be seen to go back to Jesus himself.

2 Cleopas and his companion told the disciples 'how he had been recognized by them at the breaking of bread'. This has been the experience of the Church ever since at the Eucharist.

10.23 The Evening of Easter Day (Luke 24:36–46)

The invitation of Jesus to touch him was to demonstrate that it was he Himself and not an apparition. All the resurrection stories are about a Jesus who is no longer limited to a physical body as he had been.

10.24 The Appearance in Galilee (Matthew 28:16–20)

Matthew says that the eleven made their way to a mountain in Galilee where Jesus had told them he would appear to them. All the other appearances were in Jerusalem.

The disciples were commissioned to preach the Gospel to all nations and to baptize converts in the name of the Father, Son and Holy Spirit. According to Acts, the first Christians were baptized in the name of the Lord Jesus (Acts 2:38) but by the time Matthew was written, the Trinitarian formula was used.

10.25 The Ascension (Luke 24:47–53)

Jesus had appeared to them many times during the forty days after the Resurrection. They never knew at what moment he might appear, and this taught them that at every moment they were living in his presence: he had eventually to wean them of the necessity of his physical presence. He was to go away from the few that he might be present to all men. Had he not done so, his presence would always be localized in Palestine. When they were ready for it, he gave them this dramatic acted parable to make it clear to them that they would see him no more, but that he would be with them always, everywhere. He went 'up' – this symbolized his entry into glory.

Luke's fullest account of the Ascension is in his second volume, The Acts of the Apostles.

11 A Selection of Important Teachings

11.1 Prayer (See Unit 6.9)

1 The Sermon on the Mount (Matthew 6:5–15)

(a) Do not make a show of piety. 'Go into a room by yourself' rather than try to impress others with your goodness.

(b) The hypocrites have their reward already in the admiration of men: by implication, it is the only reward they will have, for they do not impress God. If the object of prayer is to impress men, it is wrongly directed and cannot be called prayer at all.

(c) Do not indulge in meaningless repetition, for saying prayers is not the same as praying. Do not imagine that the more you say the more likely you are to be heard, for 'Your Father knows what you need before you ask Him.'

(d) The Lord's Prayer (Matthew 6:9–13) (See Unit 6.10). Note that this, the model Christian prayer contains four basic elements, Adoration, Confession, Intercession and Petition.

2 Three Important Parables

(a) The Friend at Midnight (Luke 11:5–8) (See Unit 7.4)

(b) The Unjust Judge (Luke 18:2–8) (See Unit 7.4)
Both of these stress the need to pray in faith, confident that a loving Father will give His children what is best for them.

(c) The Pharisee and the Publican (Luke 18:10–14) (See Unit 7.3)
The true attitude of prayer is trust in God's love and mercy, not in self-righteousness.

11.2 Acts of Charity (Almsgiving) (Matthew 6:1–4) (See Unit 6.9)

1 Do not make a show of your religion. When you perform a charitable act, do so anonymously, not as the hypocrites who advertise their generosity in order to win the regard of men. The only reward they will have is the approval of men (and by implication,) not of God. To give to those in need, in order to boost one's self-regard, is to be involved in a conflict of motives.

2 Give generously, not counting the cost to yourself. 'Do not let your left hand know what your right hand is doing.'

3 The example of the Widow's Offering (Mark 12:41–44; Luke 21:31–34). What made her tiny gift of more value to God than the large sums given ostentatiously by the rich from what they could easily spare was:

(a) she made a personal sacrifice, giving of her very livelihood;

(b) her motive was not to impress others.

It is not the intrinsic value of the gift that matters, but the sacrifice involved in the giving.

11.3 Fasting (See Unit 6.9)

The fast must be known only to God so that it might be to the glory of God and not self.

Jesus fulfilled each of these duties of prayer, acts of charity and fasting, but drew attention to the fact that those who perform them in order to bolster up their own pride have their reward (the good opinion of their neighbours) in this world only, and do not impress God.

11.4 Anxiety (Matthew 6:25–34) (See Unit 6.11)

Look at the birds. They do not sow or reap or store their food in barns, yet your father feeds them. You are worth more than the birds. The flowers do not work or spin yet 'even Solomon in all his splendour was not clothed like one of these'. If God looks after the grass, will he not care for you? It is a counsel of 'first things first'. Trust in God and all other things fall into place. There are enough troubles in the present without worrying about what might happen in the future. However, it is important to note that it is not sensible foresight that Jesus warns against (e.g. the family man who takes out adequate insurance to provide for his dependants in the event of his death). What Jesus does condemn is:

(a) Worrying about matters over which we have no control. Thus, 'Is there a man of you who by anxious thought can add a foot to his height?'

(b) 'Anxious thought', that is, the worry which becomes an obsession. This is counter-productive because it inhibits us, hindering us from getting on with the work of the Kingdom, and indicates a lack of trust in the promises of God.

11.5 Judging Others (Matthew 7:1–6) (See Unit 6.12)

When we condemn others we console ourselves that at least we are not as bad as they are. This inflates our ego and makes us self-righteous: exactly what Jesus condemns in the Pharisees.

Note that in the parable of the Pharisee and the Publican (Luke 18:10–14) the Pharisee thanked God that he was 'not like the rest of men, greedy, dishonest, adulterous, or for that matter, like this tax gatherer'. His was not a prayer but a boastful catalogue of his own virtues, which, he felt, entitled him to condemn the tax gatherer. Yet it was the penitent tax gatherer, who relied only on the love and forgiveness of God, whose prayer was accepted.

In the Sermon on the Mount we are told, 'Pass no judgment', yet common sense tells us that it is often necessary to make judgments on others, in the usual meaning of the word, for if we hope to help other people we have to make a judgment of what is the case. An illustration from medical practice may help us. When a doctor tells us we have an illness he is not condemning us for having a disease: he has rather, made a diagnosis on the evidence of the symptoms, which will make it possible for him successfully to treat the ailment. A Christian's judgment of another person should be similar to a diagnosis (like a doctor, he would not intrude on anyone's privacy unless it was essential), and the object must be in order to be of maximum help, not to condemn. A Christian has to bear in mind that he, too, is a sinner, and may well have a plank in his own eye when he sets about trying to remove the speck from his brother's eye.

What Jesus means is that Christians should not judge others in the sense of 'passing judgment' or condemning them. They must be generous in their attitude to others, because only God knows the full facts of the case. Only those who know their own weakness, who forgive, and are charitable to others, are in the right state to receive God's forgiveness.

11.6 Jesus and the Sabbath (See Unit 1.6)

Jesus seemed to go out of his way to break the sabbath regulations. In fact, he broke only the letter of the law when it hindered the fulfilment of its spirit. He taught that, 'The Sabbath was made for the sake of Man and not Man for the Sabbath' (Mark 2:27).
Important incidents which illustrate his teaching are:

1 The incident in the cornfield (Mark 2:23–28; Luke 6:1–5; Matthew 12:1–8)

2 The man with the withered hand (Mark 3:1–6; Luke 6:6–11; Matthew 12:9–14)

3 The man with the dropsy (Luke 14:1–6)

4 The crippled woman (Luke 13:10–17)

11.7 Jesus and the Law (See Unit 6.7)

Jesus upheld the Law, saying that 'not a letter or stroke' would disappear from it. He regularly attended synagogue worship and commanded a leper he had healed to have his cure certified by the priest and to offer the sacrifice required by the law. Yet he felt free to criticize the Law, to re-interpret it, and even to amend it.

The key to this apparent contradiction is that he drew a distinction between the ceremonial law and the moral law. He was not against ritual in principle, but saw that its observance could become a formality which took the place of genuine religion. He called the people whose religion consisted mainly in the performance of formal rites 'hypocrites', a Greek word which means 'someone acting a part'.

Above all, he criticized the ritual law when its performance stood in the way of human need. Thus he did not hesitate to heal on the sabbath day, even in the synagogue (Mark 3:1–6).

Jesus taught that the Law was to be observed in accordance with its original purpose, duty towards God and also towards men. The work of God and also human need claimed priority over the observance of the Law. He reinterpreted the moral law by revealing its deeper meaning. He said, 'You have learned that our forefathers were told. . .' and quoted a particular law; he then continued, 'But what I tell you is this'. He then proceeded not to deny the Law, but to draw out its deeper meaning. He went on to give five examples to show how his teaching fulfilled the Law of Moses. (See Unit 6.8)

1 Murder (Matthew 5:21–26)

2 Adultery (Matthew 5:27–28)

3 Divorce (Matthew 5:31–32)

4 Oaths (Matthew 5:33–37)

5 The law of 'An eye for an eye' (Matthew 5:38–42)

11.8 Wealth

Jesus took the attitude that while there is nothing evil about money or possessions in themselves, they can preoccupy men to the extent of becoming idols which prevent them from surrendering the whole of their lives to God. To possess wealth is not wrong, but there is always a danger that it will possess us.

Three important incidents illustrate his teaching:

1 The Rich Young Man (Mark 10:17–31; Luke 18:18–30; Matthew 19:16–30)

Jesus saw that the young man was so preoccupied with his wealth that nothing less than complete renunciation would do, just as an alcoholic has no alternative but to abstain completely if he is to be cured. In the event, it was more than the young man could do.

2 Zacchaeus, the Tax Gatherer (Luke 19:1–10)

Jesus approved of the promise of Zacchaeus, 'I give half my possessions to charity; and if I have cheated anyone I am ready to repay him four times over.' The young man was asked to give up everything, but Zacchaeus' offer of only half his goods is accepted.

3 The Brothers and their Inheritance (Luke 12:13–21)

The message is, 'You can't take it with you'. A man of great wealth may be a pauper in the eyes of God.

A saying recorded in the Sermon on the Mount (Matthew 16:19) is of relevance: 'Do not store up for yourselves treasure on earth where it grows rusty and moth eaten. . . Store up treasure in heaven. . . for where wealth is, there will your heart be also.' See also the Parable of the Rich Man and Lazarus (Luke 16:19–31), a plea for the responsible use of wealth.

11.9 Marriage and Divorce (Mark 10:1–12; Luke 16:18; Matthew 5:27–28, 31–32, 19:1–12)

There were two schools of Jewish thought. While it was agreed that divorce was lawful, the circumstances under which it was permitted were in dispute. According to the Law (Deuteronomy 24:1) a man could divorce his wife if 'she does not win his favour because he finds something shameful in her', but the 'shameful' thing is not defined. The school of Rabbi Shammai taught that it was adultery and nothing less serious, but the school of Rabbi Hillel taught that it could be a number of lesser offences.

Jesus agreed with neither, saying that a man and his wife were one and inseparable, and that in the days of Moses divorce was permitted because the people were then 'so unteachable'. He taught that marriage was indissoluble.

The Jewish Law allowed remarriage after divorce, but Jesus taught that this was to commit the sin of adultery by all parties: his teaching was thus more strict than the Jewish Law.

Matthew 19:9 introduces a much disputed exception to the rule, that of a wife's unchastity. (In Jewish Law a wife could not divorce her husband.) This would seem to contradict the teaching of Jesus about unlimited forgiveness, and also to make even a single act of adultery more serious a sin than persistent cruelty, whether physical or mental.

Some manuscripts use the Greek word for fornication rather than adultery both here and in the Sermon on the Mount (Matthew 5:32). This could be interpreted to mean that if a man puts away his wife because of fornication, i.e. unchastity before marriage, the issue is not one of divorce, but nullity, because the Law decreed that fornication made subsequent marriage null and void. The Christian Church allows marriage after a decree of nullity on the grounds that the original 'marriage' was not a marriage at all.

Another note of controversy is raised by Matthew 19:10–12, which could mean that Jesus admitted that his teaching was an ideal to be striven for, but was too difficult for many to achieve.

The earliest statement on marriage and divorce in the New Testament is in 1 Corinthians 7:10–17 and is to the effect that marriage between Christians is indissoluble. Paul insists that this is not his ruling, but 'the Lord's'. The subject is still highly controversial, but what is beyond dispute is the high value Jesus placed on the marriage bond.

Part II The Old Testament

12 Genesis

12.1 Historical Background and Religious Ideas

There are various modern scientific theories of how the universe came into being, but it would be unreasonable to expect the writers of Genesis living more than 2500 years ago to have knowledge in advance of their own time. Their chief interest was not in how the universe happened but rather in stating their conviction that it did not happen by chance: they saw the **whole natural order as the purposeful creation of God**. Some people take the creation stories of Genesis literally to be true and discount the scientific evidence of biologists, geologists and astronomers, but the vast majority of Bible scholars believe that the purpose of the authors was to tell about God and His purpose for mankind. Their message was conveyed within the framework of an ancient religious myth; confusion arises when we take the myth for the message.

Genesis makes no attempt to prove the existence of God. For the writers the reality of the Being of God was not a conclusion reached by logical argument, but a conviction born of their daily experience of Him. They saw behind the universe the mind and purpose of the Supreme Being they had come to know from their own experience. So the essential point of the story of the first chapters of Genesis is that **God is the self-existing Creator of everything in the universe**. This means that God did not create it out of already existing material, but out of nothing (creatio ex nihilo); there was only God when He summoned the universe into existence. It follows from this that there is a distinction between the Creator and His Creation: thus man as a created being can never become God, and when he tries to be God, which he sometimes appears to do, he can only bring disaster upon himself. This is a principal part of the message of the story of Adam and Eve. We as human beings are not in this world by right, but by the grace of God: our life is a gift to be accepted in gratitude and in a spirit of responsibility.

12.2 The Image of God

Genesis 1:27 states that **God made man in His own image**, which means that man is distinct from the animals and enjoys a special relationship with God. It does not mean that God looks like man, but larger; rather that man shares some of the qualities of God which animals do not. Man has the ability to know what is false and what is true, what is right and wrong, and what is beautiful or ugly. In other words he has power to reason, to make moral decisions, and aesthetic judgments. Psalm 8 says of man:

> Thou hast made him little less than a god,
> Crowning him with glory and honour
> Thou makest him master over all thy creatures;
> Thou hast put everything under his feet:

In his own sphere man is like a god. He is the supreme among all created things. If man is made in God's image it follows that the personality of man is of infinite value and this is reflected over and over again in the insistence of the Bible that man must deal justly with his fellow-man.

The image of the monarch upon a coin is a representative symbol, and there is a sense in which man, since he is made in the image of God, is the representative of God to the rest of the Creation. Man has been given authority by God over all created things, and with that authority goes accountability for the way in which he discharges his stewardship. Man's exploitation of the rest of the Creation is an offence against God.

Adam and Eve were free to obey or to disobey God. This tells us that God created man as a person, not a robot, one who is free to respond and find his true purpose and fulfilment in God, to be indifferent to the claims of God, or to rebel. This could not be otherwise because man could not respond to the love of God if he were not free. Love cannot be forced: it has to be a free response. On the evening of the sixth day God looked on His completed work of creation and 'God saw all that he had made, and it was very good' (Genesis 1:31). The important point being made here is that the material things of life are God's gifts and are to be enjoyed, not disdained; they can be wrongly used but they are themselves good. It is for man to use them responsibly, justly, and without waste, remembering both the claims of the needy, and his accountability to the Creator.

12.3 Genesis 2:4–3:24

The second creation story is very different from the first. It begins with the creation of man and ends with the creation of woman, whereas the first account begins with the creation of light and ends with the creation of man. The order in which things are created is quite different and it is clear that the author is not the same as the one who wrote the first account. This and a great deal of similar evidence, has prompted scholars to conclude that Genesis is a compilation of accounts derived from several different sources. The second creation story is derived from a source dating from the 9th century BC which is generally known as 'J' from the name Jahweh by which it calls God, while the first account derives from a much later source (5th century BC) called 'P', the priestly source. Genesis 2 is thus 400 years older than Genesis 1. Another source, not involved at this point is called 'E' after the name it gives to God – 'Elohim'. Genesis is the work of an editor rather than an original author and since he was co-ordinating sacred material which he could not lightly alter, this would account for the inconsistencies sometimes found in the text. Different as these stories are in detail, the underlying assumptions are the same, namely that God is the self-existing Creator of everything that is, and man is a created being whose existence in this world is not by right but by the grace of God.

12.4 The Fall

The story of Adam and Eve displays a profound understanding of human nature. We see man placed in God's Creation and given the responsibility of caring for it. It is an idyllic existence which Adam and Eve are meant to enjoy; they are forbidden only to eat of the fruit of the tree of the knowledge of good and evil. They are free to choose and in the event they use their freedom to flout the authority of God. The fact that the fruit of the tree was tempting to the eye was an immediate but secondary temptation, the real temptation was that by eating the fruit they would be 'like gods'. They were not content to be creatures: **they challenged the authority of God**.

Genesis does not give us an historical account of the origin of sin, but an inspired analysis of its meaning. The prophets, as we shall later see, had been led to understand that sin is essentially disobedience to God; all the injustices and cruelties inflicted upon the nation were the direct result of having turned away from a just and righteous God. The prophets saw that the root of sin is disobedience, rebellion against God. Thus, although the account is dressed in the garb of inspired myth, it is in essence not the story of the first man's fall, but of every man's. We do not need to ask if it is true; we know it to be true from our own experience.

The result of their disobedience was loss of innocence – they discovered that they were naked and tried to hide themselves from God.

The message is that when man puts himself at the centre of the universe in place of God, disaster follows. Man is made in the image of God but the ways in which he uses his freedom tarnish that image, frustrate God's purposes, and bring upon himself suffering and death. The first consequence of the Fall was to break the relationship between man and God and the second was to destroy man's relationship with his brother. (The story of Cain and Abel, Genesis 4:1–16.)

12.5 The Flood (Genesis 6:5–9:17)

The story probably originates in an ancient Babylonian myth written in the Gilgamesh Epic. The similarities are so close that there can be little doubt of this. However, the Hebrew version is very different in its religious and moral content. Archeological evidence indicates that about 4000 BC there was a disastrous flood in Mesopotamia, then the most advanced area of civilization, but there is no evidence to suggest that it was more than a local occurrence. The Biblical account may well go back to this event, but the total flooding of the world is not history. Inconsistencies in the story are due to the interweaving of the two sources, J and P.

Man was so sunk in wickedness that God was sorry he had created him. The point of the story is that God cannot condone sin, and the flood is the symbol of God's judgment upon mankind. Though deserving to be destroyed, mankind is saved from extinction because of one righteous man. (Compare Genesis 18:32 where God promises not to destroy Sodom if Abraham can find ten good men in it.)

12.6 The Covenant with Noah (Genesis 8:21–9:14)

The sign of the covenant, or solemn agreement, between God and Noah is the rainbow. It is to serve to remind God as well as Noah of the agreement. This is of course one of the many anthropomorphisms to be found in these early stories; for example in the creation story God walks in the garden in the evening, and here God smells the odour of Noah's sacrifice and makes the rainbow an aid to memory. (Anthropomorphism is to conceive of God as having human form, personality, or character.)

As in all the Bible covenants, that with Noah is an expression of the love and mercy of God in spite of the frailty of man. On His part God accepts obligations and in turn the articles of the agreement lay responsibilities upon man.

The terms of the covenant are:

1 God gives to man every animal for use as food, just as He at the beginning gave him 'all green plants' (Genesis 9:3).

2 The flesh must not be eaten with the blood still in it. (This reflects the primitive idea that the blood was the life. Man could eat the flesh, but the blood, the life, belonged to God.)

3 Man was no longer to live without the rule of law. 'He that sheds the blood of man, for that man his blood shall be shed' (Genesis 9:6).

4 Never again will God drown the earth.

12.7 Abraham (Genesis 11:31–12:9)

Abram came from Ur and with his father, Terah, and nephew Lot, set out for Canaan. But when they reached Harran in what is now south-eastern Turkey they settled there.

The call of Abram

God told Abram to leave his father's house and journey 'to a country that I will show you', promising to bless him and make of him a great nation. Abram trusted God, who led him to the land of Canaan, telling him that He would give this land, (the promised land) to his descendants. It was already occupied by the Canaanites, but Abram, believing in the promise of God, formally claimed it by building altars and offering sacrifice to the true God at the sacred terebinth tree of Moreh and in the hill country near Bethel. He then journeyed by stages southwards towards the Negeb and was eventually driven by famine to Egypt.

Abram prospered in Egypt as a result of his wife Sarah being taken into Pharaoh's harem, a rather discreditable episode, and he later returned with Lot to the place in the hill country where he had previously set up an altar. There he and Lot separated because the land could not support their numerous livestock. Lot, given the choice, settled in the plain of Jordan, a rich land before the destruction of Sodom and Gomorrah. Abram remained in the promised land, settled by the terebinths of Mamre at Hebron and 'there he built an altar to the Lord' (Genesis 13:18). Abram became embroiled in a local war between four kings and the kings of Sodom and Gomorrah. He was forced to intervene when the four kings plundered Sodom and Gomorrah and carried off Lot as a captive. On Abram's return in triumph he was met by Melchizedek, King of Salem, who offered gifts of food and wine (Genesis 14:18–20). Here we have the **first reference to Jerusalem**, which centuries later was to be the City of David and the site of the Temple.

The covenant with Abraham (Genesis 15:1–21 and 17:1–14)

The covenant with Noah was between God and all mankind but the covenant with Abram was between God and a special people within mankind, Abram and his descendants. Genesis records two accounts of this covenant, the first and earliest from 'J' and the second from 'P'. God appeared to Abram in a vision and spoke with him. The terms of the covenant were:

1 'Your name shall no longer be Abram, your name shall be called Abraham, for I make you father of a host of nations' (Genesis 17:5).

2 'I will fulfil my covenant between myself and you and your descendants after you, generation after generation, an everlasting covenant, to be your God' (Genesis 17:7).

3 'As an everlasting possession, I will give you and your descendants after you, the land in which you are now aliens, all the land of Canaan' (Genesis 17:8).

4 This covenant was to be sealed by the rite of circumcision (Genesis 17:10ff).

The covenant rite (Genesis 15:9–17)

The symbolism of the sacrifice which was to ratify the covenant is very profound and its meaning lies in the primitive idea that the life of an animal resided in its blood. The sacrificial victim was regarded as a bridge between God and man. In this case the animals, having been killed, a heifer, a she goat, a ram, a turtledove and a fledgeling, the carcasses were cut in half and each piece was placed opposite its corresponding half. Normally, each party to the covenant passed between the pieces and were thus symbolically bound together by the life of the sacrificed animal, but in this case it was only God who passed between them since the parties to the covenant were unequal. God is represented as 'a smoking brazier and a flaming torch passing between the divided pieces' (Genesis 15:17). This is an excellent example of Biblical imagery, fire being symbolic of the presence of God.

The son of the promise (Genesis 18:1–15)

The Promise could not be fulfilled until Abraham's wife should have a son. Abraham would have been happy for Ishmael, his son by Sarah's serving woman, to be his heir but this God would not allow. Sarah was past the age of child bearing and Abraham was very old but God appeared to him by the terebinth trees of Mamre and promised that within a year Sarah would bear a son.

The sacrifice of Isaac (Genesis 22:1–18)

The story is told with great sensitivity. Abraham had lived at Ur in his youth and must have been familar with human sacrifice at the huge Ziggurat, the Sumerian Temple, and to be called upon to sacrifice his son may not have been as unreasonable or horrifying a request as it appears to us: there are **numerous Biblical examples of human sacrifice**, for example Judges 11:29–40 (Japhthah's vow). The modern view that the story is intended to teach that human sacrifice is repugnant to God, is probably an attempt to make the story more palatable, and the real point of the story is that Abraham accepts with utter obedience the demand that God makes for the surrender of his most precious possession. If he had in fact killed his son, God's promise would have been broken, but Abraham's trust was so great that even in this extremity he believed God's promise would not fail.

The death and burial of Sarah (Genesis 23)

By purchasing the cave at Machpelah for the burial of his wife, Abraham acquired a legal right to a portion of the land which God promised to his descendants. The reluctance of Ephram to sell may be an example of the polite but devious method of oriental bargaining, but it could be a mark of the owner's genuine reluctance to allow a stranger to have a legal right to any Hittite land. He preferred to give Abraham the use of the cave rather than to sell it.

12.8 Jacob

The Bible turns from the story of Abraham, a man prepared to give utter obedience to what he believed to be the will of God, and whose integrity and unwavering faith fitted him to receive the promise of God, and to be the revered father of the People of God, to Jacob, his grandson, grasping, underhanded, deceitful, disloyal, a man of low cunning, in many ways thoroughly contemptible. His brother, Esau, seems a much more attractive character, open, honest, straightforward, generous and forgiving; yet it was Jacob who was chosen to serve the purposes of God. The Bible invites us to see something of ourselves in this selfish man, and shows us what God can do with such an unlikely prospect. We are invited to look not only on what Jacob was, but what in God's hands he became.

The birth of Jacob and Esau (Genesis 25:19–26)

As in the birth of Isaac, God takes a direct part in the conception, for Rebecca was childless.

That Jacob was born grasping his brother's heel presages his later desire to supplant him, and to be the one who would inherit the promise made to Abraham.

The Bible gives us a picture of a very poor family situation in which there is disloyalty between husband and wife, while the children are the main sufferers. Favouritism on the part of both parents has appalling consequences.

The birthright (Genesis 19:27–34)

There is no disguising the fact that Jacob's behaviour is contemptible, but the point that the Bible is making is that Esau thought so little of his birthright, which included principally the promise made by God to Abraham, that he was prepared to sell it for a trifle. Such a man was obviously unfit to inherit the promise. This is underlined by Esau's marriage to Hittite wives (Genesis 26:34–35), which caused both Isaac and Rebecca bitter grief. Jacob by contrast sought out a wife from the family of his mother's brother.

The blessing (Genesis 27:1–45)

The words of the blessing (28–29) make it clear that material prosperity rather than spiritual gifts, is being bequeathed. It is clear also that such blessing is being underwritten by God and that once given it cannot be withdrawn.

The treacherous scheme of Rebecca is craftily employed by Jacob, who passes himself off as his brother, and the blind old man is at length completely deceived. He plumbs the depth of hypocrisy by accounting for the speed at which he had provided the savoury dish by saying 'It is what the Lord your God put in my way.' It is very close indeed to blasphemy.

Neither the indignation of Isaac nor the bitter disappointment of Esau can undo what has been done.

Esau's vow of vengeance makes it advisable for Jacob to fly for safety, to his mother's relatives at Harran and there is no record that she ever saw him again, a heavy price to pay for her duplicity.

Jacob's dream at Bethel (Genesis 28:10–22)

Jacob is now a fugitive and an exile from the land of promise: fear of his brother's revenge made it impossible for him to return to the land of his inheritance. It was at this crisis of rejection, and possibly of guilt, that God called him. Strange as it may seem, there were qualities in this wicked man for which God had a purpose, but Jacob needed to be disciplined, to learn the hard way, before God's high purpose for him could be fulfilled.

Like Abraham he trusted in God in spite of all indications to the contrary, for nothing seemed less likely than that he would inherit the promise, but God confirmed it at Bethel. **Jacob and his descendants would inherit the land** and be 'countless as the dust upon the earth': Jacob would be blessed and protected until God's promise to him was fulfilled. Jacob made a vow to take the Lord as his God and to offer a tenth part of all that God should give him.

Other points of significance in the story are that (a) heaven is only a ladder away from us: God is near at all times and (b) the part played by the stone. Students of primitive religions see here an excellent example of animism. Jacob took the stone for a pillow and saw the dream to be the result of contact with the stone. When he awoke he set up the stone as a pillar, pouring oil upon it as an act of worship to the deity who dwelled within it. The stone is a god's house, a beth-el.

Jacob's wives and children (Genesis 29)

The penniless Jacob agreed to work seven years for Laban's younger daughter, Rachel, only to be cheated by his father-in-law who substituted the elder daughter, Leah, at the marriage. When the week of marriage ceremony was over he was allowed to marry Rachel, whom he loved, but had to promise to work for a further seven years for her.

Leah, whom Jacob neither loved nor wanted, bore him six sons, Reuben, Simeon, Levi, Judah, Issacher and Zebulon, while Rachel's slave girls, who in those days were allowed to act as proxies, bore him Dan and Naphtali. Leah's slave girls bore him Gad and Asher, and finally Rachel conceived and bore Joseph and Benjamin, once again after the direct intervention of God. **These sons were the founders of the Twelve Tribes of Israel.**

Jacob's meeting with Esau (Genesis 32 and 33)

Laban has prospered by Jacob's management of his flocks and herds and is unwilling for his son-in-law to leave his employ, so Jacob is finally forced to gather together his own property and run away with Rachel and Leah and the children. Laban pursues them in anger but God again intervenes by speaking to Laban in a dream, warning him not to interfere. When Jacob reaches the promised land, he hears that Esau is on his way to meet him with a large company and he has every reason to fear for his safety. In his prayer he confesses his own unworthiness but remembers the promise of God to protect him – 'thou didst say, I will prosper you, and will make your descendants like the sand of the sea'. Having sent a handsome peace-offering to his brother, and placed his family and possessions across the river Jabbock, Jacob spends the night alone.

Jacob at Peniel (Genesis 32:22–32)

In this meeting in the darkness of night with a stranger we see in picture language the spiritual struggle of a man tormented by guilt. Jacob wrestled all night with the man, and because he could not throw him the man struck Jacob in 'the hollow of his thigh', dislocating the hip, but Jacob would still not loose his hold until he received a blessing. He asked Jacob for his name and when he answered said 'Your name shall no longer be Jacob but Israel, because you strove with God and with men, and prevailed' (Genesis 32:27–28). Jacob called the place Peniel, because, he said 'I have seen God face to face and my life is spared.' When day came, Israel limped across the river, a changed man, now fit to be ranked with Abraham and Isaac. The meeting with Esau was to effect a sincere and lasting reconciliation, and Jacob settled in Canaan.

12.9 The Story of Joseph (Genesis 37)

The key to the story of Joseph is found in his words to his brothers when, as ruler of the land of Egypt, under Pharaoh, he made himself known to them, 'Do not be distressed or take it amiss that you sold me into slavery here; it was God who sent me ahead of you to save men's lives.' (Genesis 45:5).

Joseph is seen in the story as an **instrument in the hand of God** to ensure that the promise shall not fail. It is the dominant theme of the whole of the Bible – God's plan to save the world from its own selfishness and the consequences of its disobedience, by choosing a people, a community which would make His ways known, and be a blessing to all mankind. Whatever evil fortune befell Joseph, God was with him, and 'meant to bring good out of it' (Genesis 50:20).

Jacob, however, seems to have learned little from the circumstances of his upbringing and the painful consequences of being the spoiled darling of his mother; he made no secret of the fact that Joseph was his favourite son. He made 'a long sleeved robe', for the apple of his eye.

Joseph's conduct was consistent with his status: he told tales about his brothers, and his dreams, which he was careful to recall to his brothers, disclose the arrogance of a spoiled child. The main examples are in the dreams about the sheaves of corn in the field (Genesis 37:5–8) and the sun, moon, and stars (Genesis 37:9–11). The latter even Jacob could not stomach. It is not surprising that the brothers came to hate this highly unpleasant young lad; there was a streak of the old Jacob in him.

The plot against his life was, however, an extreme reaction, which was inexcusable, especially in view of the suffering it brought to their father. His loyalty to his employer, Potiphar, brought imprisonment as its reward, but this proved to be the occasion of rapid promotion. The Lord was with him and having won the favour of the governor of the prison he was put in charge of all the prisoners.

Joseph interprets the dreams of the butler and baker (Genesis 40)

1 The vine with three branches

2 The three baskets of white bread

Pharaoh's dreams (Genesis 41)

1 The seven fat cows, and the seven lean cows

2 The seven ears of corn, full and ripe, and the seven thin and shrivelled ears

Joseph claims that God has enabled him to interpret the dreams and he is bold enough to tell Pharaoh what must be done. A shrewd and intelligent man must be put in charge of controllers who will ensure that a reserve is set up during the good years against the years of famine. Pharaoh appointed Joseph and gave him authority subordinate only to his own, in the land of Egypt.

13 The Exodus

13.1 Historical Background

The Book of Exodus deals with the most significant events in the whole history of Israel, events which down to the present day are a living memory. For the Jew they are not merely events in history, but part of the eternal present, still potent, still experienced in the life and worship of the 'chosen people' of God. They provide the faithful with undeniable proof that **God chose Israel**

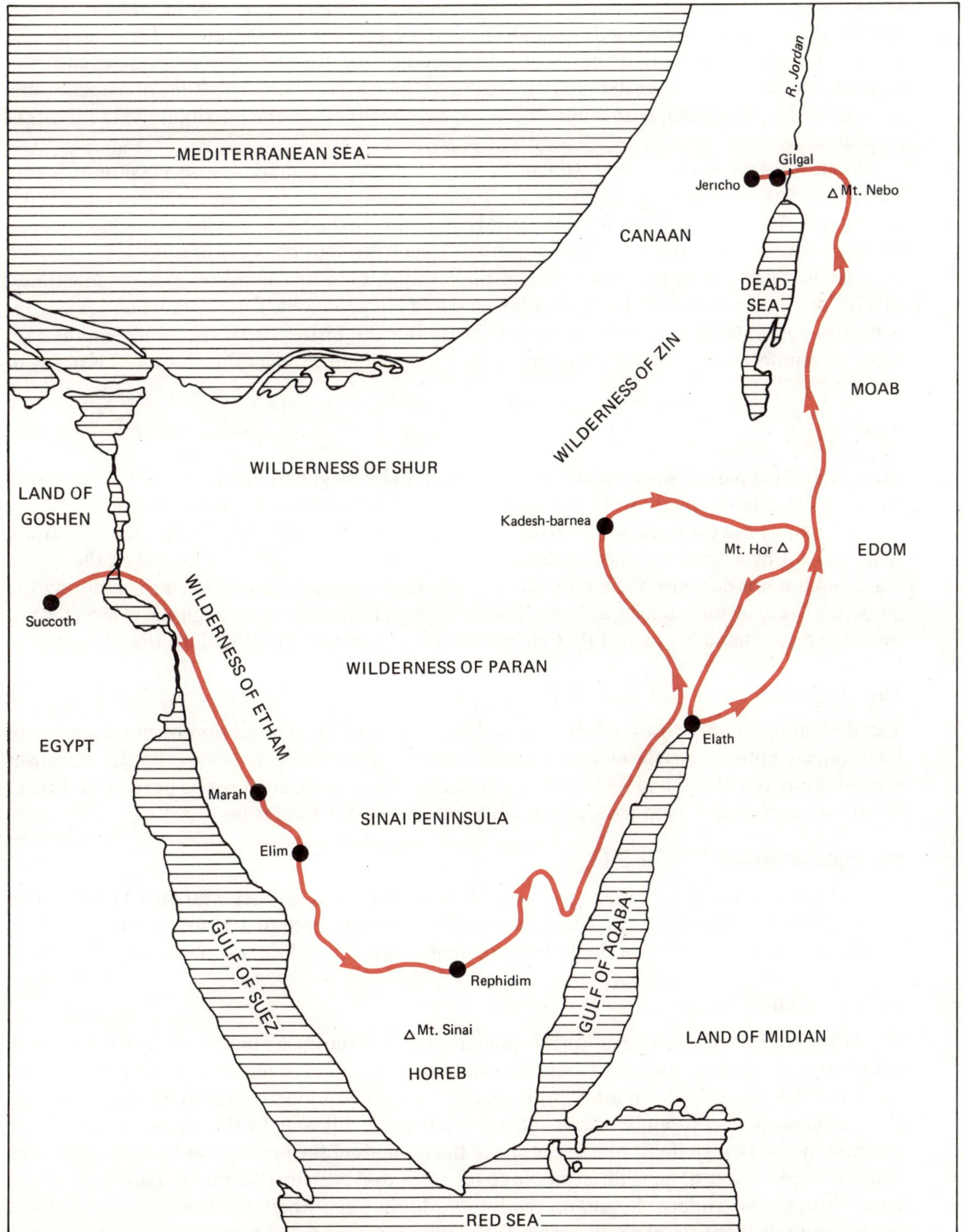

Fig. 13.1 The Exodus: route taken by the Israelites from Egypt to Canaan

for His own. Thus at Passover the Jew does not merely recall the epoch-making events of the Exodus: he brings them out of the past into the present so that they live; he does not merely remember Moses, he is with Moses in all those graphic events which are the foundation of the nation's history.

The only evidence for the Exodus is the Bible itself: there are no Egyptian records to confirm any of the story, but this is not surprising in view of the relatively small numbers involved. We can only guess as to when it happened and dates are variously given between 1450 and 1250 BC, but that it happened seems indisputable in view of the tradition so deeply embedded in the Bible and in the souls of the Jewish people. It is unlikely to have been invented because no proud nation would advertise its origin as a nation of slaves with no will to free itself from its oppressors. Ever since, that nation has gloried in the fact that it was delivered by God from a hopeless situation in which it was powerless to extricate itself.

It would be quite unfair to judge the credibility of the account by the standards of a modern historian. We are dealing with very ancient material in which fact, legend and theological interpretation are all inextricably bound up, and we are unable, therefore, for example, to account for the tradition that they crossed the bed of the Red Sea dry shod, or even if the location was the Red Sea, or the Gulf of Aqaba, or the Bitter Lakes. The Biblical name for this stretch of water is yam suph, meaning the reed sea, which widens the possibility of the location considerably.

Whatever actually happened we can be sure that only a remarkable succession of highly unusual events could have given rise to the tradition, but what is of supreme importance is the interpretation Israel placed upon them. Surely no nation in the history of the world has suffered so much for so long, and what has preserved Israel through all its sufferings has been the unshakable faith that in the experience of the Exodus, God first unmistakably demonstrated that Israel was chosen for His high purposes, and that His promise will never fail. For our study it is best to concentrate not on an attempt to tease out the history from its entanglement with legendary material, but to look at the story as it stands, and to discover the message it is trying to convey.

13.2 Moses

Many years had passed since Israel came to Egypt and new generations had risen, prospering and increasing in numbers, so that a significant minority lived on the main invasion route to Egypt. Possibly as a measure designed to increase state security, the Pharaoh decided to exercise strict control over them by putting them to forced labour on public works and in the fields. Hardship did not decrease their numbers and Pharaoh decided on direct action, ordering the midwives to kill all male babies at birth. This ruse being unsuccessful through the evasiveness of the midwives, Pharaoh ordered that all male babies were to be thrown into the Nile.

The birth of Moses (Exodus 2:1–10)

The determination of Moses' mother to save her son, and the compassion of the daughter of Pharaoh, are both seen as the means by which God operates in order to ensure that His purposes are not thwarted. Moses had a key part to play in the divine plan and must be preserved. Life in the royal court must have provided valuable experience for the future.

The flight of Moses (Exodus 2:11–22)

The incident of the killing of the Egyptian and its discovery made it necessary for Moses to run for his life to the Midian desert where he lived in the household of Jethro, priest of Midian. Eventually he married Zipporah, Jethro's daughter, who bore him a son, Gershom.

The call of Moses (Exodus 3:1–15)

We should not try to find some rational explanation of how the bush burned with fire and yet was not consumed: such an exercise would be speculative in the extreme. It is more profitable to recognize that the story is an attempt to describe a deeply religious experience by the use of picture language. Deep emotions can only be described in this way, by the use of simile or, as in this case, metaphor. In the Bible, **fire is one of the symbols of the presence of God** and there are many examples – the 'flaming torch' at the covenant with Abraham, the pillar of fire that guided Israel through the wilderness, and on the top of Mount Sinai when 'the Lord had come down upon it in fire'. What the story describes is the personal encounter between Moses and God. The upshot of the incident was that Moses was commissioned to return to Egypt, tell the Israelite slaves of his meeting with Jehovah and of God's plans for them to return to the land of their inheritance. He was to negotiate with Pharaoh for three days' release, so that they might journey into the wilderness to sacrifice to their God.

God disclosed his name to Moses: he was to tell the Israelites that 'I am' had sent him (Exodus 3:14). In the next verse God says that His name is Jehovah a word related to the Hebrew word for 'I am' and He reveals that He is no strange God but the God of Abraham, Isaac and Jacob, their forefathers. The call of Moses and his subsequent mission to the Israelite slaves was to be the beginning of a new chapter in the self-revelation of God to His people, truths about Himself which up to now had been only dimly grasped.

Moses was understandably reluctant to accept this enormous undertaking and made the excuse that the Hebrews would not believe his story that God had appeared to him, and that Pharaoh would not be persuaded. The answer, if the story is to be taken literally, is to be a display of the magic arts – the staff of Moses becomes a serpent and his arm leprous, while Nile water is turned into blood. Moses attempted further to excuse himself by pleading lack of the necessary eloquence, but God deputised Aaron, his brother, to be the mouthpiece.

Moses returns to Egypt (Exodus 4:19–5:22)

As predicted Pharaoh would not be persuaded, and accused Moses and Aaron of distracting the people from their work and encouraging them in idleness. Pharaoh's response was to lay extra burdens on the people by making them supply their own straw for brick-making (hence the metaphor - 'to make bricks without straw'), and they remonstrated with Moses for making their lot not better but worse. Moses returned to Sinai to report to God and to complain about the inadequacy of his support. (Note the anthropomorphisms.)

The call of Moses repeated (Exodus 5:44–6:13)

God reassured Moses:

1 'Though I appeared to Abraham, Isaac and Jacob I did not tell them my name as I have done to you.'

2 'I have heard the groanings of my people Israel and have called to mind the covenant I made with their forefathers.'

3 'I will rescue them from slavery, adopt them as my people, and become their God.'

4 'I will lead you to the land I promised your forefathers.'

13.3 The Plagues of Egypt (Exodus 7:14–10:29)

The explanation of the plagues is a matter about which we can only speculate. Most of them are likely to have been natural occurrences which may have been more remarkable than usual and in unusually rapid succession. These disasters are taken to be signs of God's determination to set His people free and to fulfil His promises to Moses.

13.4 The Passover (Exodus 12:1–13:16)

The Passover meal is generally regarded by scholars as being of much earlier origin than the Exodus. It was probably part of the spring festival of nomadic tribes and involved a sacrifice to the gods and the sprinkling of the blood on the tents to ward off evil spirits. The unleavened bread originates in the custom of the Canaanite people who once a year break the cycle of leavening the bread with sour dough. A small piece of dough from each mix of flour was left to ferment and was used in the next batch. Some contamination was inevitable in this process, and its effects were reduced by breaking the chain. At the harvesting of the first sheaf of the new corn the cycle was broken and unleavened bread was eaten until the new dough had begun to ferment of itself. This was an agricultural ritual rather than a nomadic one; in the ritual of the Passover, both are brought together.

Here again speculation is tempting, but is unlikely to produce firm conclusions, and it is best to take the Passover narrative as it stands and accept the interpretation laid upon the events by the writers themselves.

Details of the meal are as follows:

1 A yearling male lamb or kid, without blemish, must be selected and killed between dusk and dark.

2 Some of the blood must be smeared on the doorposts and lintel of each house in which the lamb is eaten.

3 The whole of the lamb is to be eaten roasted, not boiled or eaten raw.

4 Any left-over must be destroyed by fire.

5 It must be eaten with unleavened cakes and bitter herbs.

6 Out-door clothes must be worn – belt, sandals, and with staff in hand – and the meal must be eaten in haste.

7 The day is to be kept as a festival of remembrance for all time.

8 When the children ask the meaning of the rite, they are to be told how God 'passed over the houses of the Israelites in Egypt, when he struck the Egyptians, but spared our houses' (Exodus 12:27).

The death of the first born in Egypt induced Pharaoh to let Israel go. The Egyptians now urged them out of the country offering them gifts, as God had told Moses.

Later writers had to account for the route they took south eastward into Sinai rather than the very much shorter route along the coast. At the time of writing the war-like Philistines occupied the coastal area, so the writers presumed that this must be the obvious reason. However, the Philistines did not inhabit the area until long after the Exodus. It has been suggested on the evidence of the description in Exodus 19:14–20 that Mount Sinai was a volcano, so that Israel was guided to the Holy Mountain by the ascending smoke by day, and the glow of the eruptions by night, reflected in the column of cloud. If we take the story literally, which we are entitled to do if we are concerning ourselves only with the motives of the writers, we see in the cloud and fire **two powerful symbols of the presence of God**, recognizable in many of the most significant passages in the Bible.

The number of the people must be a gross exaggeration for 600 000 plus dependants would give an approximate total of about 2 million. The 70 odd people who came to Egypt with Jacob would have to have multiplied 30 000 times. The two midwives found adequate to deal with Hebrew births at the time of the birth of Moses indicate a very much smaller figure. The wilderness could never have supported such a vast crowd, in addition to their flocks and herds, for the long period indicated by the figure of forty years.

13.5 The Crossing of the Red Sea (Exodus 14)

As it stands the story is pure folk lore rather than history, but there can be little doubt that if the substance is inextricably entangled in legend, substance there must be. In this case it is tempting to speculate. The Hebrew 'yam suph' is wrongly translated Red Sea: it is rather 'reed sea', which indicates a shallow inland lake rather than deep sea water. Pharaoh, when he realized that Israel had escaped, took the only course possible and followed not with infantry, but cavalry and fast moving chariots. Israel moved through the marshy waters of the lake but when Pharaoh followed, God 'clogged their chariot wheels' making the fast, but now immobile, unarmoured vehicles easy prey to a numerically superior enemy. The strong east wind which had made a passage on foot possible now abated and the defenceless Egyptians were overwhelmed. There are more fantastic explanations. However, the writers emphasize that this action is the work of God. The Egyptians perceive that **'It is the Lord fighting for Israel against Egypt'** (Exodus 14:25) and 'When Israel saw the great power the Lord had put forth against Egypt, all the people feared the Lord, and they put their faith in Him and in Moses His servant' (Exodus 14:31). The ancient fragment of the Song of Miriam, the sister of Moses and Aaron, testified to the same conviction, that Israel has been saved by the hand of God alone (Exodus 15:21).

13.6 The Journey to Mount Sinai (Exodus 15:22–17:16)

The euphoria engendered after the crossing of the Red Sea was not to last. As soon as the hardships of the journey began to be experienced, Israel complained to Moses, first about the bitterness of the waters of Marah and then about the shortage of bread and meat. Moses appealed to the Lord who showed him how to sweeten the water by throwing a log of wood into it (a secret which would be more than valuable today). The point was then made that obedience to God would ensure their safety. Food was provided by the 'bread from Heaven', the manna, which is believed to be the honey-like substance excreted from the tamarisk tree by the action of parasites, and from flocks of migrating quails. Later Moses produced water by striking a rock with his staff; surprisingly, similar occurrences have happened in recent times. The point of these stories, however, is to stress that God is with them to protect and guide: He will always supply His people's needs. What is required of them is obedience and faith.

13.7 War with Amalek (Exodus 17:8–16)

The Jewish attitude of prayer is standing with arms lifted up towards heaven. When Moses, fatigued, lowered his hands Amalek prevailed, so Aaron and Hur held up his hands. What is being taught is that united waiting on God in prayer is the means by which success will be ensured.

13.8 Jethro, Priest of Midian (Exodus 18)

The visit of Jethro to Moses at this point of the story is of great interest to scholars. Some, noting the almost exaggerated reverence by which Moses receives him, and the fact that Jethro appears to preside at the sacrificial meal which follows, conclude that Yahweh (Jehovah) was the god of Sinai, that Jethro was his chief priest and that Moses had become a convert at the Burning Bush. It is also suggested that Moses was induced by Jethro to adopt the judicial system of the Midianites. These ideas cannot be dismissed as mere speculation, but need not divert us from our main course.

The important point is that we see here a delegation of authority. The role of Moses was to be the **'people's representative before God'** and they were to bring their disputes to him (Exodus 18:19); he was also to teach them what they must do. He could not continue to be the arbiter over every minor dispute, so God-fearing, honest and incorruptible men were appointed as leaders of units of various sizes. They were to sit as a permanent court to decide simple cases. We see here the **origin of the Sanhedrin**, the Great Council and ruling body in the days of Jesus.

13.9 The Covenant at Sinai (Exodus 19:1–25)

The covenant with Noah was between God and mankind, the covenant with Abraham was with a family chosen by God to be the instrument for saving the world from itself, and the covenant at Sinai was a renewal of the covenant between Abraham and God, but now with the children of Abraham, the founding father of the nation. They were to become God's special possession, a kingdom of priests and a holy nation. What was demanded of them was obedience and faith.

The details of the story display some very primitive ideas about God belonging to the earlier stages in the development of Hebrew religion. Thus God revealed Himself in the violent forces of nature, in thundering and lightning, earthquake and awesome spectacle associated with volcanic eruption, fire and explosion and dense cloud. Then again the people were to wash their clothes and not set foot upon the mountain, on pain of death. Barriers were set up to mark the area of tabu. God came down from heaven to meet and talk with Moses on the mountain top, a crude anthropomorphism to the literally minded. This, however, is to miss the point, for we have here a highly symbolic picture, full of Biblical imagery, expressing the aweful holiness and power of God and the reality of the gulf set up by the sin of man. In this dramatic way it is stressed that the covenant is not between equals, for man is a creature of his Creator and by the nature of things must remain so.

13.10 The Ten Commandments (Exodus 20)

The terms of the covenant are the Ten Commandments, and its preface (Exodus 20:1) lays down the fundamental basis 'I am the Lord your God', by implication putting man in his proper place in the order of things. The covenant marks an astonishing advance in religion, for it bound the Israelites, not only to worship Yahweh to the exclusion of other gods, but also to live the kind of life that was acceptable to Him. This was in marked contrast to the religion of Canaan which consisted of ritual and sacrifice but had no moral content. For the Hebrew, religion and morality were inextricably bound up.

1 *'You shall have no other God to set against me.'*
For centuries the greatest danger to Israel, religiously and politically, was the temptation to fall for the attractions of the gods of surrounding nations and the local fertility gods of the land of Canaan. At the time of the Exodus the Hebrews worshipped Yahweh as their tribal God in opposition to such gods as Dagon of the Philistines and Chemosh of the Moabites, whose existence was not in dispute. It was not until the great prophets of the eighth, seventh and sixth centuries BC that the Hebrews came to believe that Yahweh was the one true God of the whole world.

2 *'You shall not make a carved image . . . you shall not bow down to them or worship them for I the Lord your God am a jealous God . . . But I keep faith with thousands, with those who love me and keep my commandments.'*
The Jews alone among ancient peoples, made no images of their God, on the grounds that Yahweh was so great and holy that no earthly representation of Him could be less than blasphemous. The classical statement of **Israel's contempt for idol-worshippers** is found in Isaiah 44:12–20 – the idol maker uses part of a piece of timber to warm himself and cook his food, and with what is left he makes a god, prostrates himself before it, and prays to it for salvation.

The basic sin of idol-worship consists in worshipping a false representation of God, and this applies equally to having a wrong or distorted mental image. Because God is love, we must not imagine that He is an indulgent father who is prepared to overlook or condone the selfishness of His spoiled children. This is what is meant by God being a 'jealous God'. A truly loving father cannot evade the duty of guiding and disciplining his children.

It is also a sad fact of history that the 'sins of the fathers' are sometimes visited upon the children in the sense that the injustices of even the distant past can bear bitter fruit in the present. It is of the nature of things that this should be so. While none of us could be responsible for the cruelties of the slave trade of hundreds of years ago, we share its consequences and suffer its bitter legacy. Later in its history, Israel came to see the wrongness in principle of the saying, 'The fathers have eaten sour grapes, and the children's teeth are set on edge': God willed that each man should accept the consequences of his own individual sin.

3 *'You shall not make wrong use of the name of the Lord your God; the Lord will not leave unpunished the man who misuses His name.'*
The Hebrews held, and still hold, the name of God in greatest respect. It is so sacred that it is not allowed to be spoken, and in the public reading of Scripture the word Adonai, 'the Lord', is substituted for Yahweh. The commandment calls for a proper respect for God, and worship is man's expression of this respect. The psalmist states this beautifully in saying, 'Give the Lord the honour due unto his name. Worship the Lord with holy worship'.

The commandment may also be relevant to the swearing of oaths. To swear by the name of God was a solemn oath indeed.

4 *'Remember to keep the sabbath day holy . . . Therefore the Lord blessed the sabbath day and declared it holy.'*
The institution of the sabbath was of inestimable importance in both the development and the survival of Judaism. During the exile in Babylon there was no Temple at which they might worship, so the sabbath took the place of the Temple as the focal point in the practice of their faith. They met on the sabbath day to study their Scriptures and to pray. Throughout their subsequent history **sabbath observance has been the key to their survival**, for Judaism is a religion of the home rather than the synagogue, and it was possible even in times of severe persecution to continue to practise their religion in the privacy of their homes by observing the sabbath rituals.

We may need to be reminded that Sunday, the Christian holy day, is not the sabbath. The day of the Resurrection, the first day of the week, has been observed by Christians in its place soon after the Church ceased to be a sect of Judaism. The fourth commandment does apply in principle, though not in detail, for being human and therefore fallible, if we do not organize God *into* our lives he is organized out. Sunday is not properly observed by Christians if they have not spent at least some of the day witnessing at public worship.

By tradition, the first table of the Law ends at this point, Moses having written the duty to God on the first of the two tablets of stone. (There is no Biblical evidence of this, however.)

The second table deals with our duty to other people. What is morally wrong is not always a crime, but the emphasis here is that it is always a sin. An offence against our fellow men is also an offence against God.

5 *'Honour your father and your mother that you may live long in the land which the Lord God is giving you.'*
The Jews have a strong sense of the importance of family life, which has made an inestimable contribution to their survival. Judaism is much more a religion of the home than of the synagogue, and this has enabled them to practise their religion in times of persecution in the privacy of the home. The Jewish mother has a special role in the religious education of the children and the rituals peculiar to home life.

Every home, Jewish or gentile, is the primary influence in the spiritual and moral development of the children: we all enter into a legacy which we have not created, and the obligations and also the values of the community are handed on. The commandment reminds us of the debt we owe to those who have been in authority over us, especially our parents.

6 *'You shall not commit murder.'*
This commandment obviously applies to the crime of murder and not to capital punishment or war, both of which were allowed. Because man is made in the image of God, he is of infinite value in the sight of God and no one has the right to take the life of another. (The matter of suicide and mercy killing will be examined later.) It is surely logical to deduce from this that it is murder to allow someone to die by default on our part. The more affluent countries of the world cannot escape responsibility for allowing the deaths of countless millions from hunger and disease in the underdeveloped countries of the Third World.

7 *'You shall not commit adultery.'*
This is a plea for the maintenance of high standards of sexual morality. Many of the sanctions attached to sexual promiscuity in the past seem by many to have been eroded, such as illegitimacy and venereal disease, but statistics show that this is not the case. For whatever reasons, illegitimacy is on the increase while the incidence of venereal disease has reached epidemic proportions. To adopt a puritanical attitude is not helpful, and to treat sex as something dirty is positively dangerous. Sexuality must be seen as one of the greatest gifts of God and is therefore to be treasured and not prostituted: its proper sphere is within the bond of marriage. The essence of the sin of adultery is that it involves the exploitation of a fellow human being: it means treating a person as a thing, as a means of gratifying an appetite, rather than someone created in the image of God. Few sins strike more deeply into the moral fibre of the individual.

8 *'You shall not steal.'*
To steal is to exploit one's fellow man, to appropriate to ourselves the rewards of the labours of others. We have no right to what is not our own. We can use euphemisms like winning, scrounging, liberating, fiddling, but whatever softer word we use it is still stealing. Not to give a good day's work for a good day's pay is stealing, as is not to give a good day's pay for a good day's work. It is stealing to travel on public transport without paying, or to 'fiddle' the income tax or the expense account – the examples are endless. All the forms of stealing come under condemnation.

9 *'You shall not give false evidence against your neighbour.'*
The commandment is specifically aimed at perjury in the courts, but includes also the duty at all times to have respect for the truth, allowing for the fact that it is sometimes uncharitable, cruel, and even unnecessary to tell it. Sometimes an evasion or a 'white lie' is permissible when the naked truth could not be borne, but conscience has to decide. In principle we must acknowledge that when trust and confidence die the whole fabric of society is in peril.

10 *'You shall not covet your neighbour's house; . . . or anything that belongs to him.'*
This was the sin of Adam and Eve – they coveted the status of God. This is the only one of this catalogue of sins which looks inward into the heart and mind. Jesus said, 'Wicked thoughts, murder, adultery, fornication, theft, perjury, slander – these all proceed from the heart: and these are the things that defile a man' (Matthew 15:19). These words of acute insight illuminate this commandment. We have all coveted something which other men enjoy, but the commandment is not against a legitimate desire to want to better ourselves, but rather the greed and consuming envy which begrudges the success of others. We see something of this even in Shakespeare, surely the greatest literary genius of all time, when he speaks of 'Admiring this man's art and that man's scope, With what I most enjoy, contented least.' That he had grounds for the envy of anyone seems incredible, but like us all, he was only human.

13.11 The Covenant Rite (Exodus 24:1–18)

Moses stood at the top of the mountain, enveloped in the cloud which indicated the presence of God, while Aaron and seventy elders of Israel stood at a distance down the mountain side, the people standing at the foot. Then Moses came down and told the people all the words of God.

The next day an altar was erected at the foot of the mountain with twelve sacred pillars, one for each of the tribes of Israel. Bulls were sacrificed and Moses placed half the blood in basins and the other half he flung against the altar. He then read aloud the 'book of the covenant' and the people gave their assent. Moses then flung the blood in the basins over the people saying, 'This is the blood of the covenant which the Lord has made with you on the terms of this book' (Exodus 24:8).

The same symbolism as in the covenant rite of Abraham is employed here: half the blood is flung over the altar, which represents Yahweh, and the other half is flung over the people of Israel. The parties to the covenant, or solemn agreement, are bound together by the life of the third party, the sacrificial victim.

14 The Settlement in Canaan

14.1 Joshua and Judges

These books cover the same ground but differ radically in their account of the manner of the 'conquest'. They agree substantially on two points only:

1 Israel is destined by God to inherit the promised land;

2 Israel's failure to keep the covenant often made the hold on the land precarious.

The Book of Joshua describes a fairly rapid conquest of the land in which the Canaanites were either killed or made into slaves, while Judges gives a quite different picture. Israel was at first able only to subdue the hill country in the centre of the land, the grazing land to the south centred around Hebron, and the land east of Jordan. The fertile lands remained in the control of the Canaanites, cutting the Israelite settlements into three main divisions: Jerusalem became Israelite only when captured by David. However, Israel brought a new vigour to the occupied territories and gradually spread into the arable lands through peaceful penetration and inter-marriage. Their status probably improved because of their military prowess, which soon came to be exercised against the traditional enemies of the land. It is significant that of all the judges of Israel only Deborah and Barak have to fight a Canaanite enemy; the others fight against those who are as much the enemies of the Canaanites as the Israelites themselves – invaders from the east. Contact with a more advanced agricultural people made inevitable a change from the life of the wandering shepherd to that of the settled farmer. Gradual as the change was, it had a profound effect on the religion of Israel.

Israel at this time had no concept of Yahweh as the one true God, the God of the ends of the earth. That there were other gods was not denied, but Yahweh was their national God who had delivered them from bondage in Egypt. The land which they now occupied belonged to the Baals, the gods who had been worshipped there for centuries, and who controlled its fertility. They knew Yahweh as a God of the desert who dwelt on a distant mountain, and the question was did he know how to grow corn or understand the cultivation of vines and olives? It would take only a bad harvest or two for the doubts of some to be resolved – the gods on the spot must receive due reverence, the Baals must be served if disaster was to be avoided.

Israel learned her mistake in the bitter experience she suffered between the occupation and the establishment of the monarch. The Baals were local gods, powerful only in their own city or locality, exercising no authority and claiming no allegiance outside it; **the Canaanite religion was therefore a disintegrating force**, fostering local rather than national loyalties. The result was that the plundering hordes from the east encountered little effective resistance and destroyed their victims piece-meal. When Israel entered Canaan, the twelve tribes were united not only by blood, but by their allegiance to Yahweh, and when that allegiance was eroded, no united resistance could be offered to an enemy. The judges were men and women who, under the inspiration of Yahweh, called the people to return to their God, and when they did so they threw off the yoke of their enemies.

Thus we see in the Book of Judges a regular cycle of events – Israel turns from God and serves the Baals, and Yahweh gives them over to the hand of their oppressors. After years of suffering they repent and call on Yahweh, who gives them a judge who leads them to victory. After the death of the judge, Israel turns to evil ways and the cycle of events is repeated.

The final struggle was against the war-like Philistines from the sea-coast and this led eventually, under Saul and David, to a real unity of the people of Palestine for the first time.

Joshua

After the death of Moses, Joshua was appointed to lead Israel into the promised land. The spies sent across the river Jordan were given refuge by Rahab, the harlot, and her safety was guaranteed when Jericho was destroyed. The Jordan was crossed by a miracle reminiscent of the crossing of the Red Sea, and the Ark of the Covenant, the symbol of the presence of God, played a prominent part in the story. When the priests carrying the Ark put their feet in the river it divided and they then took their stand in the middle of the river bed while the people passed over. The Ark was displayed in this way so that the people would know that the living God was among them and would lead them to victory. Twelve stones were taken out of the river bed and set down in the camp as a memorial of the miraculous river crossing. At Gilgal all those men who had not been circumcised in the wilderness were now circumcised, and the Passover was celebrated with bread made from the corn of the land; the manna then ceased.

When Joshua came to Jericho he was met by the captain of the Lord's army. He was told, 'Take off your sandals: the place where you are standing is holy' (Joshua 5:15), an incident which recalls the story of the Burning Bush. The Ark is again prominent in the story of the fall of Jericho; it was carried in procession around the foot of the walls, and the collapse of the ramparts is attributed to the miraculous intervention of God.

Joshua's farewell at Shechem (Joshua 24)

Having assembled all the tribes at Shechem, Joshua declared the words of the Lord. The message consists of a recounting in detail of all that God had done for His people from the call of Abraham to the settlement in the promised land. Joshua then exhorted the people to adhere to the God of their fathers who had done such great things for them and warned them of the consequences if they turned to other gods. The people solemnly swore to be faithful and Joshua then made a covenant with them, writing the terms in the book of the Law of God. A great stone was set up as a witness.

The Book of Judges

When the initial force of the invasion of Canaan was spent, we see the twelve tribes scattered thoughout Palestine, settling down side by side with the native population and yet maintaining a certain distinction. While each tribe maintains its independence each is bound to the others by blood and a common religious allegiance to Yahweh and his covenant. In times of danger each rallied to the defence of the other.

As we have seen, the fundamental bond of unity was continually being threatened by the temptation to practise the religion of the land. Israel's military success gave them some social status, but as a desert people they soon discoverd they had a great deal to learn about the skills of farming from the indigenous population. Yahweh was all-powerful in the desert, but it was the gods of the land who ensured or denied fertility, and the temptation to make the best of both worlds was very real, to divide allegiance between Yahweh and Baal. Unfortunately, the Canaanite religion could not be accommodated within the religion of the stern desert God whose character was known as holy and morally good. Canaanite worship included human sacrifice and ritual sexual intercourse with religious prostitutes, both of which were consistent with the character of the Baals, whose behaviour was immoral and arbitrary in the extreme. This division of allegiance would probably have led to the complete assimilation of Israel, but for the fact that in time of danger, Yahweh, who had proved His power time and time again in war, became Israel's strength and stay. **In danger Israel was recalled to its old allegiance**.

The book of Judges does not present us with a continuous history of the period but rather with representative incidents. When Israel is subjugated to a cruel enemy, God raises up a 'judge' who leads them first to renew their allegiance, and then to overthrow their enemy.

It is probable that the '**Song of Deborah**' (Judges 5) dates from about this time, and gives us considerable information about the conditions that prevailed. Israel had chosen new gods and consorted with demons, and was now so completely subjugated that there was no weapon among them: 'Not a shield, not a lance was to be seen.' The victory is attributed entirely to Yahweh. The tribes which respond to the call of rebellion are praised and Reuben, Dan and Asher who failed to act, are spoken of with contempt. **Jael**, wife of Heber, by guile killed **Sisera**, the enemy leader, and the gory details are gloated over, as is the grief of the mother of Sisera.

In the next episode Israel again turns from Yahweh and this time it is an enemy from the east who conquers. God raises up **Gideon** to free Israel. The insidious nature of the spread of the Canaanite religion is seen in the fact that Gideon's own father had an altar of Baal and an asherah, a wooden pillar representing the presence of Baal. Judges 6:11–8:27 tells the exciting story of Gideon's exploits in battle, including a memorable surprise attack by night against vastly greater numbers. As a result there was peace for many years, and though Gideon is, by implication, praised for refusing to accept the kingship on the grounds that 'the Lord will rule over you' (Judges 8:23), he made an image out of the gold of the booty, which Israel worshipped.

The next great warrior hero is **Jephthah**, raised by God to deliver them from the Ammonites. Prominent in the story is the account of Jephthah's vow (Judges 11:29–40) which seems to indicate that human sacrifice was still practised in Israel. There is in the story of the last great judge, **Samson**, little to contribute to religious or moral ideas (Judges 13:1–16:38) and he is more like a popular hero than a religious leader. Of interest, however, is the fact that Samson is a **Nazarite**, that is, he belonged to a group of people who took a vow, lifelong in early days, to use no razor, and to abstain from any product of the vine, especially wine. The reason behind this was that the faithful of Israel looked back on the experience of the desert for their inspiration; it was the time when their religion was at its purest and most austere. The vine was looked on as the most typical product of the settled agricultural life in contrast to the nomadic existence of the wilderness.

The vine, since it takes years of careful cultivation before it yields fruit, and crops for many years thereafter, was seen as the symbol of the permanent, settled existence among the Canaanites and the consequent falling away from the high standards set by Yahweh. The Nazarite stood for the old ways and by his vow made a public protest against the new ways.

Of interest also is that the Philistines made their first entrance on the stage of Hebrew history. Unlike the other invaders they came from the western seaboard, not from the desert.

The confusion and complexity of this stage in the history of Israel is well summarized in the last verse of the book of Judges: 'In those days there was no king in Israel and every man did what was right in his own eyes.'

14.2 Israel Becomes a Monarchy

By the time of Samuel a merging had taken place between the Hebrews and their racial cousins, the native population, through inter-marriage and trade. Yahweh had become the dominant God but many of the religious practices of the land had been incorporated into His worship. The Hebrews had become farmers and Yahweh's function could no longer be to guide and protect His people through the wilderness. His role as a fertility God was new, and it was natural that if He were to take over the duties of the Baals, some of the features of their worship would be relevant and appropriate. The observance of the three great agricultural feasts of Israel, the **Feast of Unleavened Bread** (later incorporated into the Passover), the **Feast of Weeks** (the first fruits of the harvest) and the **Feast of Ingathering** at the end of the harvest, are evidence of syncretism, that is, a mixing of two religions.

The form of testing by which Israel was to learn more about the meaning of their calling as the People of God, His chosen instrument, had now changed. In Joshua and Judges their temptation was to abandon the austere religion of Sinai for the sensual and morally degenerate polytheistic religion of Canaan, but in Samuel and Kings the testing takes the form of the dangers attending the rise of a monarchy and the consequent exposure to the perils of power politics (See Unit 14.4). The story of the monarch is told from the point of view of the prophets who exercised the right to criticize even the kings when they failed to observe the terms of the covenant.

14.3 The Prophets

The Hebrews learned from the Canaanites the practice of ecstatic prophesy. The prophet was a person possessed by the god in such a way that his personality became for a time, taken over. Thus 'the spirit of the Lord took possession of Gideon' (Judges 6:34), and 'the spirit of the Lord suddenly took possession' of Saul (1 Samuel 10:9–10) and 'God gave him a new heart'. It is plain that the prophet was regarded as speaking the words of God, that his will was taken over: the content of the message could not be foreseen.

There were two classes, the **seer** and the **ecstatic**, though in later years their functions merged. The **typical seer is Samuel** (1 Samuel 9:1–10:16). Saul and his servant, after an unsuccessful search for some lost asses, found themselves near Ramah, where Samuel the seer lived. The servant suggested that they consult this 'man of God'. Details of the story disclose the nature of the role of the seer.

1 As a seer, Samuel was in direct communication with God. He actually heard the words of God as well as experiencing visions. God spoke to him 'in these words' (1 Samuel 10:15).

2 The seer was to some extent in control of his powers because people went to him to ask questions about matters concerning themselves, which might be either important or trivial, like finding lost articles; they were confident of an answer. For this service a fee was normally charged (1 Samuel 9:8).

3 Samuel held a position of authority and honour in the town. At the sacrificial meal at the shrine, the people waited until he pronounced the blessing. Samuel was a man of high standing before God and the people.

We have more detailed evidence for the **role of the ecstatic**. It appears that the ecstatic became conscious of a world other than, but just as real as, the world in which men ordinarily live, as though a veil had been withdrawn so that the secret world of the gods was revealed to him. He claimed to have direct contact with the deity in these experiences. When in the ecstatic state the whole body might be affected: it might become rigid, accompanied by trance, or subject to violent convulsive behaviour, often presenting the appearance of a frantic dance. Sometimes there would be a babble of strange unintelligible sounds and the body might become insensitive to pain, the dancers cutting themselves until the blood gushed out. (The prophets of Baal on Carmel 1 Kings 18:26–29.) The state of ecstasy could be induced by music: the ecstatics encountered by Saul were 'led by lute, harp, fife and drum and filled with prophetic rapture' (1 Samuel 10:5).

These ecstatics gathered together in bands because ecstasy was readily communicated. Thus when Saul met a company of Nebiim (ecstatics) at Gibeah (1 Samuel 10:10) he too had an ecstatic experience, to the amazement of all. The incident indicates that the social status of the ecstatics was very low, for Saul's friends were surprised at his association with such disreputable people. Nevertheless, the words and symbolic acts of the ecstatics were regarded as coming from God.

While in early times there was a strong contrast between the seer and the ecstatic, by the time of the monarchy the distinction had disappeared, hence 'what is nowadays called a prophet used to be called a seer' (1 Samuel 9:9). Groups of prophets came to be attached to the sanctuaries of Israel and also to the king's court, but later we see the rise of prophets who had an independent existence. Thus Elijah was independent of the guilds of prophets, though it is clear from 2 Kings 2:3–7 that they recognized that he was a true prophet, for two separate bands of prophets came to meet Elijah and prophesied that he would be taken away from Elisha, his disciple, just before he was taken up to heaven in the whirlwind. The prophet Micaiah was also independent, for he had the courage to stand forth and contradict the unanimous prediction of the band of court prophets, that King Ahab would be victorious in battle (1 Kings 22:10–28).

It is difficult to associate the great prophets of Israel's later history with men of this kind, but their origin may be traced to these early prophets. The experiences of the ecstatics were refined as time went on, and their direct descendants, the Great Canonical Prophets, became the first to realize the necessary connection between morality and religion and that God was not only all-mighty, but just and holy.

14.4 The Monarchy

The life of the nomad by its very nature lays far greater stress on the value of the personality than does that of the farmer or townsman. This is because of the simplicity of the way of life. Goods have to be readily portable and there is little room for luxury; the animals tend to belong to the community rather than the individual, the tent to the family, and as a result there is little in the way of social strata. Some men may gain status through respect for their age, experience, or their performance in times of crisis or war, but on the whole, men have a common status as 'brethren'. Each man enjoys a large degree of personal liberty, limited only by the common interests of the community and the general will. These features of desert life, which we can roughly call democratic, were taken for granted, and preserved to a large extent in the promised land. It is natural that Yahweh, the God of the desert, should be associated with this way of life, so that anyone trying to impose personal authority over the nation would be held in breach of a religious law. Thus when Israel demanded that Samuel should give them a king, God's reaction was 'it is I whom they have rejected, I whom they will not have to be their King' (1 Samuel 8:7), with the strongest implication that the true king of the nation is God Himself, and that no earthly king can ever take His place. Even the king is under the rule of God.

By contrast, society in most of the rest of the world had a rigidly pyramidal structure, with a king at the summit, whose exercise of authority was thoroughly arbitrary: the king was responsible to no one and owed allegiance to no power except that of the gods whose actions were equally arbitrary.

The monarchy in Israel was an attempt at a compromise between these two social structures. **The agitation of the people for a king** came as a result of the territorial ambitions of the war-like Philistines, who occupied part of the Mediterranean coast. It was not an unrealistic demand, for archeological evidence has revealed that the Philistines had over-run a large part of the land, destroying a number of towns, notably Shiloh, the central sanctuary. There was a need for a soldier-king with authority to organize resistance. In Saul's appointment and in the early events of his reign, we see the familiar pattern of the judges: as a result of military pressure from outside, a leader is thrown up who, possessed by the spirit of God (1 Samuel 10:10), leads God's people to victory. The difference is that eventually, as a result of his constant struggle with the Philistines, **Saul** built up a more complete organization of the nation than ever before, and imposed his authority over a much larger area of land than any of his predecessors. If he began as little more than a judge, he died a king.

David is a different case. Unlike Saul he does not owe his authority to prophetic possession, but rather to his outstanding personal qualities, his qualities of leadership, personal bravery, charm of manner, and political shrewdness. He won his kingdom, and he was the only King of Israel to hand on to his successor a united realm stretching from Dan to Beersheba. Even so, his reign was based on a covenant between all the elders of Israel and himself, 'sworn before the Lord' (2 Samuel 5:3).

Solomon's reign marks the beginnings of a decline. He earned a great reputation for magnificence but Samuel's warnings to the people when they demanded a king, were vindicated during his reign. Four months of every man's time were claimed for forced labour on his various projects, and this was the most resented of all his violations of the rights of free men, as the Israelites understood it. The national discontent was, as expected, given voice by the prophetic party, the true guardians of the nomadic tradition. Solomon was too strong to be overthrown, but on his death the prophet Ahijah inspired a successful revolution which led to the dividing of the Kingdom. The principle of a 'limited monarchy' seems to have been maintained in the Northern Kingdom until it was again challenged when Ahab married the foreign princess Jezebel, who tried to introduce an oriental despotism. She realized that the main obstacle to absolute rule was the power of the prophets, so she made their elimination her prime objective, and was so successful that Elijah only was left of all the prophets of the Lord (1 Kings 19:10). In the end Yahweh was triumphant.

Of great significance is the story of Naboth's vineyard, for when the freeholder refused to give up the land of his ancestors, it never occurred to Ahab either to compel him to do so, or to gain possession by deceit. He accepted the situation, though he did not like it. Jezebel, however, brought up under a quite different theory of royal authority, had no such scruples, yet even she had to proceed through the appearance of the forms of justice, perpetrating a judicial murder. (False witnesses charged Naboth with blasphemy and treason; he was stoned to death and his property was then forfeit to the crown, 1 Kings 21:1–14.)

It can be seen that Israel's concept of kingship was unique among the nations.

15 The Book of Samuel

15.1 Samuel the Prophet

The prophet dominates the book. He exercised authority over the king himself and while he lived, overshadowed Saul whom he had anointed to be the first King of Israel.

The book opens with an account of Samuel's birth and as in the case of Sarah, Rebecca and Rachel, his mother, Hannah, conceived by the help of God. From birth Samuel was dedicated to the priesthood of Yahweh; it is significant that 'no razor shall ever touch his head' (1 Samuel 1:11) – like Samson he was to be a Nazarite. Note that the Song of the Virgin Mary (Luke 1:46–55) is strongly reminiscent of Hannah's Hymn of Thanksgiving.

The story of the bad behaviour of the sons of Eli the priest (1 Samuel 2:12–20), gives us incidentally an insight into the form of worship at Shiloh which was then the central sanctuary of Israel. It was of greatest importance that the ritual of sacrifice should be scrupulously followed, and their failure in this respect was a serious sin. Eli was the religious head of the tribes and the priesthood at Shiloh was hereditary: the death of Eli and his sons made it possible for Samuel to succeed to the priesthood.

The life and work of Samuel

The young Samuel grew up in the Temple 'in the presence of the Lord' (1 Samuel 2:21), and, in contrast to Eli's reprobate sons, in favour with God and the people. 'In those days the word of the Lord was seldom heard, and no vision was granted' (1 Samuel 3). In other words, two essential constituents of the experience of the seer were missing – auditory and visual. The boy Samuel who slept in the Temple (by implication a building of some sort rather than a tent, as in the desert) heard the voice of God calling him by name. When he replied 'The lord came and stood there' (1 Samuel 3:10) and in this way Samuel's experience came through the seeing eye as well as the ear, so that having been chosen, he became a seer, in direct communication with God. God pronounced a judgment against Eli and his sons because of the abuses of the sacrifices and offerings at the shrine. When Samuel grew older all Israel recognized that he was a prophet of the Lord and his word had authority throughout the land after the death of Eli.

The Ark is captured (1 Samuel 4)

When Israel was defeated in battle, they sent to Shiloh for the Ark of the Covenant, the sacred symbol of the presence of God, and set it up in the camp, but the result was the opposite of that expected, and the Ark was taken. The two sons of Eli were killed. The news of the capture of the Ark caused the death of Eli the old priest.

Meanwhile the Ark was placed in the Philistines' temple at Ashdod. A series of disasters ensued in each of the cities to which the Ark was sent and after seven months the Philistines returned it with an offering of gold in restitution. The Ark eventually was housed at Kiriath-jearim, Shiloh, its original home, having been destroyed by the Philistines.

Samuel becomes the undisputed leader of Israel

The defeat of Israel at Ebenezer, and the destruction of the principal shrine at Shiloh, led to Samuel taking up residence at Ramah and from this centre he went on circuit to the main religious shrines Bethel, Gilgal and Mizpah (1 Samuel 7:16). The history of the Book of Judges seems to have repeated itself, for after the defeat of Eli's sons and the capture of the Ark, there was a movement away from Yahweh, but after twenty years of Philistine supremacy the people again returned. As evidence of their sincerity, Samuel exhorted them to banish the Baals and the Astaroth, promising that God would then give them victory. Samuel assembled Israel at Mizpah to make intercession to God in sacrificial worship. (War was regarded as a religious act, and God's help was needed to overcome the gods of the Philistines.) Water was poured out in a drink offering, the people fasted, confessing their sin against God, and Samuel offered a lamb as sacrifice. Their prayers were answered, the advancing Philistines were thrown into confusion by a violent storm, and as they fled were pursued by Israel and slaughtered in great numbers. The occupied territories were restored and Israel's supremacy was maintained as long as Samuel lived.

15.2 Israel Demands a King (1 Samuel 8:1–21)

Samuel had appointed his sons as judges to assist him, but they proved to be corrupt and there is a suggestion that Samuel may have contemplated setting up an hereditary office of judge which, in the circumstances, Israel would not tolerate; they demanded a king. Samuel was displeased, feeling he had ben rejected by those he had served so well, and when he prayed to the Lord he found that his feelings were shared by Him. The message he received was that by making this request they were really rejecting the God who had brought them up from Egypt. Samuel was charged to tell them of the heavy price they would pay for having a king (1 Samuel 8:11–18), but they would not hear.

15.3 Saul

The visit of Saul to the shrine at Ramah had been foretold to Samuel the day before. When Saul came, seeking news of his father's lost asses, Samuel invited him to a feast at the shrine and then gave him hospitality at his house for the night. The next morning he accompanied Saul to the end of the town, took a flask of oil and poured it over Saul's head, anointing him King over Israel. Samuel prophesied that Saul would be given three signs to confirm what had taken place:

1 two men at the tomb of Rachel who would tell him that the asses had been found;

2 three men going up to Bethel to worship would give him two loaves, part of their sacrificial offering;

3 at the hill of God he would meet a company of prophets coming down, and the spirit of the Lord would take possession of him; he too would be filled with prophetic rapture, and become another man.

All these signs were given, and Saul went home, telling no one that he was now King.

Samuel presents Saul to Israel (1 Samuel 10:17–25)

It appears that two different traditions are amalgamated in the story of Samuel. This second account of Saul's appointment is quite different from that of Chapter 9. Here Israel was summoned to come before the Lord at Mizpah and Saul was selected by sacred lot; he was then presented to Israel as King and was accepted.

His first test was not long delayed. 'The spirit of God suddenly seized him' when he heard of the Ammonite threat to Jabesh-Gilead, and hewing in pieces the oxen with which he had just been ploughing, he sent the portions to each tribe of Israel with a demand that they should take the field or their oxen would suffer the same fate (1 Samuel 11:6–9). (Compare the similar action of Gideon.) The result was the utter defeat of the Ammonites and Saul was formally invested as king at Gilgal. It seems that his kingship was confirmed by the fact of his possession of the spirit of God and his subsequent victory.

The rejection of Saul (1 Samuel 13:7–14)

This passage may have been inserted by an editor, in order to account for Saul's rejection presuming that it must have been due to a serious sin. The Philistines mustered to attack Israel in force and Saul's ill-equipped army began to melt away, some even fled across the Jordan. Saul however remained firm at Gilgal, waiting for Samuel to come to offer the necessary sacrifice before the army could go into action. (War was a religious act.) He waited for seven days and when Samuel did not arrive, in desperation, since the army was deserting him, Saul offered the sacrifice. Just as he finished sacrificing, Samuel arrived, and was furious at Saul's 'disobedience'. No excuse could be accepted for Saul's action. (Apparently a major defeat would have been preferable. Since he had performed a rite lawful only for a priest, it was regarded as a sin against God.)

Probably Samuel was not alone in viewing with apprehension the rise of the monarchy, on the ground that it marked a deviation from Israel's former dependence solely on God as its king. Israel was unlike all other nations because of its **unique relationship with God**. Later writers, with the advantage of hindsight, knew that the monarchy had been Israel's downfall.

The second rejection of Saul (1 Samuel 15)

Samuel the prophet declared to Saul 'the very word of the Lord of Hosts' commanding him to destroy the Amalekites because (several hundred years before) they had attacked Israel on their way up from Egypt. Every living thing as well as the property was to be destroyed. This action the law demanded (Deuteronomy 20:16) as proper vengeance on the enemies of Yahweh; it was the law of Cherem or the 'Ban'. The enemies of a nation were also the enemies of its god, and a battle would be fought by the command of the god or at least in his name. A condition of victory might be that all the spoils of war should be 'devoted' to the god and become his property; that being so it became dangerous to appropriate any of it, and it was all destroyed. Not to carry out the ban was a serious sin: it was the offence of Achan (Joshua 7) who took for himself some of the booty of Jericho. He, all his family, and all that he had, were destroyed.

After massacring the Amalekites Saul kept the best of the sheep and cattle to sacrifice them at Gilgal. Samuel was not impressed – 'Obedience is better than sacrifice' he said, and that is what God requires. Because of this sin God would reject Saul and give his Kingdom to another. Then Samuel took Agag the captive King of the Amalekites and hewed him in pieces before the Lord.

This kind of savagery was normal practice among the nations. The famous Moabite Stone describes a similar event when the King of Moab, at the command of his god Chemosh 'devoted' the Israelite town of Nebo, massacring all its 7000 inhabitants in honour of the god.

The rift between Saul and Samuel was now complete and Samuel began to look around for Saul's successor.

15.4 David

A very large part of the narrative from 1 Samuel 16 to the end of 2 Samuel is taken up with the story of David. The main incidents only need be studied in depth; it is important that the text should be well known in these instances.

David is anointed King (1 Samuel 16:1–13)

Note that after his anointing 'the spirit of the Lord came on David and was with him from that day onward'.

David becomes Saul's harpist and armour-bearer (1 Samuel 16:14–23)

David and Goliath (1 Samuel 17:1–58)

There is evidence of the use of two separate traditions, for Saul would hardly have failed to recognize his armour-bearer and private harpist (1 Samuel 17:55–58).

David's rapid promotion and popularity (1 Samuel 18:1–5)

The estrangement between Saul and David (1 Samuel 18:6–29)

This was caused by the welcoming but tactless refrain of the women celebrating the victories of David. Saul in a fit of madness tried to kill David with a spear and, failing, then tried to engineer his death at the hands of the Philistines. Saul became afraid of David 'and was his enemy for the rest of his life' (1 Samuel 18:29).

Saul plots against David (1 Samuel 19)

Jonathan, David's friend, interceded for David and Saul promised to spare him. However, 'an evil spirit from the Lord' came upon Saul and he tried again to spear David. Next Saul tried to arrest him but Michal, Saul's daughter, the wife of David, warned him and saved his life. David fled to Samuel at Ramah. When Saul sent his men to arrest David they found Samuel at the head of a band of ecstatic prophets and the spirit fell also on them. When Saul came in person, he too, possessed by the spirit, fell into prophetic rapture.

The friendship of Saul and Jonathan (1 Samuel 20)

Jonathan helped David to escape. (The incident of the arrows shot at a mark.)

David escapes to the cave of Adullam (1 Samuel 21)

On his way David called on Ahimelech the priest at Nob who, unaware of David's fall from favour, gave him five loaves of the Bread of the Presence which had been taken out of the sanctuary prior to replacing them with fresh ones. Jesus referred to this incident (Matthew 12:3) in support of his teaching about sabbath observance. Ahimelech also gave David the sword of Goliath. Doeg the Edomite observed what happened and informed Saul who took revenge by killing 'every living thing in Nob'. Abiathar alone escaped and brought the news to David at the cave of Adullam.

David spares the life of Saul at En-gedi (1 Samuel 24)

Saul went into the cave in which David was hiding, to relieve himself. David was tempted to kill him but instead cut off a piece of his cloak. When Saul emerged from the cave David showed him the evidence that he could have killed him; he could not bring himself to kill 'the Lord's anointed'.

David spares the life of Saul at Ziph (1 Samuel 26)

David entered the camp at night and took the spear and water-jar from beside the head of the sleeping Saul.

Saul and his sons are killed on Mount Gilboa

Note the famous poem 'David's lament over Saul and Jonathan' (2 Samuel 1:17–27). David is made King of Judah (2 Samuel 2:1–6), probably with Philistine permission, for he had ingratiated himself with them when forced by Saul to flee into Philistine territory (1 Samuel 27:1–6).

Saul's son Ishbosheth, with Abner the commander of his father's army, claimed the allegiance of the other tribes. There was war between David and 'the house of Saul' and eventually, with the murder first of Abner and then of Ishbosheth, David became King of all Israel. He made a covenant with the people before the Lord. David captured Jerusalem from the Jebusites, a Canaanite tribe whose strong fortress had until now withstood Israel (2 Samuel 5:6–8) and made it his capital. Later the Ark was removed to the city and placed in a tent, a return to the tradition of the Exodus.

David showed kindness to Mephibosheth, Jonathan's son (2 Samuel 9), restoring to him all the estates of Saul and giving him a permanent place in the court.

David and Bathsheba (2 Samuel 11)

Chapters 9–20 are regarded by scholars as being of very great value because they are either a contemporary record or that of a writer with access to contemporary accounts. They are therefore among the oldest writings in the Old Testament. Careful attention should be paid to the passages selected.

It is astonishing that in an age when court chroniclers were concerned to depict their kings as men of virtue and military genius, glossing over all faults and failings, David is painted like Cromwell, 'warts and all'. All his actions good and bad stand under the judgment of a righteous God, and this is a reflection of the fact that Israel had a covenant relationship with God to which even the kings, however highly they might be regarded, had to conform. It cannot be over-stressed that **this relationship between God, king and people is unique**.

So the account moves from the story of one of David's most generous and loyal features, the restoring to favour of the son of Jonathan, to one of the shabbiest deeds of the reign, namely David's adulterous relationship with Bathsheba, the wife of Uriah the Hittite, and the treacherous arrangement for the killing of the brave and noble soldier. David's subsequent communications with Joab his general, plumb the depths of hypocrisy.

Nathan's Parable (2 Samuel 12:1–14)

Yahweh could not be deceived and Nathan the prophet was sent to pronounce judgment with the parable of the 'little ewe lamb', one of the best loved in the whole of the Bible. The punishment was that the child of Bathsheba would die.

That a prophet could so rebuke a king, and bring him to repentance is astonishing, and the story illustrates both David's allegiance to the covenant, and the power that the representatives and messengers of God exercised in Israel.

The character of David

He is the greatest of the kings of Israel and an inspiration to the nation in times of trouble. His impact was great enough to give rise to the vision of another David, a Messiah, who would arise to restore the former glories of the nation.

1 He was loyal to Yahweh and the covenant.

(a) He twice spared the life of Saul, at Engedi (1 Samuel 24) and at Ziph (1 Samuel 26) because he was the Lord's anointed, the King of Israel. For the same reason he killed the man who had dispatched Saul (2 Samuel 1:1–16).

(b) He placed the Ark in Jerusalem, making it the Holy City and chief shrine of Israel. He planned to build a Temple.

(c) He accepted the rebuke of Nathan the prophet and the punishment of God (2 Samuel 12).

(d) He ruled subject to the authority of God.

2 He was a **brave man**. He killed Goliath in single combat (1 Samuel 17).

3 He was a natural leader of men, inspiring the kind of loyalty that would make men risk their lives to fulfil his slightest whim: for example, the three men who brought him water from Bethlehem (2 Samuel 23:15–17).

4 He was a **brilliant commander** of men. He enlarged the boundaries of Israel further than ever before.

5 He could be **deeply affectionate**: for example his friendship with Jonathan.

6 He was **politically shrewd**. His choice of Jerusalem as capital, a city captured from the Jebusites, was very wise, because until that time it had never been part of Israel. As former neutral territory, its choice avoided the jealousy which would have arisen if he had chosen a city in any of the tribal lands (2 Samuel 5:6–10).

There are several blots on his record

1 His **adultery** with Bathsheba (2 Samuel 11).

2 His arranged **murder** of Uriah the Hittite (2 Samuel 11).

3 His **cruelty** to conquered enemies. (This, however, must be seen in the light of the times, as normal practice in war.)

4 His **weakness** and indulgence in his relations with his worthless son Absolom, which caused great suffering in Israel.

16 The Books of Kings

16.1 Solomon

Typical of the methods of an oriental ruler, Solomon began his reign by removing all possible opposition. He killed his elder brother Adonijah who aspired to the throne, and Joab his father's commander-in-chief. Abiathar the priest he dismissed from office, appointing Zadok in his stead. Zadok 'took the horn of oil from the Tent of the Lord and anointed Solomon' (1 Kings 1:39).

Solomon asks for wisdom (1 Kings 3:4–15)

God appeared to Solomon in a dream and invited him to ask for whatever he wished; the King asked for wisdom to rule God's people well, and his request was granted, and because he had not asked selfishly for long life or wealth, or the lives of his enemies, these gifts also would be given.

Solomon and the two mothers (1 Kings 3:16–28)

An example of Solomon's wisdom immediately follows in the famous incident of the two mothers, each of whom laid claim to the same child. It is an example of shrewdness rather than profound thought.

The building of the Temple (1 Kings 5, 6 & 7)

Solomon concluded a trade agreement with Hiram King of Tyre for the supply of building materials for the Temple, in exchange for corn and oil. Large numbers of labourers were conscripted for felling trees, quarrying stone and transporting them to Jerusalem.

The building was of stone, lined with cedar and consisted of two chambers, the inner one containing the Ark was overlaid with gold. On the day on which the Ark was brought into its shrine, a cloud filled the Temple, as in the old days in the wilderness the presence of God came down upon the Tabernacle. 'The glory of the Lord filled his house' (1 Kings 8:12).

Unlike Herod's Temple, it had only one court, and was one of a complex of buildings which included the royal palace. Herod's Temple was much larger.

The character of Solomon

1 He had a reputation in Israel far greater than warranted by the record of his deeds. By the convention of Bible editors of ascribing authorship of anonymous works to well-known people, he has been given credit for such works as Proverbs and the Wisdom of Solomon, which he could not possibly have written. However, his **reputation for wisdom** must have some basis, though it is not borne out by the record.

2 He lived a life of ostentatious **extravagance**, and while much of this was financed by his own skills in fostering trade among neighbouring nations and by his foreign political alliances, he oppressed his people, subjecting them to severe forced labour to supply his needs.

3 He cemented his alliances by marrying the daughters of neighbouring kings. The figure of 1000 wives and concubines must be an exaggeration, but it is evidence of his **reputation as a lover of women**.

4 He had a **large standing army** with a considerable force of cavalry which was a running drain on the nation's resources. This however did not prevent him from losing Edom and Syria, parts of David's kingdom.

5 His administrative organization was better integrated than that of David, but its main purpose was to maintain the king's household in luxury.

6 He allowed his foreign wives to seduce him away from Yahweh to the **worship of other gods**. This damaging blow to national unity contributed to the division of the Kingdom.

7 While everything Solomon did was on a grand scale, his policies resulted in the **division of the nation soon after his death**. He failed to appreciate the 'democratic' temper of his people which they had inherited from the years in the wilderness, and while his subjects were powerless to oppose his **autocratic rule** while he lived, his successor could not contain a popular revolt.

He probably owes much of his reputation to the fact that his priestly historian was influenced by **Solomon's establishment of the Temple in Jerusalem**, an event of such immense importance in the history of Israel that it outweighed his many failings.

16.2 The Divided Kingdom

Jeroboam

The prophecy of Ahijah regarding Jeroboam (1 Kings 11:26–40) Jeroboam was appointed by Solomon to be in charge of the labour gangs in the tribal area of Joseph. He was travelling alone when he met the prophet Ahijah who took off his new cloak, tore it into twelve pieces and gave them to Jeroboam. This is the **first symbolical act of a prophet** to be recorded, a device which was later to become common. This was an acted parable, conveying a message from God to the effect that, after Solomon's death, the Kingdom would be divided: ten tribes would be ruled by Jeroboam and the two remaining, Judah and Benjamin, by Solomon's heir. Jeroboam was forced to flee to Egypt for safety.

After the death of Solomon, his heir, Rehoboam, went to Shechem to have his succession confirmed. He received a request from the people to lighten the heavy burdens Solomon had laid upon them as a condition of their loyal support. He took the advice of his luxury-loving young friends rather than that of the elders of the people, and refused the request saying 'My little finger is thicker than my father's loins . . . My father used the whip on you; but I will use the lash' (1 Kings 12:10 and 11).

The result of this ill-advised attitude was that the ten northern tribes (**from this point onwards referred to as Israel**) seceded from the house of David, and took as their king, Jeroboam, who had returned from Egypt.

The sin of Jeroboam

Jeroboam rebuilt Shechem and made it his capital in opposition to Jerusalem. He saw that if the people had to go up to Jerusalem to sacrifice there was a danger that they would be weaned from their allegiance, so he made Bethel in the south and Dan in the north, the two chief sanctuaries of his Kingdom. The Ark of the Covenant was in Jerusalem so he placed in each shrine a golden calf as the symbol of God's presence.

His action would certainly have been considered sinful in the days of Josiah, three centuries later, but hardly so at this time, for Jerusalem had been part of Israel only from the days of David, and it was not at this time considered to be the only true sanctuary in the land. The local sanctuaries or high places had been used since the days of the patriarchs, and Gideon, Samuel, and Elijah had offered sacrifice in them. Bethel had strong associations with Jacob and was an ancient and revered holy place, while Dan had certainly been served by a priesthood descended from Moses rather than Aaron and may well have still been served by it. Again, the second commandment prohibiting 'graven images' could not have been strictly interpreted at this time because there were figures of cherubim, lions and bulls in Solomon's Temple (1 Kings 7:29). In spite of the partition of the Kingdom, there was still only one people, both bound by the same covenant, both serving the same God.

The history was written from the point of view of the Temple priests at Jerusalem living many centuries later when a quite different attitude to the Law obtained. Jeroboam's sin was looked upon as unforgiveable, and the very existence of the Northern Kingdom was looked upon as an offence against God. As a result, according to the chronicler, all its kings, including the great Omri, 'did what was wrong in the eyes of the Lord'; and 'followed in the footsteps of Jeroboam' (1 Kings 16:25–26). In spite of the fact that the Temple and the Ark were in Jerusalem, Rehoboam himself frequented the High Places.

16.3 The Prophet Elijah

Elijah, Ahab and Jezebel (1 Kings 16:29–22:38)

Omri, a distinguished King of Israel of the 9th century, who impressed even the war-like Assyrians, is dismissed in only a few verses, because however successful his reign had been from a secular point of view, it appeared to have little to contribute religiously. By the same token his far less important son Ahab is given a great deal of treatment because his marriage to Jezebel, a princess of Tyre, resulted in a serious confrontation between the monarchy and Elijah, one of the greatest prophets of Israel. The issue at stake was Yahweh or Baal-Melkart, the chief god of Tyre.

Elijah was the greatest of the prophets since the days of Samuel and like his illustrious predecessor had the power to read the signs of the times and to speak with the authority of God. He warned and rebuked without fear or favour and was prepared to admonish even the king. He is the only prophet whose appearance is described – 'A hairy man with a leather apron around his waist' (2 Kings 1:8). Compare John the Baptist (Mark 1:6).

Ahab's marriage to Jezebel was a disaster, for she was fanatically devoted to Baal-Melkart, the god of Tyre, and was determined to introduce his worship into the Kingdom. He, on the other hand, was weak and vacillating, and soon became dominated by his ruthless wife.

Elijah the prophet declared there would be a year of drought, Yahweh's punishment for Ahab's unfaithfulness. Forced into hiding he eventually went down to Zarephath, a village of Sidon where Yahweh told him that a widow would look after him.

1 Kings 17:1–16 tells the story of how the **ravens fed Elijah** when he was in hiding east of Jordan, and how miraculously the widow of Zarephath's jar of flour and flask of oil remained undiminished throughout his stay with her, though there was no food in the land. In 1 Kings 17:17–24 we have the story of how he brought her son back from the dead. (The details of both of these stories should be learned.) In the third year of the famine Yahweh told Elijah that he must show himself to the king and that the drought would end. As soon as Ahab saw the prophet he said 'Is it you, you troubler of Israel?' Elijah replied, 'It is not I who have troubled Israel, but you and your father's family.'

It appears that when Ahab allowed his wife to introduce the worship of Melkart, the prophets had protested and, furious at this, Jezebel massacred a large number of them. Obadiah, the king's comptroller, had saved a hundred of the prophets of God by hiding them in a cave. Elijah now demanded a show of strength between himself and the prophets of Baal, 450 of whom been introduced by Jezebel. All Israel was assembled on Mount Carmel and Elijah issued the challenge, 'How long will you sit on the fence? If the Lord is God, follow Him; but if Baal, follow him' (1 Kings 18:21).

They answered not a word.

The account of the contest on Mount Carmel is most dramatically told (1 Kings 18:22–46). There are many points of interest but note:

1 The behaviour typical of the ecstatic prophets, the wild dancing, the gashing of themselves with swords (in their frenzy they would be insensitive to pain), their loud cries of invocation.

2 God answered by fire as He had on Mount Sinai.

3 When the prophets of Baal had been killed, God's honour was vindicated, so the rain came.

Elijah had to flee for his life and came at last 'to Horeb, the mount of God' – Horeb is Sinai – (1 Kings 19:9) and entered a cave. There God spoke to him and told him to 'stand on the mount before the Lord'. Elijah expected God to speak to him in storm and earthquake and fire as He had to Moses, but He spoke not in the wind, nor the earthquake nor the fire, but in 'a low murmuring sound' or 'a still small voice' (Authorized Version) or literally, 'a sound of thin silence'.

He was given a new vision of God, as one who speaks in the stillness and silence to those who wait patiently on Him. Then came the question 'Why are you here Elijah?' The prophet offered his excuses, but they could not be accepted. What Elijah learned was that:

1 He was not alone as he imagined, for there were thousands in Israel who were still loyal to God.

2 God had an immediate task for him. He was to return by way of the wilderness to Damascus and anoint Hazael to be King of Aram: Jehu was to be anointed King of Israel and Elisha was to be commissioned as prophet in his place. The reign of Ahab and Jezebel would thus come to an end.

Naboth's vineyard (1 Kings 21:1–26)

This is another beautifully told story. Some points to note are:

1 Naboth was not unreasonable to refuse the generous offer of the king, because there was a strong prejudice against surrendering the family heritage. His piece of land was part of the family portion of the promised land held in trust from God. Thus 'Every Israelite shall retain his father's patrimony' (Numbers 36:9).

2 From the beginning of the monarchy there was the strongest opposition to absolute rule: there was a limited monarchy because the true King of Israel was God, and each monarch was expected to rule by the principles of the covenant. Therefore it never occurred to Ahab either to gain the land by compulsory purchase, or by deceit. His wife, brought up under a very different theory of monarchy, took a very different line, yet even she, with no scruples to hamper her, found it necessary to go through the appearance of justice. A judicial murder was better than an assassination. False witnesses were procured and Naboth was put to death for blasphemy and treason; his property was forfeit to the king.

3 As in the case of David's arranged murder of the innocent Uriah the Hittite (2 Samuel 11) the deed could not be hidden. As Nathan the prophet rebuked David, so Elijah fiercely pronounces judgment on Ahab: disaster will fall upon his house and 'Jezebel shall be eaten by dogs by the rampart of Jezreel' (1 Kings 21:23).

The time came for Elijah to be taken up to heaven (2 Kings 2). He is accompanied by Elisha who had been called by God to succeed him (1 Kings 19:19–21). Elijah divided the waters of Jordan by striking the river with his cloak; they then passed over. Elisha asks for a double portion of the spirit of Elijah and is promised that if he sees Elijah being taken away, his request will be granted. In the event there appeared chariots and horses of fire and Elijah was taken up to heaven in a whirlwind. When Elisha saw it he said 'My father, my father, the chariots and the horsemen of Israel' (2 Kings 2:12).

The importance of Elijah

1 It was he alone who prevented the designs of Jezebel to root out the religion of Yahweh and to substitute the worship of Baal-Melkart. If he had failed, the Hebrews would have sunk into paganism. He laid a firm foundation for the later prophets of Israel.

2 In the tradition of Nathan, the great prophet of David's reign, Elijah upheld the truth which was the essence of the covenant, that religion must not be divorced from morality. This is seen most clearly in the incident of Naboth's vineyard.

16.4 Elisha

Many of the stories about him closely resemble those of Elijah.

1 Having picked up the fallen mantle of his master he divided the waters of Jordan as Elijah had done.

2 He replenished the widow's flask of oil in a miracle similar to the one performed by Elijah at Zarephath (2 Kings 4:1–7).

3 He restored the Shunammite woman's son to life in a manner very similar to that of Elijah at Zarephath (2 Kings 4:34–47).

4 At the end of his life we hear the identical words used by Elisha at his parting from Elijah; 'My father, my father, the chariots and horsemen of Israel' (2 Kings 13:14).

Elisha cures Naaman the leper (2 Kings 5)

This is another beautifully told story, notable for the revealing request of the restored Naaman. The prophet refused to take payment, but agreed to Naaman's request for 'two mule loads of earth' in order to sacrifice to the Lord (2 Kings 5:17). He can only worship Yahweh on land which belongs to the jurisdiction of Yahweh. At his home in Syria he can spread out the load of Yahweh's earth, build an altar on it, and then offer him due service. The prophet raised no objection to Naaman's intention to go through the form of worshipping Rimmon the Syrian god on his return home while swearing life-long allegiance to Yahweh. We find evidence in this incident that **Israel had not yet arrived at the concept of Monotheism**, that is the belief that there is but one supreme God. Yahweh was the God of the nation, whose writ ran within the national boundaries only; Rimmon enjoyed similar authority in Syria. The prophet would presumably have strongly objected to the worship of Rimmon in Israel.

The account contains a number of instances of imitative magic. Elisha purified the polluted spring of Jericho by putting salt in it (2 Kings 2:19–22). A broth was being prepared by some of the prophet's followers and one of them, in ignorance, put a poisonous herb into the pot; Elisha put some meal into the pot and the broth became harmless (2 Kings 4:38–41). A man dropped an iron axe head into the water and Elisha cut a piece of wood, threw it into the water at the place and the iron floated (2 Kings 6:5–7). The principle is that the salt and the meal are harmless, so the water and the broth become harmless: the wood floats, so the iron in imitation does the same.

In 2 Kings 6:24–7:20 we have an account of the siege of Samaria by Ben-hadad, details of which are quite horrifying. Elisha promised that the siege would be raised the next day, and the besieging army hearing a sound like horses and chariots in the night, fled in haste, convinced that a powerful force was coming to relieve the city.

Elisha wielded considerable political power behind the scenes. He inspired a revolution in Damascus which resulted in the murder of Ben-hadad by Hazael who succeeded him (2 Kings 8:7–15) and another palace revolution in Israel by which Jehu became king (2 Kings 9:8–13) and killed both Jehoram, King of Judah, and Ahaziah, King of Israel (2 Kings 9:14–28).

One of the most dramatic stories in the Bible is about the **death of Jezebel**. She was thrown down from a window near the gate and her body was trampled by the horses of Jehu (2 Kings 9:30–37). Jehu eventually killed all the descendants of Ahab in both Samaria and Jezreel, as well as the nobles and priests. He then proceeded to kill all the prophets and priests of Baal and destroyed his temple. He thus stamped out the worship of Baal in Israel (2 Kings 10). All these actions were as much religious as political and the inspiration for it came from Elisha. The revolution is sometimes called the 'prophetic revolution'.

17 The Minor Prophets

17.1 Introduction

An important stage in the evolution of the prophet came when one of the band or guild dissociated himself from the message given by the others and delivered a message of his own. The first person in the Bible to do this was Micaiah (1 Kings 22:5–28).

The Kings of Israel and Judah, Ahab and Jehosaphat, had joined forces with the intention of attacking Damascus and, as was the custom, decided first to consult the prophets about the prospects of victory. The band of prophets, under their leader Zedekiah foretold a victory. Jehosophat required a second opinion and insisted that Micaiah should be consulted in spite of

Ahab's advice that the prophet's message was always unfavourable. In reply to Ahab's question Micaiah agreed with the other prophets but the king sensed that this was not the truth. On being pressed further the prophet foretold disaster. For his pains Micaiah was put in prison. In the event the allies were defeated and Ahab as prophesied was killed (1 Kings 22:5–28). A century later it was the message of the single independent prophet that was relied upon, rather than that of the band of ecstatics.

Amos, Hosea and Micah are among the 'Minor Prophets'. This is not a reflection of their importance as prophets, for they are very important indeed: it means that their writings are less extensive than such 'major' prophets as Isaiah, Jeremiah, and Ezekiel.

17.2 Amos

Historical background

The long reign of Jeroboam 11 (788–747 BC) had given Israel renewed prosperity. Taking advantage of the weakness of Syria he had recovered much of the territory lost since the days of David and the trade routes between Asia and Africa had come under his control. He could now levy tolls on a considerable proportion of world trade and the wealth of the country was no longer derived principally from the products of the land. This led to an economic boom and the rise of a wealthy and powerful middle class of merchants, whose style of life stood in stark contrast to that of the majority of the population. The privileged classes lived in ostentatious luxury and controlled the rest of society by a system of heartless social injustice. The expansion of commerce led to the rise of money-lending as a profession, and the small farmer who failed to redeem his mortgage, lost his land and became the tenant of the new owner, paying a large part of his produce as rent. Justice could be bought, sometimes for the price of a pair of shoes.

The luxury of the rich stood out in striking contrast to the misery of the poor, and religion was not a restraining influence; indeed it had become an instrument of oppression. The Law stated that if a man's outer garment were taken in pledge it had to be restored at nightfall, in order that he might sleep in it protected from the cold, but if the money-lender could say that he needed the article for a religious ceremony, there was no obligation to return it. There was no lack of religion, for the people flocked to the shrines at Bethel, Gilgal and Beersheba, but it was **concerned only with outward form** and had no spiritual or moral content. The pure religion of the wilderness no longer held sway.

Amos, as an outsider, a man of Judah, saw the rottenness of the situation and fearlessly spoke out against it, in the name of God. He was brought up, as was Elijah, in the simple life of the hill country. The influence of the religion of Canaan on the religion of Israel had not been uniform, and it had made least impression in the semi-desert regions of the south and east: there they retained the simpler and purer standards of the pastoral life. Tekoa, where Amos was brought up, was a small village south of Jerusalem in marginal land not far from the Dead Sea. He earned his living as a shepherd and a gatherer of 'sycamore-figs', a coarse kind of fig eaten by the

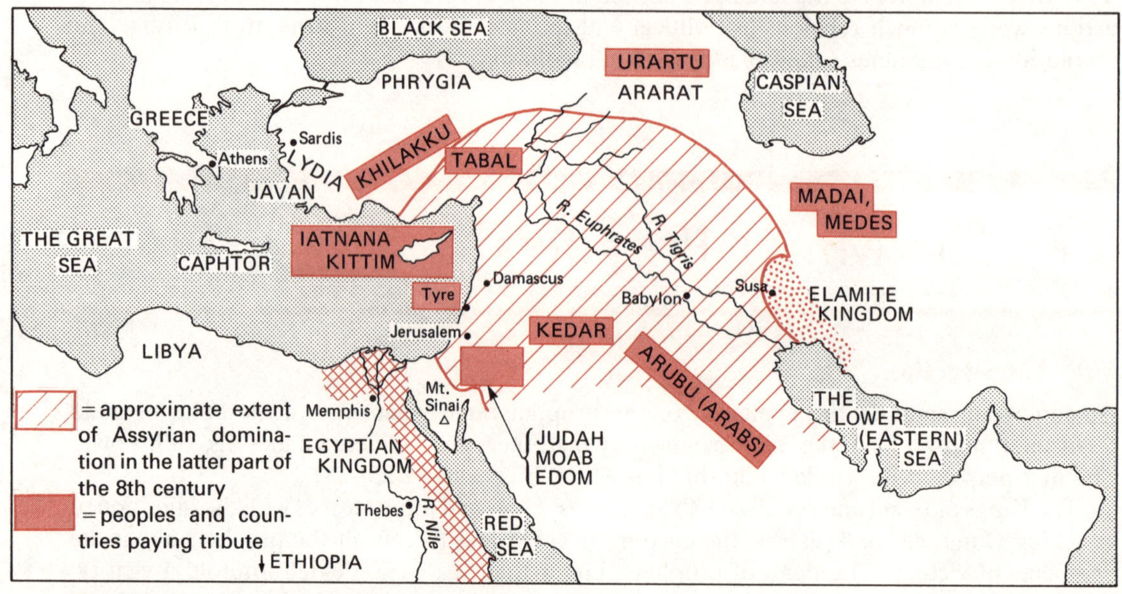

Fig. 17.1 The Assyrian Empire

peasants. Year by year Amos set out for the great cities of the north to sell his wool and fruit, and what the inhabitants of 'civilization' could accept as normal, shocked and disgusted him. He saw the artificiality, shallowness and licentiousness of their way of life, and saw too the need for someone to speak for God against it. Eventually the conviction that he was called by God came upon him with overwhelming certainty. 'The Lord took me as I followed the flock and said to me "Go and prophesy to my people Israel"' (Amos 7:15).

The call could not be resisted: 'The lion has roared; who is not terrified? The Lord God has spoken; who will not prophesy?' (Amos 3:8).

The teaching of Amos

1 The universality of God This was not yet to declare belief in monotheism, that is, that there is one God, but Amos proclaimed the truth that Yahweh is supreme: there may be other gods, but they are of inferior status and Israel must have nothing to do with them. It is:

He who made the Pleiades and Orion,
who turned darkness into morning
and darkened day into night,
who summoned the waters of the sea
and poured them over the earth. (Amos 5:8)

2 Yahweh is the Creator of the heavens and earth God is the Lord of all nations, a belief which follows from the concept of the universality of God. Though He had chosen Israel, He was also concerned with other nations, 'Did I not bring Israel up from Egypt, the Philistines from Caphtor, the Aramaeans from Kir?' (Amos 9:7) Thus the great racial migrations such as the Philistines from Crete were as much the work of God, as Israel's Exodus from Egypt. Every nation was subject to the supreme moral Law of God and for this reason Amos declares the judgment of God on all the neighbouring nations in Chapters 1 and 2.

3 The universal moral law The neighbouring nations worshipped false gods but they are not condemned for this so much as for their crimes against humanity. The message of Amos is that the Lord of the nations wills that justice should prevail among men and nations. Injustice and cruelty are a denial of the true nature of man and are sins against God. When Amos turned to the judgment on Judah (Amos 2:4–5) the principle is that much should be expected from those who had received much, so the condemnation is the more severe for 'they have spurned the Law of the Lord and have not observed His decrees' (Amos 2:4).

The judgment on Israel was equally severe, basically for the same reason. The prophet pointed out what the nation had forgotten, that it owed its very existence to God, for

It was I who brought you up from the land of Egypt,
I who led you in the wilderness forty years,
to take possession of the land of the Amorites. (Amos 2:10)

He reminded them of their special relationship with God, 'For you alone have I cared among all the nations of the world' (Amos 3:2) but that position of privilege carried the responsibility to do the right. If she fails to do this she will lose the only reason for her existence, and God Himself will bring about her ruin.

4 God demands righteousness The necessity of social justice. The few lived in ostentatious luxury and fattened on the many who lived in poverty. Amos records a catalogue of their impositions. They 'hoard in their palaces the gains of crime and violence' (Amos 3:10); 'They give short measure and sell the sweepings of the wheat' (Amos 8:6); they build up large estates from lands of the peasants, and erect great houses for both winter and summer; the women make increasing demands on their husbands to provide for luxuries and are referred to as 'You cows of Bashan who live on the hill of Samaria, you who oppress the poor and crush the destitute' (Amos 4:1). The process of justice was corrupted by bribery: 'they sell the innocent for silver and the destitute for a pair of shoes' (Amos 2:6). 'You have turned into venom the process of law, and justice itself into poison' (Amos 6:12).

5 God required pure worship Since God is just and holy, the worship of corrupt and evil people is not acceptable: 'I hate, I spurn your pilgrim feasts, I will not delight in your sacred ceremonies.' (Amos 5:21). The people must realize that God could not be bribed by sacrifice and their worship would not be acceptable until they were prepared to 'Let justice roll on like a river and righteousness like an ever-flowing stream' (Amos 5:24).

6 The complacency and self-satisfaction of Israel was misplaced

(a) Their reliance on the covenant. Amos also accepted the covenant as the foundation of the nation's religion, but the people interpreted their special relationship with God to mean that no matter what the circumstances, God would always be on their side (Amos 5:14). Amos taught that they were called not to privilege but to service. The covenant relationship made their responsibility greater and their punishment for failure the heavier:

> For you alone have I cared
> among all the nations of the world;
> therefore will I punish you
> for all your iniquities. (Amos 3:2)

In the same way the privileges of wealth carried with them responsibilities for the welfare of their fellow men.

(b) The popular belief in 'the Day of the Lord'. This was that God would one day intervene directly in human affairs to deliver His people from oppression and establish a kind of Golden Age. Amos saw that far more important than the special relationship sealed by the covenant, was the fact that Yahweh was a moral God. The manner of His intervention would be related to the moral standing of the nation, so what could they expect?

> Fools who long for the day of the Lord,
> What will the day of the Lord mean to you?
> It will be darkness, not light . . .
> A day of gloom with no dawn. (Amos 5:18–20)

This is the first reference in the Bible to the longed for Day of the Lord which was later to be prominent in the message of the prophets.

(c) The inviolability of the capital city, Samaria. As a result of continued fortification over several reigns, it was immensely strong. More important, it was the sanctuary of God, and God would not allow it to be destroyed. Amos foretold its fall.

The visions of Amos

Amos denied that he was a professional prophet attached to any guild of prophets, but the way in which his message from God was received was consistent with the experience of the ecstatics. They heard the word of God or saw a vision. These visions seem to be induced by meditation on an object in the world of sense – the real world for most people – and through this to become conscious of a world other than, but equally real, in which there is direct contact with God. Three visions of Amos are recorded:

1 The Plumbline (Amos 7:7–9) He saw in a plumb-line placed against a wall to test its uprightness, the message that God was testing the 'uprightness' of Israel and found that she was badly out of true. The wall was bound eventually to fall: Israel was not fit to survive.

2 The Basket of Summer Fruit (Amos 8:1–3) Amos saw a basket of ripe summer fruit, in Hebrew 'kais' and from this is suggested 'kes' – the end.
'The time is ripe for my people Israel.' (Amos 8:2)

3 The Plague of Locusts (Amos 7:1–3) Amos observed them hatching out as the corn was beginning to sprout and saw in this a message that the nation would be destroyed.

The judgment on Israel

The nation ignored the signs sent to warn it, drought, famine, plague, pestilence, war and earthquake (Amos 4:6–11). (The reference to the plague 'I sent plague upon you like the plagues of Egypt' (Amos 4:10) is of interest in that it may help us roughly to date the prophecy: a widespread outbreak of plague, according to Assyrian records, occurred in about 763 BC.)

In spite of all these warnings Israel did not return to God and the sentence of doom was pronounced upon them, 'Israel, prepare to meet your God' (Amos 4:12).

There was of course a means of avoiding destruction, even at this late moment:

> Seek good and not evil,
> that you may live,
> that the Lord the God of Hosts may be firmly on your side,
> as you say he is. (Amos 5:14)

Destruction will be complete, 'I will smash them all into pieces, and I will kill them to the last man with the sword' (Amos 9:1). The prophecy of doom was fulfilled when in 721 BC the Assyrians conquered Israel and deported its people who were never to return.

Amos is rejected

He had dared to prophesy against Israel at Bethel, 'the king's sanctuary, a royal palace'. His message that the nation would be deported and Jeroboam killed, was to Amaziah the priest as dangerous as it was impudent, and Amos was reported to the king. 'Amos is conspiring against you in Israel; the country cannot tolerate what he is saying' (Amos 7:10). He was open to a charge of treason, but was protected by the fact that his ecstasy made it clear that he could claim the authority of God and therefore must not be harmed. To the sneer of Amaziah that he was a professional prophet Amos replied, 'I am no prophet, nor am I a prophet's son; I am a herdsman and a dresser of figs. But the Lord took me as I followed the flock and said to me "Go and prophesy to my people Israel."'

He was forced to return to Tekoah. There his oracles were written down, probably by his followers. He is 'the first writing prophet'.

Amos the social reformer

To think of him merely as the first great social reformer is to look at him from the wrong perspective. He is first and foremost a man of God and his concern for social justice springs from his devotion to God. In the covenant, religion and morality are not separate, but inextricably entwined, whereas in the Canaanite religion they were separate: religion was a matter of ritual ceremony, not morality, and this was the danger to Israel when it came into contact with the immoral worship of the fertility gods. The **message of Amos** was a call for a return to the articles of the covenant, insisting that a righteous God demands justice, so that worship and morality must go hand in hand. For Amos social justice was not an extra to the religious life, but the very heart of it.

It is important to understand that he did not offer Israel a new programme for the reorganization of society. He did not denounce the system and offer a new one, but said rather that the current system must be subject to the principles of the covenant. The Nazarite and Rechabite believed that civilization was inherently evil and that a return should be made to the ways of the wilderness, but the habits of the desert were not appropriate to a rich and fertile land, and new social structures were bound to develop under the new conditions. What Amos saw was that the social and political life of Israel was in imbalance with its spiritual life, and that the cure was not to return to the old ways of the wilderness, but to apply the unchangeable moral and spiritual principles forged in the desert, to the new conditions.

There was always hope for the nation if it would turn back to God.

> Hate evil and love good;
> enthrone justice in the courts;
> it may be that the Lord the God of Hosts
> will be gracious to the survivors of Joseph. (Amos 5:15)

However, the words 'it may be' suggest that Amos had little hope that the nation would reform. Many scholars hold that Amos 9:11–15 represents an attempt by a later hand to contrive a happy ending and is not part of the original prophecy.

17.3 Hosea

Historical background

The degeneracy of Israelite society described by Amos had in no way been checked, and in the days of Hosea there was the added factor of political insecurity. The long reign of Jeroboam II (788–747 BC) had given Israel economic prosperity and political stability, but at his death the strong hand was removed, and there follows a period described in 2 Kings 15:10, of palace intrigue and murder; one man after another lays violent hands on the crown, 'King after king falls from power, but not one of them calls upon me' (Hosea 7:7).

In 745 BC the throne of Assyria was seized by an able soldier, Tiglath-Pileser, who had military designs on Egypt. The kings of Israel played the diplomatic game, now submitting to Assyria and paying tribute, now flirting with Egypt, 'Ephraim is a silly senseless pigeon, now calling upon Egypt, now turning to Assyria for help' (Hosea 7:11).

The Assyrians took Damascus in 734 BC and in 724 BC Samaria was besieged: it fell in 721 BC and with its fall the Kingdom of Israel was destroyed. Its people were transported to Assyria (2 Kings 18:11); the ten tribes ceased to exist.

Only one prophetic voice comes from Israel in the eighth century BC, that of Hosea. Like Amos, he was a layman, and while each had a passionate love for God, they were in all other respects very different. Amos was a man of the country and an outsider of Judah, while Hosea was a townsman who lived among his own people, identifying himself with their way of life, and sorrowing over their sins as though they were his own. Above all Hosea's call was very different, indeed unique, for he saw in the deep personal tragedy of his personal life, his faithless wife, a revelation of the character of God.

Hosea and his wife

He had married a wife named Gomer and there is in the account of this an immediate problem of interpretation. The message from God is 'Go, take a wanton for your wife and get children of her wantonness; for like a wanton this land is unfaithful to the Lord' (Hosea 1:2). Taken literally this can only mean that Hosea was called to take to wife a woman he knew to be a wanton. Taken together with the second account in Hosea 3:1–3, which is that of the prophet himself, rather than in the third person, we may interpret it to mean that God had called him to love a woman who was destined to become an adulteress. In the Hebrew idiom this could be stated as 'love a woman, loved by another man, an adulteress' (Hosea 3:1). It was from his experience of a wife who became unfaithful to him, that he learned of the love of God for a faithless people, Israel.

Gomer bore him three children to each of whom he gave prophetic names, and then she deserted her husband and children. Though the Law permitted him to divorce her he found that he could not bring himself to do so, for he loved her still; even in his suffering he finds he cannot help loving her. Then the message came from God:

> Go again and love a woman
> loved by another man, an adulteress,
> and love her as I the Lord love . . . the Israelites
> although they resort to other gods. (Hosea 3:1)

This was a flash of inspired insight, marking a tremendous step forward in man's understanding of God, who goes on loving man in spite of man's unfaithfulness. Eventually Gomer's wealthy lover abandoned her and she became a Temple prostitute. This was not as reprehensible as it seems to us, because at this time sacred prostitution was an institution Israel had borrowed from the Canaanite religions, and was quite respectable. Such women were regarded as sacred, and it is indicative that one of the Hebrew words for harlot is literally 'holy woman'. After a search he found her, bought her back, and brought her home. He told her that at first they would not live as man and wife but there was hope that eventually they could make a fresh start. There is nothing in the text, however, to suggest a happy ending.

Faithless Israel and God

There is no observable sequence in the prophetic utterances and the same themes often recur. This makes the understanding of the work quite difficult. We can only pick out extracts which give the general tenor of Hosea's message.

While Amos saw that a righteous God demanded righteousness of His people, Hosea saw that Israel's sinful way of life was not merely breaking God's law, but wounding His heart of love. God, for Hosea, is not only righteousness but also love. Just as Gomer had been unfaithful to him, Israel had been unfaithful to God, and he expresses this thought in three powerful similes:

1 It was God who had brought them out of Egypt, and cared for them in the wilderness: 'When Israel was a boy, I loved him; I called my son out of Egypt' (Hosea 11:1).

2 It was Yahweh who gave them the richness of the promised land: 'It is I who gave her corn, new wine and oil, I who lavished upon her silver and gold' (Hosea 2:8). To credit other gods with this was a sin against love.

3 Israel is the son of Yahweh, and Hosea employs the simile of the father teaching the child to walk:

> It was I who taught Ephraim to walk,
> I who had taken them in my arms;
> but they did not know that I harnessed them in leading strings
> and led them with bonds of love –
> that I had lifted them like a little child to my cheek. (Hosea 11:3–4)

Israel, like Gomer, deserved to be abandoned, but as Hosea could not forget Gomer, so God could not forget Israel.

> How can I give you up, Ephraim,
> How surrender you, Israel? . . .
> I will not turn round and destroy Ephraim;
> for I am God and not a man. (Hosea 11:8ff)

Out of the bitterness of his personal tragedy Hosea learned that in love there is suffering. God's righteousness cannot condone sin, but because God loves he cannot hate the sinner: he can only suffer and go on suffering if need be. Hosea called his first son Jezreel (Hosea 1:4). This was the city where Jehu at the command of Elisha slaughtered the house of Ahab. Hosea with his deeper insight into the nature of God condemned such butchery. The house of Jehu will be punished 'for the blood shed in Jezreel' (Hosea 1:4).

Hosea's second child was a daughter – Lo-ruhamah – 'no mercy', for it seemed inconceivable that God should forgive the nation. The third child, a boy, he called Lo-ammi – 'not my people': 'for you are not my people, and I will not be your God' (Hosea 1:9). This notion he was later to reverse (Hosea 2:23).

The manner of Israel's unfaithfulness:

1 In politics

'They make kings, but not by my will' (Hosea 8:4). The gaining of the throne by conspiracy and murder was not the will of God. Power based on treachery could command no respect. 'Samaria and her King are swept away like flotsam on the water' (Hosea 10:7).

Israel played the political game, flirting with one powerful nation and another. The only hope for Israel was to put her trust in God.

2 In social life

> There is no good faith or mutual trust,
> no knowledge of God in the land,
> oaths are imposed and broken, they kill and rob;
> there is nothing but adultery and licence,
> one deed of blood after another. (Hosea 4:1–2)

3 In religion

All the evils of the nation stemmed from the nation's desertion of Yahweh for the Baals, the Canaanite fertility gods. The great agricultural feasts were orgies of drunkenness and immorality:

> for they have forsaken the Lord
> to give themselves to sacred prostitution.
> New wine and old steal my people's wits. (Hosea 4:10–12)

The priests were as bad as the rest:

> like robbers lying in wait for a man,
> priests are banded together
> to do murder on the road to Shechem. (Hosea 6:9)

Idol-worship was commonplace and Hosea attacks the bull worship of Shechem which had been instituted at the time of the division of the Kingdom, when the commandment against graven images was not as strictly interpreted:

> Your calf-gods stink, O Samaria . . .
> For what sort of a god is this bull?
> It is no god,
> a craftsman made it. (Hosea 8:5ff)

This condemnation paved the way for the clearer message against idol-worship, to come later. Hosea was in this also, far in advance of his time.

Prophecy of doom

God could not deny his own righteousness, and judgment was bound to follow. 'Israel sows the wind and reaps the whirlwind' (Hosea 8:7) but punishment would not be for revenge but reform. The nation will surely fall 'Samaria will become desolate because she has rebelled against her

God' (Hosea 13:16). She 'and her king are swept away' (Hosea 10:7). There will be a new captivity 'Ephraim shall go back to Egypt or in Assyria they shall eat unclean food' (Hosea 9:3). for, 'My God shall reject them because they have not listened to Him' (Hosea 9:17).

There is however a strong ray of hope founded on God's unchanging love for His people. In spite of the disasters which were bound to follow, the end of the Kingdom in blood and violence, the nation dispossessed and carried into captivity, there was the hope that Israel would learn the hard lesson and return to God in sincere penitence for its past misdeeds. Israel would then again be restored. Hosea expresses with delicate sensitivity God's love for His fallen but penitent Israel, in the language of courtship:

> I will woo her, I will go with her into the wilderness
> and comfort her . . .
> she will answer as in her youth,
> when she came up out of Egypt. (Hosea 2:14–15)

The quality of Hosea's own love for Gomer is revealed in the deeply moving lines, 'On that day she shall call me "My husband", and shall no more call me "My Baal"' (Hosea 2:16).

The meaning of the word Baal is 'master' or 'owner', but God is here claiming a far more intimate relationship than an owner. He is the beloved husband of a wife, not the owner of a chattel.

Israel will be God's new sowing in the land and its fruitfulness will show that the nation is forgiven for the evil of Jezreel. 'Lo-ruhamah' – 'no mercy' will be shown love, and 'Lo-ammi' – 'not my people' will become 'You are my people' and he will say 'Thou art my God' (Hosea 2:21–23).

Hosea was the last great prophet of the Northern Kingdom.

17.4 Micah

The first chapter tells us that Micah prophesied in the reigns of Jotham 740–735 BC, Ahaz 735–720 BC and Hezekial 720–692 BC, Kings of Judah. He is thus a contemporary of Hosea in the north and Isaiah in the south, prophesying before and after the fall of Samaria in 721 BC.

In contrast to Isaiah who lived all his life in Jerusalem and came from an aristocratic family, Micah was a countryman living in Moresheth, a country village near the Philistine border. While there is no account of his call, Micah was quite certain that God had called him, and that he spoke the words of the Lord.

The sins of Israel

Like Amos before him he condemned Israel: God has 'a case' against it. In Chapter 6 Amos pleaded that case by reference to all God's mercies to the nation in times past: Israel had forgotten them all.

We find the same social injustices which Amos condemned: 'Your rich men are steeped in violence' (Micah 6:12), the judges 'give judgment for reward' (Micah 7:3). Traders use 'the infamous short measure, the accursed short bushel' (Micah 6:10). The result is a general fall in moral standards, for if the leaders of the people are corrupt it is not surprising that their example is followed. Honesty is no more and no confidence may be placed in either family or friend:

> Trust no neighbour, put no confidence in your closest friend;
> seal your lips even from the wife of your bosom.
> For son maligns father,
> daughter rebels against mother,
> daughter-in-law against mother-in-law,
> and a man's enemies are his own household. (Micah 7:5–6)

Disaster is bound to befall Israel: 'I will make Samaria a heap of ruins in open country' (Micah 1:6).

The sins of Judah

Micah saw, to his dismay and horror the same evils prevalent in Judah, and knew that the same doom would come upon it, 'What is the crime of Jacob? Is it not Samaria? What is the hill-shrine of Judah? Is it not Jerusalem?' (Micah 1:5). With vivid imagination he imagines the progress of the Assyrian armies through the land, village after village falling to the invaders. God Himself is coming in judgment:

> the Lord is leaving His dwelling-place
> down He comes and walks on the heights of the earth . . .
> and all for the crime of Jacob and the sin of Israel. (Micah 1:3–5)

It was not domestic or foreign politics which concerned Micah but rather the plight of the poor of the countryside. The economic and social changes which had first affected Israel had now spread to Judah: the money-lenders, traders and property speculators were now battering on the peasantry of Judah:

> They covet land and take it by force:
> if they want a house they seize it:
> they rob a man of his home
> and steal every man's inheritance. (Micah 2:2)

They lie in bed planning the wicked deeds they will do in the day, knowing they have the power. His most powerful and bitter metaphor is of the butcher who first skins his victim, strips off the flesh and even breaks the bones to extract the marrow. So the rich exploit the peasantry to extinction.

Micah continued his diatribe against the religious authorities, who, far from condemning these injustices, tolerated them and were themselves corrupt. 'Her priests give direction in return for a bribe, her prophets take money for their divination'. (Micah 3:11) they were prepared to 'promise prosperity in return for a morsel of food (Micah 3:5), and to curse those who would not do so.

The inviolability of Jerusalem

Just as Israel believed that Samaria could not fall because it was under the protection of God, it was popularly believed in Judah that no harm could come to Jerusalem beçause God's Temple was within its walls. Micah fearlessly prophesied that this belief was false:

> Zion shall become a ploughed field,
> Jerusalem a heap of ruins,
> and the temple hill rough heath. (Micah 3:12)

These very words were quoted in defence of Jeremiah more than a hundred years later when he was on trial for his life, accused of prophesying the doom of the city (Jeremiah 26:18ff). It was argued that King Hezekiah had heeded the words of Micah, and the reforms he then instituted were the means of saving Jerusalem from the Assyrian invaders. God had relented, and revoked the disaster He had promised. Micah had not been killed by his king, and it was argued that to kill Jeremiah for conveying the same message from God, would be to invite disaster. The words of Amos and Hosea evoked no response, but those of Micah bore fruit in a religious reformation.

Three great passages

1 A definition of true religion

In order to approach God it is not of prime importance to offer him animal sacrifice, still less should a man offer his children as a sacrifice for his sin, for:

> God has told you what is good;
> and what is it that the Lord asks of you?
> Only to act justly, to love loyalty,
> to walk wisely before your God. (Micah 6:8)

For the sheer beauty of the poetry, rather than accuracy, the Authorized Version may be preferred:

> 'He hath shewed thee O man what is good ; and what doth the Lord require of thee, but to do justly, and to love mercy, and to walk humbly with thy God?'

2 The Messiah (Micah 5:2–5)

We find in Micah the first reference to the hope of a Messiah whom God will send to deliver His people. As David did, He will come from Bethlehem. He will be a shepherd to the people and 'his greatness shall reach to the ends of the earth, and he shall be a man of peace' (Micah 5:4ff).

3 The Age of Peace (Micah 4:1–4)
In God's good time all the nations of the world shall look to Jerusalem to know the ways of God and walk in His paths:

> They shall beat their swords into mattocks
> and their spears into pruning-knives;
> nation shall not lift sword against nation
> nor ever again be trained for war,
> and each man shall dwell under his own vine,
> under his own fig-tree, undisturbed,
> For the Lord of Hosts himself has spoken. (Micah 4:3–4)

18 The Major Prophets, Isaiah

18.1 Isaiah of Jerusalem

The book of the prophet Isaiah well illustrates a feature common to all the prophetic books, in that it consists of isolated items often of different authorship and date, collected over a long period of time and finally written down. Some of the prophetic books, for example Amos, are more or less the work of one man, but Isaiah is the work of at least three different prophets living at different times and in very different situations.

The Isaiah who gave the book its name lived in the eighth century BC, and was roughly a contemporary of Amos and Hosea. He is known as Isaiah of Jerusalem and his work is spread over Chapters 1–39. Chapters 40–55 are the work of the Second, or Deutero-Isaiah, an unknown prophet who lived about two centuries later when Israel was in exile in Babylon. Chapters 56–66 are the work of a Post-Exilic prophet living in Palestine, who is unknown, but is called Third or Trito-Isaiah. A further and major difficulty for us it that we seldom find a consecutive narrative of any length or any chronological sequence in the prophecies. We can do no better than to extract from the writings the most prominent features of their message.

18.2 Historical Background

After Jeroboam, the last strong King of Israel died in 767 BC, the throne was seized by a succession of murdering conspirators until the Kingdom fell in 721 BC. Uzziah or Azariah, King of Judah 786–744 BC, was able during this period of instability to free himself from dependence on the Northern Kingdom, and to bring prosperity to Judah. His successor, Jotham, reigned for only three years and in 741 BC the weak and corrupt Ahaz came to the throne.

Assyria, which had been in a period of decline, set out on a policy of expansion, threatening its less powerful neighbours and eventually making them tributaries. In 738 BC Menahem, King of Israel, was made a tributary vassal of Assyria, so Judah with the border of the Assyrian empire only thirty-two kilometres from Jerusalem felt herself to be directly threatened. There were three possibilities open to her:

1 to take an initiative in submitting to Assyria;

2 to join in alliance with Egypt, which was too weak to oppose Assyria on its own, and risk Assyrian reprisals;

3 neutrality. This had the drawback of alienating the neighbouring states which were combining against Assyria.

18.3 The Call of Isaiah (Isaiah 6:1–13)

This is a classic example of prophetic vision – **Isaiah saw and heard**. Standing at the threshold of the Temple he saw all the externals, the walls, the furniture, the altar of incense, but saw much more. Through these externals he saw the thin veil which stands between this world and that of the spirit raised, and he 'saw' the Lord, high and lifted up, seated on a throne, the train of his robe filling the Temple. About him were the seraphim, angelic creatures, each having six wings: two of their wings covered their faces for they could not look on God, with two they covered their feet, a sign of humility, and with two they flew, ready for the service of God. They called ceaselessly one to another 'Holy, holy, holy, is the Lord of Hosts: the whole earth is full of His glory' (Isaiah 6:3).

The vision of God's holiness brought a double reaction, first a conviction of the unworthiness of both himself and the nation, and then a deep desire to serve. Yet how could one so unworthy be of service to the Holy Lord? Only God could make him worthy: a seraph carried a live coal from off the altar and touched the prophet's mouth saying: 'See, this has touched your lips; your iniquity is removed' (Isaiah 6:7). Then he heard the call of God 'Whom shall I send? Who will go for me?' and the response was 'Here am I; send me' (Isaiah 6:8).

From this vision, Isaiah drew his own unique revelation of the being of God – his holiness. **Amos saw God as Justice, Hosea as Love,** and **Isaiah as Holiness**: this vision was the main spring and inspiration of all his teaching. Isaiah was not a social reformer nor a politician, but a prophet whose attitude to social injustice and political strategy, stemmed from his overwhelming sense of the holiness of God. The idea of the holiness of God was not new, for it is a concept necessary to all religion. The root meaning is 'set apart'. Thus food set apart for God is holy (c.f. the Shew Bread, the loaves set before God in the Temple, which it was not lawful for anyone to eat, except th priests), and those who by rites of purification set themselves apart are holy, but Isaiah introduced a new dimension: God is holy because unlike men there is in Him no stain of sin: 'the Lord of Hosts sits high in judgment and by righteousness the holy God, shows himself holy' (Isaiah 5:16).

Because God is a holy God, a moral God, His demands upon His chosen people are moral. He is 'The Holy one of Israel', 'high and exalted', far removed in this sense from sinful humanity. The technical term is 'transcendent'. However, though God is apart in His holiness He is also near, 'Cry out, shout aloud, you that dwell in Zion. For the Holy one of Israel is among you in majesty' (Isaiah 12:6). He is Immanuel, 'God with us' (Isaiah 7:14). He is therefore not only 'transcendent' but also 'immanent', present with his people.

18.4 The Universality of God

Like Amos before him, Isaiah believed that Yahweh was the creator of the whole world, and though Israel enjoyed a special relationship, all nations were therefore God's care, for 'The whole earth is full of His glory' (Isaiah 6:3). Yahweh was then the God of history, and Isaiah saw in the power politics of the nations, the hand of God. Assyria, like Judah, was an instrument of God, and He was using this powerful nation to punish her for her sins.

Yahweh, the God of History

Whereas Amos was a herdsman living in the semi-desert and Micah was a peasant from the country, Isaiah was born of an aristocratic family and lived all his life in Jerusalem. He was part of the court circle, and intimate enough with the kings of Judah to be privy to their councils. He was a man of influence and the great statesman of his age.

Because of God's universality, Isaiah could not confine himself to a narrow nationalism. Most of the neighbouring nations, including Philistia, Moab, Damascus, Tyre and Egypt are the subjects of prophecies of invasion and defeat. He has nothing but contempt for the rulers of Egypt but this is because he sees their powerlessness compared with the might of Yahweh. It is foolish to trust in Egypt and neglect the Lord of all the nations.

He believed that if Judah was central to God's plan for the world she could not perish: she would be punished, but not destroyed:

> You must not be afraid of the Assyrians, though they beat you with their rod and lift their staff against you as the Egyptians did; for soon, very soon, my anger will come to an end, and my wrath will all be spent. (Isaiah 10:24)

Assyria is an instrument of God, to punish evil nations, including Judah: 'He is the rod that I wield in my anger, and the staff of my wrath is in His hand' (Isaiah 10:5). But Assyria will not be allowed to destroy Judah:

> 'When the Lord has finished all that He means to do on Mount Zion and in Jerusalem, He will punish the King of Assyria for this fruit of his pride and for his arrogance and vainglory. (Isaiah 10:12)

The advance of the conquering Assyrian army terrified the citizens of Jerusalem, but Isaiah remained unmoved. No earthly power could destroy a city which was the sanctuary of the Lord of Hosts. In their panic the people of Judah examined every possible expedient to save themselves, but the advice of Isaiah was to stay calm, and to trust in God for He alone controlled the destinies of nations, 'Come back, keep peace, and you will be safe; in stillness and in staying quiet, there lies your strength' (Isaiah 30:15).

18.5 Isaiah and Ahaz

In 735 BC Pekah, King of Israel, and Rezin, King of Syria, formed an alliance against Assyria and needed the support of Ahaz of Judah to protect their right flank, and form a solid line of resistance. When Ahaz refused they marched on Jerusalem and Judah was panic stricken (Isaiah 7:1–3). Isaiah went to meet Ahaz and took his son Shear-jashub ('A remnant shall return') with him, as a living symbol of God's message that Judah would be overrun but that her people would not be utterly destroyed. This was hard for Ahaz to accept and he ignored it. Ahaz was then told that God would give a convincing sign if he asked for it, but the king said he would not tempt God. Isaiah then said, 'The Lord himself shall give you a sign: a young woman is with child and will bear a son, and will call him Immanuel' ('God with us' Isaiah 7:14). Before the child will be old enough to know the difference between good and evil, Israel and Syria will be destroyed, and Judah inundated by the enemy.

His word was still ignored but Isaiah decided to take the word to the people. He wrote the prophecy on a large tablet 'Maher-shalal-hash-baz' ('speed – spoil– hasten – plunder') and had it witnessed by Uriah the High Priest. When later, Isaiah's wife bore him a son, he gave him this name. The Lord said, 'Before the boy can say Father or Mother, the wealth of Damascus and the spoils of Samaria shall be carried off and presented to the King of Assyria' (Isaiah 8:4).

Ahaz asked Assyria for help against Syria and Israel (2 Kings 16:8), bribing him with the gold and silver of the Temple. Assyria invaded Syria and Northern Israel in 734 BC destroying Damascus with the usual barbarity. Samaria survived for a few more years but in 734 BC Hoshea, King of Israel, rebelled against Assyria; there was a swift response: Samaria was soon surrounded and in 721 BC after a terrible siege it fell. The Northern Kingdom passed forever from the stage of history. Note that from now on the name of Israel is no longer applied exclusively to the Northern Kingdom, but is used to include all the descendants of Jacob.

18.6 Isaiah and Hezekiah (Isaiah 36 and 37)

Hezekiah came to the throne of Judah in 725 BC at the age of 25 years. At first he appears to have tried to learn the lessons of the past: he suppressed the hill shrines and centred worship in Jerusalem (Isaiah 36:7) but eventually was seduced by the intrigues of Egypt. Shebna, the king's treasurer, was the leader of the pro-Egyptian faction and Isaiah told him that by his policies he was digging his own grave. Isaiah thundered against Egypt, not only because she was not to be relied upon, but because the hope of Israel was to trust in the Lord of Hosts, the Holy One of Israel. For three years he went about naked and barefoot, an acted prophecy, (the only one in his book) of the coming captivity of Egypt. Isaiah was able to influence Hezekiah so that when Ashdad rebelled and was destroyed, Judah escaped. However, this success earned only a respite. The pro-Egyptian party won over the king who entered into secret negotiations with Egypt. The Assyrians, predictably, reacted by invading Judah, and Jerusalem was besieged. Chapters 36 and 37 provide a detailed narrative and should be read carefully. Sennacherib, the Assyrian King, sent his chief officer, Rabshakeh, to try, if possible, to frighten Jerusalem into surrender. Hezekiah sent out envoys to negotiate terms, but Rabshakeh insisted on 'open diplomacy' – the intimidating propaganda, as well as the terms, were yelled out in the Hebrew language so that the people on the walls could hear every word.

Who could the people trust?

> Do not be taken in by Hezekiah. He cannot save you. Do not let him persuade you to rely on the Lord, and tell you that the Lord will save you.
> . . . Where are the gods of Samaria? Did they save Samaria from me? . . . and how is the Lord to save Jerusalem? (Isaiah 36:11–20)

Then Hezekiah tore his clothes and put on mourning but Isaiah was full of confidence:

> This is the word of the Lord: do not be alarmed at what you heard when the lackeys of the King of Assyria blasphemed me. I will put a spirit in him, and he shall hear a rumour and withdraw to his own country; and there I will make him fall by the sword. (Isaiah 37:6–7)

The chief officer withdrew, but the King of Assyria sent messengers with a letter, threatening dire consequences if the city did not capitulate. Hezekiah took the letter to the Temple and 'spread it out before the Lord' (Isaiah 37:14) praying for deliverance. Isaiah was sent to declare God's answer to that prayer. It was that the King of Assyria:

> . . . shall not enter this city
> nor shoot an arrow there,
> he shall not advance against it with shield
> nor cast up a siege-ramp against it.
> By the way on which he came he shall go back;
> . . . .
> I will shield this city to deliver it,
> for my own sake and for the sake of my servant David. (Isaiah 37:33–35)

The Assyrian army never invaded Jerusalem: Isaiah 37:36 suggests that a disaster, like a plague, may have decimated the army, and Sennacherib was forced to return home.

18.7 Isaiah and the Religion of Judah

His book begins with the familiar prophetic judgment on popular religion. There is a condemnation of the formal worship involved in endless sacrifices.

> I am sated with whole-offerings of rams
> and the fat of buffaloes;
> I have no desire for the blood of bulls,
> of sheep and of he-goats.
> Whenever you come to enter my presence –
> who asked you for this? (Isaiah 1:11–13)

They go through the motions with their round of festivals and offer the same repetitive prayers, 'I can put up with them no longer' (Isaiah 1:14). They have rebelled against the Holy One, and 'their land is filled with idols, and they bow down to the work of their own hands' (Isaiah 2:8). They practise necromancy and 'Seek guidance of ghosts and familiar spirits' (Isaiah 8:19). There is no knowledge of God:

> The ox knows its owner
> and the ass its master's stall;
> but Israel, my own people,
> has no knowledge, no discernment. (Isaiah 1:3)

18.8 Social Injustice

Similiar to both Amos and Micah are the denunciations against those who oppress the poor. The wealthy who buy up the land and dispossess the peasants are scornfully accused by God:

> 'Shame on you! you who add house to house
> and join field to field,
> until not an acre remains,
> and you are left to dwell alone in the land. (Isaiah 5:8)

The metaphor Isaiah uses is the court of law. God comes forward to open the indictment against the elders of the people and their officers:

> You have ravaged the vineyard,
> and the spoils of the poor are in your houses.
> Is it nothing to you that you crush my people
> and grind the faces of the poor? (Isaiah 3:14–15)

The money-lenders 'strip my people bare and usurers lord it over them' (Isaiah 3:12). Justice is a commodity to be bought and sold:

> Your very rulers are rebels, confederate with thieves;
> every man of them loves a bribe
> and itches for a gift;
> they do not give the orphan his rights,
> and the widow's cause never comes before them. (Isaiah 1:23)

It was the leaders of the people who most offended. The women are singled out for their flagrant display of their ill-gotten wealth, with their 'anklets, discs, crescents, pendants, bangles, coronets . . . fine dresses, mantles, cloaks, etc' (Isaiah 3:18–23). They walk 'with mincing gait and jingling feet' (Isaiah 3:16); their men are 'mighty topers, valiant mixers of drink' (Isaiah 5:22).

To Isaiah all this stood out in vivid contrast to the moral holiness of God. Rites and ceremonies, however magnificent, however frequent, were no substitute for righteous living.

God's demand was this:

> Put away the evil of your deeds, away out of my sight.
> Cease to do evil and learn to do right,
> pursue justice and champion the oppressed;
> give the orphan his rights, plead the widow's cause. (Isaiah 1:16–17)

God told Isaiah that he was prepared to discuss the situation, to 'argue it out', 'though your sins are scarlet they may become white as snow' (Isaiah 1:18).

The forgiveness the prophet had experienced at his vision in the Temple, could be offered to a people who turned in penitence from their wicked ways.

The Parable of the Vineyard (Isaiah 5:1–7)

The message is clear: the nation had rejected God and brought a judgment upon itself. Yahweh could now only abandon her to her enemies.

18.9 The Remnant

Though Isaiah and his message were rejected he nevertheless earned the title of 'The Prophet of Hope'. He believed that regardless of how many of the nation forsook Yahweh, or how catastrophic the doom which would befall, there would always be a faithful few, a 'Remnant' through whom God's purposes would be fulfilled. At the very beginning of his ministry he called his first son Shear-jashub - 'A remnant shall return'. The nation has brought judgment on itself, but Yahweh is not only justice, but love, and justice will be tempered with mercy. Isaiah saw in the future a redeemed society purged by suffering, ready to do the will of God. 'Then those who are left in Zion, who remain in Jerusalem, every one enrolled in the book of life, shall be called holy' (Isaiah 4:3).

18.10 The Messiah

Such a nation as envisaged by Isaiah would need a leader, so Isaiah's concept of a restored Israel exhibiting God's righteousness and moral holiness was bound up with an ideal leader, a Messiah (Anointed One). A new age was to dawn inaugurated by God, and a descendant of the House of David would reign in peace and righteousness. Isaiah gives us a vivid picture of this Prince of Peace (Isaiah 9:2–7).

In Chapter 11 Isaiah goes on to expand the concept of the Messiah and his Kingdom. He shall be from the stock of Jesse (the father of David) and be endowed with many gifts, wisdom and understanding, administrative skills, and knowledge of God. He will rule with justice, protecting the poor and punishing the wicked. (In stark contrast to the kings of Judah and Israel.) It will be an idyllic age in which:

> The wolf shall live with the sheep,
> and the leopard lie down with the kid;
> the calf and the young lion shall grow up together,
> and a little child shall lead them. (Isaiah 11:6)

It is not surprising that Christian devotion has seen in these prophecies a deeper significance.

18.11 Immanuel (God with us) (Isaiah 7:10–17)

This is not a Messianic prophecy, but rather one which is relevant to a definite historical event. Ahaz, King of Judah, was being threatened by the kings of Israel and Syria who intended to depose him; he decided to defend Jerusalem and to enlist the aid of Assyria. Isaiah advised against involving Assyria, and to trust only in God, prophesying the downfall of both Israel and Syria. He asked the king to seek a sign from God and when he refused, declared that God Himself would give a sign, 'A young woman is with child, and she will bear a son and will call him Immanuel.' Before the child will be old enough to know right from wrong, Israel and Syria will be overthrown. There is no prophecy here of a virgin birth.

19 The Major Prophets, Jeremiah

19.1 Historical Background

Whatever reforms had been carried out by King Hezekiah in the days of Isaiah were undone by his successor Manasseh (696–641 BC). His reign was the longest in Judah's history and by far the worst. He rebuilt the hill shrines, set up an image of the goddess Asherah in the Temple (2 Kings 21:1–7) and 'shed so much innocent blood that he filled Jerusalem full to the brim' (2 Kings 21:16). Worst of all he practised human sacrifice and 'made his son to pass through the fire' (2 Kings 21:6). The prophetic movement was driven underground, preserving there the true faith of the nation and collecting the oracles of the great 8th-century prophets. His son, Amon, named after the Egyptian god, succeeded him, but was assassinated after only two years, and Amon's young son, **Josiah**, came to the throne in 639 BC. When he grew to manhood he was to play a great part in the revival of the worship of Yahweh.

During the reign of Ashurbanipal (668–627 BC) the Assyrian empire went into decline, and this coincided with a revival of the power of Egypt which enabled her to throw off the yoke of Assyria, now menaced by the Medes and Babylonians. An alliance with Egypt was now a practical proposition for Judah, but she was in the invidious position of a buffer state between two great warring nations engaged in the quest of world domination, Egypt and Babylon.

Jeremiah was roughly a contemporary of Josiah, and was brought up at the village of Anathoth, to the north-east of Jerusalem. It was here that Abiathar, the faithful priest of David, had lived when deposed by Solomon. Jeremiah's father was a priest who served the famous shrine at Anathoth, so that he grew up in the atmosphere of religion.

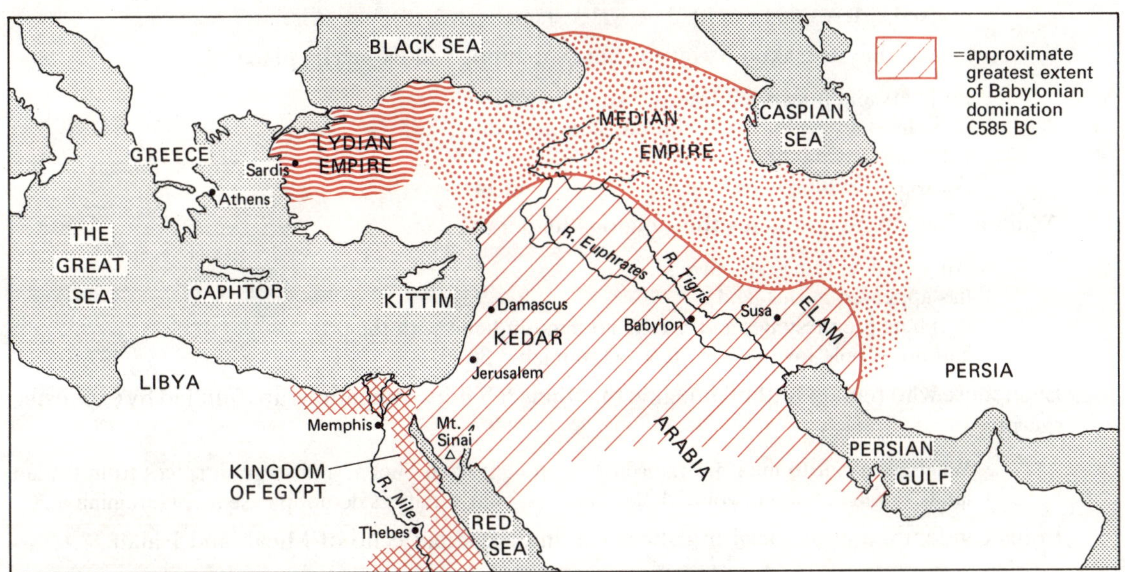

Fig 19.1 The Babylonian Empire

19.2 The Call of Jeremiah (Jeremiah 1:4–10)

He was still only a very young man when God's call came to him:

> Before I formed you in the womb I knew you for my own; before you were born I consecrated you, I appointed you a prophet to the nations. (Jeremiah 1:5)

The response of Jeremiah was quite different from that of Isaiah, who, at his call, felt instantly compelled to offer himself in God's service. Jeremiah drew back, as Moses before him had done at the Burning Bush (Exodus 4:10) – 'Ah! Lord God, I do not know how to speak; I am only a child.' But the Lord said 'Do not call yourself a child; for you shall go to whatever people I send you and say whatever I tell you to say. Fear none of them, for I am with you and will keep you safe' (Jeremiah 1:6–8). Then the Lord touched his mouth (compare the call of Isaiah (Isaiah 6:7) where the prophet's mouth is touched with a live coal from off the altar) and said 'I put my words into your mouth. This day I give you authority over nations and over kingdoms' (Jeremiah

1:10). The vision of the almond tree accompanied his call (Jeremiah 1:11). He saw an almond tree in full bloom, the first tree to flower in spring. Its branches were bare of leaves and looked quite dead but it broke into flowers. In the classical manner, in an object of this world he saw revealed a truth of the world of the spirit: God spoke to him telling him that just as the branches of an almond appear to be dead though new life is hidden within, a spiritual spring would soon take place and God would make His power felt.

A second vision is of a boiling cauldron, 'tilted away from the north' (Jeremiah 1:13). Judgment is to come from the north: Judah is to be invaded. God will state His case against His people for worshipping other gods. Jeremiah is to stand up against the people and to speak without fear:

> This day I make you a fortified city,
> a pillar of iron, a wall of bronze,
> to stand fast against the whole land,
> against the kings and princes of Judah,
> its priests and its people.
> They will make war on you but shall not overcome you,
> for I am with you and will keep you safe. (Jeremiah 1:18–19)

He was promised that his message would be received with hostility, but also that God would be with him and protect him.

19.3 The Religious and Social Evils of the Time

> Two sins have my people committed:
> they have forsaken me,
> a spring of living water,
> and they have hewn out for themselves cisterns,
> cracked cisterns that can hold no water. (Jeremiah 2:13)

Not only have they forsaken Yahweh, but worshipped idols in His place:

> Prophets and priests are frauds, every one of them;
> they dress my people's wound, but skin deep only,
> with their saying "All is well."
> All well? Nothing is well! (Jeremiah 6:13 and 14)

Again and again he castigates the religious leaders:

> An appalling thing, an outrage,
> has appeared in this land:
> prophets prophesy lies and priests go hand in hand with them,
> and my people love to have it so. (Jeremiah 5:30–31)

Even those who remain faithful imagine that their religious obligations are fulfilled by expensive sacrifices:

> What good is it to me if frankincense is brought from Sheba, and fragrant spices from distant lands? I will not accept your whole-offerings, your sacrifices do not please me. (Jeremiah 6:20)

In his condemnation of social injustices we find a distinct echo of Micah and Isaiah:

> Among my people there are wicked men,
> who lay snares like a fowler's net
> and set deadly traps to catch men.
> Their houses are full of fraud,
> as a cage is full of birds.
> They grow rich and grand,
> bloated and rancorous;
> their thoughts are all of evil
> and they refuse to do justice,
> the claims of the orphan they do not put right,
> nor do they grant justice to the poor. (Jeremiah 5:26–28)

Immorality is rife. 'Each neighs after another man's wife, like a well-fed and lusty stallion (Jeremiah 5:8). If you search the streets and squares of Jerusalem says the prophet, 'can you find any man who acts justly, who seeks the truth?' (Jeremiah 5:1). So utterly were they corrupt that the possibility seemed remote: 'Can the Nubian change his skin or the leopard his spots?' (Jeremiah 13:23).

19.4 The Reforms of Josiah 621 BC

After this first outburst of prophecy, Jeremiah was not conspicuous until 621 BC when the Book of the Law was discovered in the Temple during the major repairs ordered by King Josiah (2 Kings 22). When the book was read to the king 'he rent his clothes' in horror when he realized how far Judah had fallen from the ways of God; he determined on a reform of the nation. Gathering together all the people, high and low, before the Temple, he read to them the words of the book and in their presence made a covenant before God to fulfil all the commandments of the Law.

The main provision of the Book of the Law was that the Temple should be the only shrine where sacrificial worship could be offered. **The corrupt worship of all other shrines should be abolished**. Josiah went about his task with great energy: the pagan altars in Jerusalem and all the country shrines were thrown down and all objects associated with pagan worship were burned. The altar in the valley of Hinnom, where human sacrifice had been offered to Moloch, the Assyrian god, was desecrated; and he suppressed the hill shrines including even that at Bethel, which was in the territory of Assyria. Finally Josiah ordered that a Passover be kept as in the days of old (2 Kings 23:2–23).

19.5 The Book of the Law

Usually the book is identified with an original draft of the Biblical Deuteronomy, and it is generally held to have reflected the outlook of the 8th-century prophets, or rather that it was a compromise between the ideals of Hosea and Isaiah with their strong moral aims, and the priestly establishment with its aim of safeguarding the continuance of sacrificial worship. Its place in the religious history of Israel is therefore of great importance for it marks an effort to translate the moral ideals of the great prophets into practical life. With the demands of the ritual, and behind it, lie the considerations of the care of the weak and helpless and the protection of those unable to defend themselves – foreigners, women and slaves. All this makes the Law of Israel unique. It is believed that the original Book of the Law found in the Temple must have been the work of the prophetic schools, driven underground by the virulent persecution of Manasseh, and preserving the traditions until a better day should dawn.

Another important result was that there was now, for the first time in this age, a book which was immediately and generally accepted as having the authority of God. Up to this time the will of God had been made known through priests and prophets but this was now less necessary. Jeremiah himself witnesses to the reverence paid to a code:

> How can you say, "We are wise,
> we have the law of the Lord",
> when scribes with their lying pens
> have falsified it? (Jeremiah 8:8)

Men were saying that now they had the will of God before them in black and white they had no more need of prophets or priest to interpret it. There was much truth in this for prophet and priest were not to continue for ever, and it was as well for Israel that **for the greater part of her history she had the solid spiritual support of a book**.

19.6 Jeremiah and the Reforms of Josiah

Jeremiah could not be anything but glad about the reforms of the king. The covenant of Josiah was a return to the covenant of Moses, and Jeremiah called on the people to obey it:

> Then the Lord said: "Proclaim all these terms in the cities of Judah, and in the streets of Jerusalem. Say, Listen to the terms of this covenant and carry them out". (Jeremiah 11:6)

His support of Josiah must have made him unpopular with his family at Anathoth for their shrine was abolished in the reforming measures. Eventually Jeremiah became disillusioned with the reforms for three reasons:

1 Changes in the form of worship did not lead to a change of heart. The sacrificial system was still the main instrument of worship and this induced the people to continue with the notion that the duty of God could be fulfilled by burnt offerings. No prophet could accept that. There was no real penitence and the people were yet confident that they were right with God.

2 He ultimately saw the truth that however perfect it might be, it was imposed from without, and could therefore only be fragile. He saw that to be effective it must be written on men's hearts, a conviction which found expression in the prophecy of the New Covenant (Jeremiah 31:31–34).

3 Isaiah had proclaimed that because the Temple, the Sanctuary of God, was within Jerusalem, the city could never fall to an invader, and the reforms of Josiah which set up the Temple as the central shrine of the nation by adding to its importance confirmed the belief of the people that it could never fall. Jeremiah was given the disquieting task of correcting this misapprehension.

19.7 The Suffering Prophet: The Confessions of Jeremiah

Apart from the references to Josiah's reforms we have no words of Jeremiah from 626 BC, when he received his call, to 607 BC. What he did during this period is not known. What we do know, from the nature and disposition of the prophet, is the agony of mind and spirit he endured. He was called upon by God to isolate himself from normal human relationships: he was not to marry, he was not to join in the sorrows of a funeral or the joys of any feast, as a sign that all sounds of joy and gladness were to be denied the nation. He felt himself to be 'a man doomed to strife, with the whole world against me' (Jeremiah 15:10). His own kinfolk at Anathoth threatened to kill him if he did not cease from prophesying. 'I had been', he said, 'like a sheep led obedient to the slaughter' (Jeremiah 11:19); a simile which may well have inspired the second Isaiah's picture of the Suffering Servant of God.

As though this were not enough to bear, there was an agony of doubt: what if he were an 'enticed' prophet; what if God were deceiving him? Even at this time it was believed that God could 'put a lying spirit' in the mouth of a prophet (1 Kings 22:19–23), and there was only one infallible way in which the true prophet was to be distinguished from the false – the true prophet's word came to pass. For years Jeremiah had prophesied a doom that had not come. In 612 BC Nineveh fell to the Medes and Babylonians so there was now no danger from Assyria; Jerusalem seemed secure, and this the people took as evidence that his prophecy was false. Eventually he came to believe this himself:

> O Lord, thou hast duped me, and I have been thy dupe:
> thou hast outwitted me and hast prevailed.
> I have been made a laughing stock all the day long,
> everyone mocks me. (Jeremiah 20:7)

Yet in spite of the loneliness of his struggle, the jeers and plotting of his enemies, and worst, his own self-doubt, the word of God was a burning fire within him and he could not deny its expression:

> Whenever I said "I will call him to mind no more,
> nor speak in his name again",
> then His word was imprisoned in my body,
> like a fire blazing in my heart,
> and I was weary with holding it under,
> and could endure no more. (Jeremiah 20:9)

The full extent of the heroism of the prophet can only be appreciated when we take account of the fact that at this stage in Israel's religious development belief in eternal life had not yet evolved. For Jeremiah there could be no compensation for his suffering in the rewards of heaven. What saved him from a predicament which was more than human flesh alone could bear, was the assurance of God's presence. Isolated as he was from all human support, he poured out his trouble to God. In his despair he found that he was not alone for God was with him:

> I will make you impregnable, a wall of bronze.
> They will attack you but they will not prevail,
> for I am with you to deliver you
> and save you, says the Lord. (Jeremiah 15:20)

He had found a personal relationship, a personal faith, which nothing could shake.

19.8 Personal Religion

Out of his deep sense of isolation came, though unconsciously, the greatest contribution Jeremiah made to the knowledge of God. Religion is concerned with the relations between God and man, but in the older view the human unit was not the individual but the community. In primitive societies it is the norm that the individual is like a cell within the body, so that in isolation from the body he has no life, no real existence. In the history of Israel we see that the individual stands in relation to Yahweh through the tribe or nation; the unit of religion is the group, not the individual, and the punishment for the sin of the individual has to be borne by the whole community. Thus in Joshua 7 defeat at the hands of the men of Ai is attributed to the fact that an individual, Achan, has stolen some of the booty of Jericho which had been dedicated to Yahweh. In expiation, Achan is stoned to death, so that the nation is clean again. **Up to the time of Jeremiah the general rule is that a man does not stand alone in his relations with God**.

There is great value in this truth, but there is another side which must not be obscured, namely that a man in a very real sense also stands alone before God, and it was Jeremiah through his own bitter experience of isolation from his people, who appears first to have realized this. Isolated as he was from everyone in the community he found that he was still the prophet of the Lord, commissioned to reveal the will of God, still upheld by the power of God and safe under His protection. From this tremendous experience came the knowledge that such a personal relationship as his could be open to every man. The covenant of Moses had been concerned with the relationship between God and the nation, but there would be a New Covenant, greater than the old.

19.9 The New Covenant (Jeremiah 31:31–34)

The Old Covenant was written upon tablets of stone, but the **New Covenant would be written in the hearts of men**. It would cement a relationship not between God and the nation as of old, but between God and the individual, a personal relationship founded in personal faith:

1 The covenant will be the act of God. It is not a reward for a deserving people, but rather the contrary; it is an act of God's love.

2 Through it God will offer forgiveness of sin.

3 The covenant is personal and involves an inner relationship between God and the individual, but it is made 'with Israel and Judah'. It envisages a new community of men who have an inner allegiance to God. In spite of his rejection by them Jeremiah never lost sight of belonging to his people, 'I will set my law within them and write it on their hearts; I will become their God and they shall become my people' (Jeremiah 30:33).

19.10 Personal Responsibility

A consequence of the sense of a personal relationship with God was inevitably a sense of personal responsibility to Him. The Old Covenant stressed the responsibility of the nation but **Jeremiah taught the responsibility of the individual to God**. He recalled a popular saying, 'The fathers have eaten sour grapes and the children's teeth are set on edge' (Jeremiah 31:29), but said that this will no longer be said, 'for a man shall die for his own wrongdoing: the man who eats sour grapes shall have his own teeth set on edge' (Jeremiah 31:30).

When disaster came the people were able to say that, in accord with the second of the Ten Commandments, they were suffering the consequences of the sins of their fathers, while they themselves were innocent. In this way they ministered to their own self-righteousness, refusing to accept responsibility for their own sins. Jeremiah saw that this was unacceptable to God. The second commandment is misleading: it can be taken to imply that God has a running vendetta with certain families, but, though by the very nature of things children do sometimes suffer for the faults of their parents, it is not God's will that this should be so. While it is true that social and psychological factors as well as our heredity affect the degree of our responsibility for what we do, there is still a large area in which we have choice, and must be prepared to accept responsibility. According to Jeremiah each individual must stand face to face with God, by and for himself.

All this stems from Jeremiah's prophecy of the New Covenant, which for Christians was fulfilled by Jesus at the Last Supper.

19.11 The Reign of Jehoiakim (607–597 BC)

Nineveh had fallen in 612 BC, and in 607 BC Egypt, alarmed at the growing power of Babylon marched northward to reinforce the remnants of the Assyrian armies. Josiah went to meet Pharaoh, Necho, at Megiddo, whether as friend or foe is not stated, but Josiah was killed, a tragic end to a promising reign. Eventually, Jehoiakim, the eldest son of Josiah, came to the throne as a puppet of Egypt. About this time Jeremiah took up residence in Jerusalem and was joined by Baruch who was to record his words, and to whom we owe the greater part of the book of Jeremiah. Jeremiah was not slow to make comparisons between the new king and his father, Josiah, who had:

> . . . dispensed justice to the lowly and poor;
> did this not show he knew me? says the Lord.
> But you have no eyes, no thought for anything but gain,
> set only on the innocent blood you can shed,
> on cruel acts of tyranny. (Jeremiah 22:16–17)

The conduct of the people had not improved and he prophesies against them in the famous parable of the Potter (Jeremiah 18:1–12).

He watches the potter at work with his wheel and sees that now and then the clay is faulty and does not conform to the guidance of his fingers; the potter cannot rectify the fault and has to re-knead the clay and begin again; he fashions it into another vessel to his liking.

> Can I not deal with you, Israel, says the Lord, as the potter deals with his clay? You are clay in my hands like the clay in his. (Jeremiah 18:6)

Total disaster awaited the nation, and God would remake it only after much suffering had been endured.

A similiar message is proclaimed in the superb acted parable of the Broken Jar (Jeremiah 19:1–13). Jeremiah is told to take an earthenware pot to the valley of Hinnom (where human sacrifice had until recently been made to Baal) and there before some of the elders of the people and the priests, to prophesy a terrible disaster upon the city. God told him that:

> Then you must shatter the jar before the eyes of the men who have come with you and say to them, These are the words of the Lord of Hosts: Thus will I shatter this people and this city as one shatters an earthenware vessel so that it cannot be mended. (Jeremiah 19:10–11)

Here is an excellent example of a 'word' of God expressed in an action.

19.12 Jeremiah and the Temple

From the valley of Hinnom Jeremiah went to the Temple, stood in the open court, and prophesied disaster to it. The son of the chief priest had him flogged and put into stocks. On his release he was soon preaching in the Temple again. He prophesied that the Temple would become 'like Shiloh' (Jeremiah 26:6), the original sanctuary of the Ark of the Covenant which had been destroyed by the Philistines centuries before and was still a ruin. Thus, to his treasonable attacks on the king was added blasphemy, for ever since the days of Isaiah, the Temple, the dwelling place of Yawheh, was regarded as indestructable, being as it was under the protection of God. The Temple would not save them:

> You steal, you murder, you commit adultery and perjury, you burn sacrifices to Baal, you run after other gods whom you have not known; then you come and stand before me in this house, which bears my name, and say "We are safe". (Jeremiah 7:9–10)

He condemned the people's reliance on the efficacy of sacrifice, saying that in the wilderness Yahweh had made no demand for sacrifice. 'What I did command them was this: if you obey me, I will be your God and you shall be my people' (Jeremiah 7:23).

The last straw was when he attacked the religious leaders: he was seized and threatened with death. He was saved only because his judges remembered that when Micah the prophet had declared that Jerusalem and the Temple would be destroyed, King Hezekiah, who could have killed him, showed reverence for a prophet of the Lord. Banished from the Temple he bade Baruch, his scribe, to write down, at his dictation, all his prophecies, from the reign of Josiah to date, and to read them to the people in the Temple on a fast day. When they heard the prophecy repeated that the Babylonians would overwhelm Judah, the officers of the people, fearing for the life of Jeremiah advised Baruch to take the prophet into hiding before they reported everything to the king. When the scroll was read to the king he took a penknife and cut it into pieces, column by column, and burned it (Jeremiah 36:1–26). Jeremiah then had his words written on another scroll, with, in addition, a prophecy against the king.

19.13 The First Capture of Jerusalem 597 BC (Jeremiah 24:1)

Egypt and Assyria were utterly defeated by Babylon at Carchemish in 605 BC: Babylon now ruled the world. Jeremiah saw Nebuchadnezzar, the Babylonian commander, as the servant of God to punish Judah for her transgressions. In 603 BC Nebuchadnezzar demanded and received the tribute of Jehoiakim. The king, however, intrigued with Egypt and rebelled, and predictably the Babylonians invaded. The help offered by Egypt never came and after a siege of only a few months Jerusalem fell. In the meantime Jehoiakim died and his son, Jeconiah (or **Jehoiaachin**), succeeded him. The city was stripped of its treasures, and the new king, his court, his best soldiers and craftsmen were led captive to Babylon. The king's uncle, Zedekiah, was made king in his stead. This marked the first stage of the sixty years of Jewish exile in Babylon, and also the Diaspora, the Dispersion of the Jews throughout the world.

Jeremiah discouraged any idea that the exile would be short or that rebellion had any chance of success. God had destined Babylon to be supreme among the nations.

19.14 The Vision of the Basket of Figs (Jeremiah 24:1–10)

Jeremiah saw two baskets of figs, one good and one bad, set out in front of the sanctuary. The good figs represented the exiles of Judah to whom God would show favour:

> I will look upon them meaning to do them good, and I will restore them to their land . . . they shall become my people and I will become their God, for they will come back to me with all their heart. (Jeremiah 24:6 and 7)

He wrote them a letter telling them to 'Build houses and live in them; plant gardens and eat their produce. Marry wives and beget sons and daughters' (Jeremiah 29:5–6), so that they might 'increase there and not dwindle away'. He saw the future prospect of Israel in their hands, for after seventy years they would return and be gathered 'from all the nations and all the places to which I have banished you, says the Lord' (Jeremiah 29:14).

On the other hand the basket of bad figs represents Zedekiah and the people of Jerusalem who because of their deeds are destined to destruction (Jeremiah 29:15–23).

19.15 The Siege and Second Capture of Jerusalem 586 BC (Jeremiah 39:1–10; 2 Kings 25:1–7)

In 589 BC Zedekiah, with the encouragement of Egypt, refused to pay the tribute to Babylon and the country was soon invaded. Jerusalem was surrounded, but an Egyptian relief column compelled the Babylonians to retreat. The people were full of joy, but Jeremiah still prophesied destruction at the hand of Babylon. Infuriated by this, his enemies had him arrested for treason, accusing him of trying to desert. In prison he refused to change his message. His enemies then hit upon a plan to get rid of him: they could not kill him but they could ensure that he would not continue to live by lowering him into a deep dungeon (probably a disused water cistern), full of mud and filth. He would certainly have died but for the compassion of Ebed-melech, who pleaded for the prophet's life and was allowed to raise him up with ropes giving him rags to place under his armpits to prevent chafing (Jeremiah 38:9–13).

He was now confined in the guard house and remained there until the city fell. While there he bought a field at Anathoth from his cousin and had the deed of purchase properly drawn up and witnessed, as a sign from God that the coming disaster would not be final and that there was a future for Israel in the land. The Babylonians returned in force and the city fell. The savagery of their reprisals is recorded in 2 Kings 25:1–9. Jerusalem was utterly destroyed.

Jeremiah was well treated by the invaders, and was able to save the life of Ebed-melech. He chose to remain in the land with the peasants who had not been transported. After about three years, Gedaliah, a Jew who had been given charge over the land, was assassinated and, fearful of further Babylonian reprisals the people fled to Egypt, taking the protesting Jeremiah with them. He had advised them to remain in Judah, but to no avail.

19.16 Jeremiah's Contribution to Religious Thought

Much of his teaching is based on the foundations laid by earlier prophets:

1 God is righteous, loving and Holy;

2 wrong-doing will inevitably bring punishment. Judah therefore stands under the judgment of God.

His unique contribution to the development of religious ideas is the conviction that each individual is capable of entering a personal relationship with God, and what follows, that each individual is personally responsible to God.

The life-blood of religion is the inner experience of God. The New Covenant with God is not to be written on tablets of stone, but in the hearts of men.

20 The Major Prophets, Ezekiel

20.1 Historical Background

After the first capture of Jerusalem, in 597 BC, the Babylonians behaved in a relatively mild manner. They left a puppet king, but deported the most prominent and useful people to Babylon; among these was a young priest, Ezekiel. They expected, in spite of the warning of Jeremiah that their stay would be a short one, but when Jerusalem finally fell in 586 BC they were still in captivity. This meant to many of them that the gods of Babylon were more powerful than Yahweh, for how else could the Temple have been destroyed? It was the prophets of the exile who convinced them that God had allowed the destruction of the city as a punishment for the sins of the nation.

From the letter of Jeremiah to the exiles (Jeremiah 29:1–23), it is clear that they were allowed considerable freedom. He exhorts them to build houses, plant gardens, marry and beget children. In Ezekiel 8:1 the prophet meets the elders in his own house. The Jews lived in communities, which increased their sense of separateness from their conquerors, and made it possible for them to practise their religion with the traditional customs, as far as the new situation allowed.

The Temple had been destroyed and was in any case inaccessible, and in consequence the observance of the sabbath took its place as the focus of worship. The **main elements of synagogue worship were laid down at this time**. The rite of circumcision was strongly emphasized as a distinctive mark of a Jew for they now lived among a people who did not practise it and by this means they stressed their separateness. In religious matters they were very conscious of their superiority to their captors and another way of emphasizing this was strict observance of the food laws. The **experience of the exile led to a heightening of the consciousness of exclusiveness:** they were the people of God and therefore different, and indeed superior, to others. For a captive nation they appear to have been treated very well by the standards of the time. From the record in Ezra 2:64–69 considerable wealth had been built up by many of the people by the time they were free to return. From Ezra too we learn that they settled not only according to families, but also according to the districts they had occupied in Judah. The system of selecting elders as leaders of each community, which had lapsed during the monarchy was now restored. Their captivity by the waters of Babylon bore little resemblance to that of their ancestors by the river of Egypt.

20.2 The Ministry of Ezekiel

'In the fifth year of the exile', the call of God came to Ezekiel and it was expressed in terms of a fantastic vision in the fullest meaning of the word. Much of it is so obscure and even grotesque in detail, that it is impossible accurately to draw a mental picture of it. Out of a vast cloud come flashes of fire and brilliant light and there emerge four living creatures in human form, but each has four faces, four wings and hooves rather than feet. The faces are those of a man, a lion, an ox and an eagle; around them is a radiant fire from which come flashes of lightning. Beside each creature is a wheel, the hubs of which have a projection which has the power of sight and the rims of the wheels are full of eyes. The wheels move with the creatures.

All this leads up to a vision of God. The whirring of the wings is like the noise of a great torrent, but then the roaring ceases and there is seen above the firmament a sapphire throne with an encircling radiance, like a rainbow – 'it was like the appearance of the glory of God' (Ezekiel 1:28). Little wonder that the prophet fell prostrate on his face. (Note that there are details in the vision reminiscent of the call of Isaiah – the winged creatures, the fire, and God seated high on His throne.)

A spirit came, raised him to his feet and said, 'Man, I am sending you to the Israelites, a nation of rebels who have rebelled against me' (Ezekiel 2:3). The vision continues, and Ezekiel sees a hand stretched out holding a scroll in which is written words of woe, and the spirit says 'Man, eat what is in front of you, eat this scroll; then go and speak to the Israelites' (Ezekiel 3:1). The prophet then consumes the scroll – 'I ate it, and it tasted as sweet as honey' (Ezekiel 3:3). The vision over, he is transported by a spirit to the exiles at Tel-abib who were settled by the river Kebar.

The prophet was in a trance for seven days and then the word of the Lord came again. 'Man, I have made you a watchman for the Israelites: you will take messages from me and carry my warnings to them'. He was to speak for God even though the people would neither hear nor obey (Ezekiel 3:17). A solemn obligation is then laid upon him: if he warns a wicked man who will not give heed, that man shall die, but if the prophet does not warn the wicked one, he himself will be answerable for the death of the wicked.

Ezekiel has more in common with the early 'ecstatic' prophets than any of the classical prophets. It seems that when possessed by the spirit, he lost consciousness, his body went rigid, and he was unable to speak (Ezekiel 3:25 and 26).

The call of Ezekiel came six years before the final destruction of Jerusalem. His message was no more welcome than that of Jeremiah, if we remember that Jerusalem still stood and was little changed from the time when the first exiles left it; they expected to return to their indestructible city. The acted parables by which he conveyed God's messages were often bizarre in the extreme and even nauseatingly crude; they were probably not quite what was expected of a prophet of God.

20.3 The Doom of Jerusalem

He employed several vivid acted parables to emphasize his message:

1 He took a tile and drew on it a picture of Jerusalem; then he added all the paraphernalia of a siege, and placed an iron griddle between himself and the city. The message was that Jerusalem would be besieged, cut off from all help and destroyed (Ezekiel 4:1–3).

2 He lay on his left side for 190 days and on his right side for 40 days, each day to represent a year. The 190 days were to represent the years of the captivity of Israel and 40 days the captivity of Judah (Ezekiel 4:4–8).

3 He cut off his hair and beard using a sword as a razor, and divided it into three equal parts. One part he publicly burned, the second he cut up with the sword and scattered all around the city, and the third he threw to the winds. A few of these last he held back. The first two parts symbolized the destruction of Jerusalem and the slaughter of many of its people, the last part symbolized the scattering of the Jews, and the last few hairs that were reserved symbolized the remnant of God's people who would be faithful and form the nucleus of the future nation (Ezekiel 5:1–4).

20.4 Personal Responsibility

Jeremiah was the first to learn, from his suffering, God's desire for personal relationship with each individual, and Ezekiel developed this revelation on similar lines.

Many of the exiles had lost faith in a God who they believed had been defeated by the gods of Babylon. Others felt that they had been punished for the sins of their ancestors, and that this was unfair. Ezekiel said to them, 'You say that the Lord acts without principle? Listen you Israelites, it is you who act without principle' (Ezekiel 18:25).

He referred to the same popular saying quoted by Jeremiah, 'The fathers have eaten sour grapes, and the children's teeth are set on edge.' (Ezekiel 18:2) and told them that God forbids the use of this proverb ever again. Each individual bears responsibility for his own sins alone and 'the soul that sins shall die' (Ezekiel 18:4). God has no pleasure in punishing evil, so repent: 'Throw off the load of your past misdeeds: get yourselves a new heart and a new spirit' (Ezekiel 18:31). The call to personal repentance however did nothing to give them hope for the future. Ezekiel assured them that there is always hope for those who turn to God.

Ezekiel's teaching about personal responsibility was a big advance on the second of the Ten Commandments, which laid the sins of the fathers on the children, but there was a great deal yet to be learned. Ezekiel's God is transcendent in holiness, of infinite distance in this respect from sinful man, and He is a judging and punishing God. It is true that He takes no pleasure in punishing and that His desire is not for revenge but for repentance which will make possible forgiveness, but the second Isaiah, Isaiah of Babylon, reveals a tenderness in the love of God, a deep compassion for His people which is lacking in the God of Ezekiel.

Ezekiel's teaching about suffering is also seriously deficient. The traditional Hebrew belief was that suffering is the consequence of sin and Ezekiel upholds this view, one which is not adequate to account for the suffering of the innocent. However, his message was one which answered the need of the people at this time.

20.5 The Fall of Jerusalem 586 BC

Before the Fall, Ezekiel's beloved wife died. He buried her in dignified silence and without any outward show of mourning, no weeping and none of the normal funeral customs. It was another acted parable to say that Jerusalem would soon fall and that many of her inhabitants would die by the sword. They must bear their grief in silent restraint as he had done (Ezekiel 24:15–17). The fall of the city fulfilled the prophet's words and thus raised his authority among the exiles.

20.6 The Prophet of Hope

Before the destruction of Jerusalem, Ezekiel is the prophet of doom like Jeremiah before him; after it he is the prophet of hope. He saw in this tragic event clear evidence of the supreme power of God who had not been defeated and had not abandoned His people nor his land; God's action would be vindicated when a purified people would return: 'When they see that I reveal my holiness through you, the nations will know that I am the Lord, says the Lord God' (Ezekiel 36:23).

20.7 God, the Shepherd of Israel (Ezekiel 34)

Ezekiel laid the principal blame for the disaster upon the leaders of the nations. The shepherds 'have cared only for themselves and not for the sheep' (Ezekiel 34:8) and as a result the sheep have scattered and become the prey of wild beasts. The weak have not been cared for, and even the strong have been driven ruthlessly. God will dismiss the shepherds, and rescue the sheep: the shepherds shall 'feed on them no more.' God Himself will be their shepherd, and will search them out no matter where they are scattered, 'I will bring them out from every nation, gather them in from other lands, and lead them home to their own soil . . . I myself will tend my flock, I myself pen them in their fold, says the Lord God' (Ezekiel 34:13–16). As for the flock, God will judge between one sheep and another – 'Then I will set over them one shepherd to take care of them, my servant David: he shall care for them and become their shepherd. I, the Lord, will become their God' (Ezekiel 35:23 and 24). Ezekiel therefore had a clear conception of a Messiah, the 'Anointed One' of God from the family of David. Under his rule the people will dwell in safety in a fruitful land.

20.8 The Valley of Dry Bones (Ezekiel 37:1–14)

The people felt that as a nation they were finished and Ezekiel brought them God's promise of restoration, indeed national resurrection, in a dramatic vision. He saw a valley full of dry bones and God called on him to prophesy over them that the dry bones might live. The bones fitted themselves together and were overlaid with sinews and flesh, but there was no breath in them. Then God called him to prophesy to the wind, 'Come O wind, come from every quarter and breathe into these slain that they may come to life' and they stood up, a great army. This nation which imagined itself dead would be raised to new life and God would put his spirit into them.

20.9 The Parable of the Two Tablets (two sticks) (Ezekiel 37:15–23)

This again bore a tremendous message of hope. The prophet took one leaf of a wooden tablet and wrote on it 'Judah and his associates of Israel', and then a second on which he wrote 'Joseph, the leaf of Ephraim and all his associates of Israel.' Then he brought the two together to form one tablet. The message was that Israel and Judah, the northern and southern kingdoms, would be united again as a single nation: 'My servant David shall become king over them, and they shall have one shepherd. They shall conform to my laws, they shall observe and carry out my statutes' (Ezekiel 37:24).

20.10 A Holy Nation

Before it could be restored, the nation had to be cleansed of its sin for only then could it approach a holy God. The initiative is to come from God Himself: 'It is not for your sake, you Israelites, that I am acting, but for the sake of my holy name, which you have profaned among the peoples' (Ezekiel 36:22). God's holiness must be vindicated: it cannot be thwarted by a sinful people, so God set about bringing them to repentance. 'I will sprinkle clean water over you, and you shall be cleansed from all that defiles you; I will cleanse you from the taint of all your idols. I will give you a new heart and put a new spirit within you' (Ezekiel 36:25 and 26). Repentance was to follow, not precede, this action of God.

20.11 The Temple

Ezekiel, in the last chapters, gives us a picture of the worship of a restored Temple in Jerusalem. He describes in meticulous detail the architectural features and dimensions of the altar and its surroundings and the regulations for the sacrifices, the various rituals, the vestments of the priests. This became the pattern down to New Testament times.

The final vision is of the River of Life (Ezekiel 47). Led by his supernatural guide, the prophet stands at the gate of the Temple and sees a trickle of water flowing past the altar eastward growing deeper and stronger as it flows, finally falling into the Dead Sea. It brings abundant life to the desert and fish are able to live in the Dead Sea. The vision suggests the blessings which will flow from the restoration of Israel.

21 The Second or Deutero-Isaiah

One of the assured results of Biblical research is that the Book of Isaiah is not the work of a single author. For practical purposes the book can roughly be divided into three sections, Chapters 1–39 being the work of Isaiah of Jerusalem, Chapters 40–55 that of Second or Deutero-Isaiah, and Chapters 56–66 that of the Third or Trito-Isaiah. The reasons for this classification are based on historical background, differences in style and language, and differences in the logical content. In Deutero-Isaiah we meet an unknown author who is perhaps the greatest of all the prophets: certainly it can be said that in his teaching on the nature of God we reach the highest point in Hebrew religious belief.

21.1 Historical Background

Cyrus, the ruler of a petty principality in Media, to the east of Babylon, came to the throne in about 559 BC and by 546 BC had overthrown his overlord, taken control of Media and defeated Croesus the King of Lydia. This now powerful king of the Medes and Persians invaded Babylon and in 539 BC by the defeat of its army became ruler of the greatest empire ever hitherto known, stretching from the Indus to the Nile. It was to last for more than 200 years.

Cyrus was a most enlightened ruler: he believed in conciliating the subject nations rather than oppressing them, giving them local rulers of their own race, and offering the many exiles freedom to return to their home-lands.

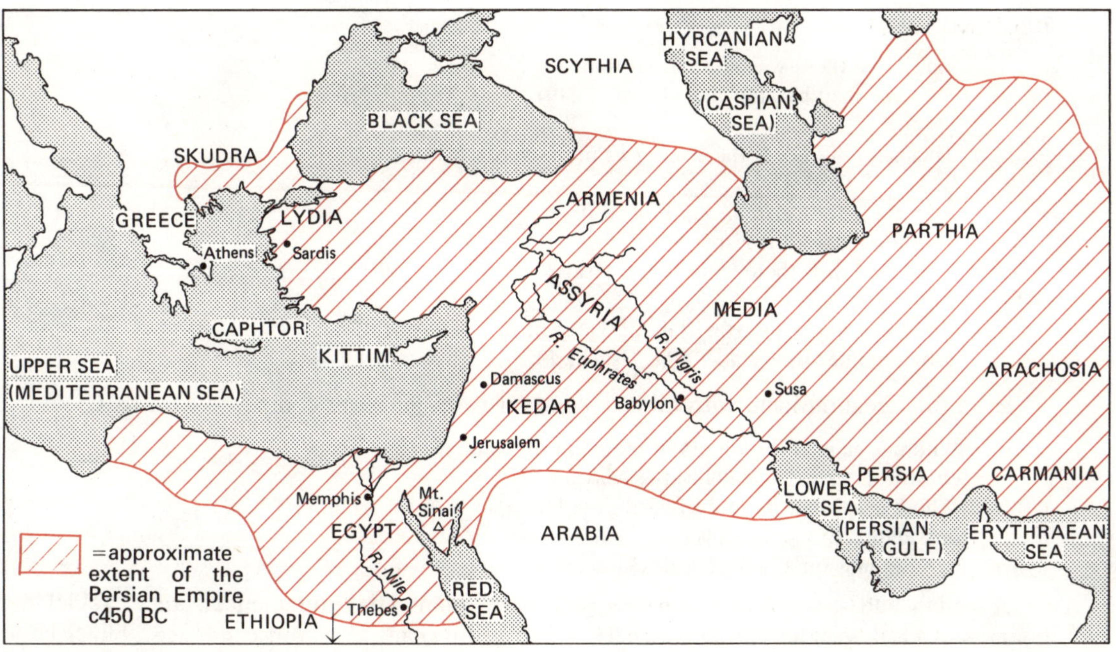

Fig. 21.1 The Persian Empire

From internal evidence of the text it can be deduced that the writings of Deutero-Isaiah belong to the period between 546 BC and 538 BC, the year of the first return of Jewish exiles to their home-land. In view of the fact that his prophecies contained condemnation of Babylonian religious practices and predictions about the fall of the city, it is not surprising that the writer kept his identity secret. Chapters 40–48 suggest that the fall of Babylon is imminent and 49–55 are consistent with the conquest of the city.

21.2 Cyrus, the Lord's Anointed

The prophet sees the victory of Cyrus over Babylon as the work of God, and Cyrus as his chosen instrument in the deliverance of Israel:

> I say to Cyrus "You shall be my shepherd
> to carry out all my purpose,
> so that Jerusalem may be rebuilt
> and the foundations of the temple may be laid". (Isaiah 44:28)

Cyrus is the anointed of God 'whom he has taken by the hand to subdue nations before him' (Isaiah 45:1). Cyrus does not know the Lord but he has been sent by him to redeem Israel:

> For the sake of Jacob my servant and Israel my chosen
> I have called you by name
> and given you your title, though you have not known me. (Isaiah 45:4)

The prophet begins his work with an incomparable poetic expression of the joy of those who have been delivered:

> Comfort, comfort my people;
> – it is the voice of your God;
> speak tenderly to Jerusalem and tell her this,
> that she has fulfilled her term of bondage
> that her penalty is paid. (Isaiah 40:1 and 2)

A voice is heard crying 'Prepare a road for the Lord through the wilderness.' and we have a superb picture of God shepherding and bringing them in triumph to their own land. From the tops of the mountains the good news is to be proclaimed, 'Your God is here':

> Here is the Lord coming in might
> Coming to rule with His right arm . . .
> He will tend his flock like a shepherd
> and gather them together with His arm. (Isaiah 40:10–11)

21.3 Second Isaiah's Concept of God, Monotheism

The belief that there is only one God had been implied in the teaching of earlier prophets, but never before had it been stressed and insisted on as in Second-Isaiah. For him God is the eternal, only God:

> before me there was no God fashioned
> nor ever shall be after me' (Isaiah 43:10)
> ....
> Thus says the Lord, the creator of the heavens,
> he who is God,
> who made the earth and fashioned it
> and himself fixed it fast . . .
> I am the Lord, there is no other' (Isaiah 45:18)
> ....
> The Lord, the everlasting God, creator of the wide world,
> grows neither weary nor faint. (Isaiah 40:28)

Even the Creation itself with all its splendour is insignificant compared with its Creator:

> Who has gauged the waters in the palm of his hand
> or with its span set limits to the heavens?
> Who has held all the soil of earth in a bushel,
> or weighed the mountains on a balance
> and the hills on a pair of scales? (Isaiah 40:12)

Jeremiah had stressed the nearness of God to man, His 'immanence', and Ezekiel the holiness of God and the gulf between Him and sinful man, His 'transcendence', but Second Isaiah's conception of God is unique in that he combined these truths and showed them to be complementary. He sees on the one hand the awful majesty and holiness of God and on the other His tender care for His people:

> Can a woman forget the infant at her breast,
> or a loving mother the child of her womb?
> Even these forget, yet I will not forget you. (Isaiah 49:15)

This is Second Isaiah's unique contribution to the knowledge of God.

21.4 Universalism

Since Yahweh is the one true God, the prophet could not restrict Him to the role of a national God. In the name of God he invites the nations of the world to draw near:

> You fools, who carry your wooden idols in procession
> and pray to a god that cannot save you . . .
> Look to me and be saved,
> you peoples from all corners of the earth;
> for I am God, there is no other. (Isaiah 45:20 and 22)

It is because of this insight that the prophet sees Cyrus as the Lord's anointed, appointed to do God's will.

Since Israel is to be God's instrument for the conversion of the gentiles it is natural that she should have pre-eminence. Some passages describe the nations as becoming subservient to Israel:

> Kings shall be your foster-fathers
> and their princesses shall be your nurses.
> They shall bow to the earth before you
> and lick the dust from your feet. (Isaiah 49:23)

The final concept is however very different: the nations will not be subservient, but will accept the leadership of Israel only because God is to be found in her. God cares for all nations and Israel has pre-eminence only because she has been chosen as the instrument for the conversion of the gentiles. Second Isaiah envisages the religion of Israel becoming a world religion.

21.5 A Purified Nation

Ezekiel had taught that before Israel could be worthy of God she must be purified, and since she had proved unable to do this for herself, God would Himself cleanse her. Second Isaiah also believed this:

> I have swept away your sins like a dissolving mist,
> and your transgressions are dispersed like clouds;
> turn back to me; for I have ransomed you. (Isaiah 44:22)

The people had not deserved to be so treated and Yahweh had performed it because His nature is love.

21.6 Idols

In Chapter 44:9–20 he shows how ridiculous is the absurdity of idol-worship. From a log of wood a craftsman makes a god: some of the wood he takes to make a fire, to warm himself and to bake bread and with the rest he makes a god, bows down to it and prays to it. In worshipping the work of his own hands 'He feeds on ashes indeed.'

21.7 The 'Servant of the Lord' Songs

Among Second Isaiah's greatest contributions is his concept of the Suffering Servant described in four short passages. Many books have been written attempting to interpret the exact meaning of the songs, but they are still the centre of keen debate. A number of solutions are offered as to the identity of the servant, none of which offers conclusive proof, but what matters most is the basic religious teaching of the songs.

1 42:1–4 In the first song God bids the nations to consider His servant 'my chosen one in whom I delight'. God has bestowed His spirit upon him, that is, given him the gift of prophecy, but unlike some of the former prophets with their loud dramatic declamations, he will be restrained and gentle:

> He will not call out or lift his voice high,
> or make himself heard in the open street.
> He will not break a bruised reed

He will preach the true religion to the nations who will look to him for guidance.

2 49:1–6 In the second song the Servant is said to have been named by God from his mother's womb (compare Jeremiah 1:6) and chosen and prepared for a future purpose. Now the moment has come: Yahweh not only purposes to bring the exiles home, but to make His Servant 'a light to the nations', to bring God's salvation 'to earth's farthest bounds.'

3 50:4–9 The third song. God has given the Servant eloquence to teach his truth and 'skill to console the weary'. He is attentive to the word of God as a pupil to his teacher, and has obeyed even though cruelty and insults were offered. He is confident that God stands by to help him so that in the end he will be justified.

4 52:13–53:12 The fourth song. It is a picture of one who suffers for the sins of others (vicarious suffering).

> He grew up before the Lord like a young plant
> whose roots are in parched ground.

An underprivileged background, it appears:

> his form, disfigured, lost all the likeness of a man,
> his beauty changed beyond human semblance.
> He was despised, he shrank from the sight of men
> tormented and humbled by suffering.

It was believed at this time that suffering was God's punishment for sin, yet the Servant is a righteous man who has done no evil; it was here that Second Isaiah put forward a new concept of suffering. The Servant is paying the price of sin, but since he himself is sinless he must be suffering for the sins of others, bearing the burden that justice demands they should bear for themselves.

> Yet on himself he bore our sufferings,
> our torments he endured,
> while we counted him smitten by God,
> struck down by disease and misery;
> but he was pierced for our transgressions,
> tortured for our iniquities;
> the chastisement he bore is health for us
> and by his scourging we are healed . . .
> the Lord laid upon him
> the guilt of us all.

It was thus part of God's plan that he should suffer, praying for others and giving his life, and in the end he is vindicated by God.

It is possible that the prophet is faintly indicating a belief in life after death. If the Servant was '. . . cut off from the world of living men, stricken to the death for my people's transgression' and '. . . assigned a grave with the wicked, a burial place among the refuse of mankind' how could he, after being dead, have 'a portion with the great' if death is indeed the end? It is not surprising that Christians believed this prophecy to be fulfilled in Jesus.

The identity of the Servant

There is no firm conclusion to be drawn:

1 It may have been a definite person, but if so he is not known to us. Some believe it is meant as a portrait of the Messiah.

2 The orthodox Jewish view is that the Servant represents the whole Jewish nation, called upon through its suffering to reconcile the whole world to God.

3 It may stand for the faithful Remnant of the nation, the purified ones, those purified by suffering to be the instruments of God.

Part III Christian Social Responsibility

22 Introduction

The scope of this subject is so wide that it is possible in a book of this kind only to deal with the more common social problems which confront people. It is also inevitable that there should be a limit to the depth at which each problem can be explored. My object is to introduce students to some of the basic facts and principles, in the hope of stimulating thought and enabling them to express an informed personal opinion on certain controversial social and moral problems.

Each subject contains areas in which moral judgments have to be made, and it is essential to understand that moral statements are unlike statements of any other kind: in particular, they cannot be shown to be true by experiment like a scientific statement, or by logical argument. In the last resort, what we regard as right or wrong depends on what we conscientiously believe to be so, in the light of our understanding of our own true nature (or for a Christian, the nature of God), so that the only valid form of a moral statement that we ourselves can make is 'I believe that this is right, or wrong.' What it amounts to is that something is right or wrong for me, because my conscience tells me so.

It is as well to remember, however, that conscience is not infallible, for someone just as honest as myself, can, in good conscience, take a stance on a moral issue which is opposite to my own. Conscience can be uneducated, insensitive, prejudiced, biased by environmental factors or past experience, and it is simply not true, therefore, to say that one opinion is as good as any other: one may be better informed, less prejudiced, better balanced and more charitable, than another.

The object of what is to follow then, is to present at least some of the issues as fairly as possible, to stimulate the student both to seek further information and to form a personal opinion which is consistent with the facts and basic principles.

23 The Family

23.1 Love

'All you need is love', sang the Beatles, but it depends on what we mean by the word. Its use is often indiscriminate and therefore confusing: 'I love walking, I love football, I love chocolate, I love dogs, I love my mother, I love my wife, I love God'. Each of these uses is proper to the word and yet they cover a vast spectrum of meaning. We need, therefore, to define the word more closely.

The Greeks used three words for love:

1 Eros, meaning erotic love or sexual passion. This is the kind of love experienced when we 'fall in love', a state of intense physical attraction and excitement. It is a matter of the emotions and one of its features is that it makes us feel that the object of our affections is the most wonderful person in the whole world, and that life without him or her would be empty. It is a condition which is very powerful indeed, an absorbing interest which has been exploited by the media, especially in films, television, books and magazines, encouraging people to believe that it is the only foundation for a happy marriage, that it will last for ever, and is an end in itself. Falling in love is an unforgettable experience, but it is a mistake to think that marriage will continue on this high level of excitement; when enriched by other kinds of love, it can, however, last a lifetime. In a film, rather more sensible than the usual, a young suitor was asking a father's permission to marry his daughter. Father said, 'Why do you want to marry my daughter?' The surprised young man answered, 'Because I love her, of course.' 'Yes', said the father, 'I know that, but do you like her?' Not such a silly remark!

2 **Philia**, meaning the love between friends. This is usually based on common interest, the enjoyment of companionship and conversation with someone who is like-minded, whom we admire for some personal qualities or skills, and with whom we know we can share. Normally this kind of love involves only part of our lives though it may grow into something deeper.

3 **Agape**. In Latin this is called 'caritas', and from this word is derived 'charity'. It is the kind of love shown by the Good Samaritan (Luke 10:25–37) involving practical help regardless of personal cost, a consistent attitude of care and concern for the well-being of another person. It is the kind of love that Jesus showed, and nearly always, at some time or another, involves suffering.

A perfect description of 'agape' is to be found in Paul's great hymn of love in 1 Corinthians 13. It is the kind of love God has for us. He loves us all equally, for our own sake, irrespective of what we deserve, and His love is not merely an emotional regard but a compassionate concern for our ultimate good, which makes demands upon us. Christians believe that 'God created man in his own image' (Genesis 1:27) which means that He has given us a share in His own nature: we are capable of love because God is love – 'We love because He loved us first.' (1 John 4:19). We cannot fulfil our true nature unless we love: we need to love as well as to be loved.

When people speak of 'making love' they mean 'having sex', but love is much more than this; it involves the whole personal relationship between a man and a woman, which includes sex. The perfect recipe for love within marriage is eros, philia and agape.

23.2 Marriage

The New Testament view of the purpose of marriage is stated in Matthew 19:5 '. . . a man shall leave his father and mother, and be made one with his wife; and the two shall become one flesh'. This idea of the fusion of personalities in marriage is also found in Paul's letter to the Ephesians 5:25–33, a passage which should be read carefully. (Incidentally, it certainly gives the answer to those who contend that Paul had a low regard for the state of marriage.) Paul says, '. . . men also are bound to love their wives, as they love their own bodies. In loving his wife a man loves himself. For no one ever hated his own body' (Ephesians 5:28–29). He goes on to take the marriage bond to be an analogy of the union that exists between Christ and his Church.

This principle, clearly laid down by both Jesus and Paul, provides an important insight into the **basic purpose of marriage**. Marriage is not first and foremost a convenient institution for the bringing of children into the world, but rather has a value in itself apart from this function. Children are a by-product of a marriage, rather than its purpose. In the Marriage Service of the Church of England Prayer Book of 1662 the 'procreation of children' is given as the primary purpose of marriage, and 'the mutual society, help and comfort that the one ought to have of the other' is third and last, whereas in the most recent revision it is 'the union of their hearts and lives' which is given prior place; this is in line with the New Testament view of the purpose of marriage.

The view that the union of man and woman in marriage has a value of its own, has an important bearing on the problems associated with sexual intercourse within marriage, making it possible to hold that intercourse has a value distinct from that of the means of producing children: it may legitimately be seen as the supreme expression of the unity of mind and heart, independently of its function of procreation.

'Charity begins at home'

Man and wife together form a family, a relationship in which the love of God and other people (*agape*) may be fostered. Children may then be born into an environment which is a training ground for growth in agape, an atmosphere of love and care between parents and children which overflows to neighbours and to the world at large. Christians see the family as the sphere in which love for God is the inspiration for learning the true value of the human personality, the practice of respect for persons, that is, to treat people as an end in themselves, rather than as a means to an end. The family is the training ground for human relationships.

23.3 The Home

Ideally speaking, every family should begin with a man and a woman falling in love and wanting to live together for life. Pop songs and fairy tales alike speak of love which is 'forever', but where love is based only or mainly on physical attraction, it is unlikely to survive the stresses of married life. Where *agape* is present, deep respect for each other and commitment to each other's happiness, marriage can more than survive, it can grow richer, even when physical attraction has waned. This is what lies behind the marriage vows, for those being married take each other 'for better, for worse, for richer, for poorer, in sickness and in health, to love and to cherish, till death us do part'.

Problems, however, come sooner or later:

1 Somewhere to live

The average age at which people marry is falling and the building programme has not kept pace with the increased demand for housing. In addition, financial restrictions have to be considered, as young people are usually at the lower end of the career and wage ladder, and building societies require a fairly substantial capital to be deposited before they will grant a mortgage. For many young people it means putting their names down on a long waiting list in order eventually to obtain rented accommodation. The only solution may be to live temporarily with in-laws, an expedient which may put the marriage under strain and perhaps make it necessary to defer having a family.

2 Economic problems

Since the war there has been a general rise in the standard of living in the sense that the 'poverty threshold' is much higher. Cars, televisions, washing machines, refrigerators, vacuum cleaners, which were formerly the prerogatives of the affluent are for many wage-earners necessities of life. Much of the drudgery of housework has been eradicated and many housewives have enough time at their disposal to take, in addition, a part-time or even full-time job. The whole pattern of married life is changing. In the early days of marriage both young people may be earning and thus be enjoying a high standard of living, but the prospect of having a baby means acceptance of a much lower standard of living and also a loss of economic independence for the woman. For some it means a choice between starting a family and keeping the car.

3 Children

The procreation of children may not be the central purpose of a marriage, but it is important where it is possible, and most married people want children. How many children to have is a matter for the couple to decide and this introduces the subject of family planning which we will examine later. For the Christian the number should not be dictated by considerations of convenience or inconvenience but rather the number consistent with the proper care of the children and the health of the mother. Most people would agree that to bring to birth so many children that they are a drain on the health of the mother and so great a strain on the family resources, that over-crowding, under-nourishment and ill-health are unavoidable, is to act irresponsibly.

4 Personal relationships

In recent years a revolution in the home has taken place. Whereas in the old days, in fact well within the living memory of the middle-aged, there were many husbands who ruled their wives and children with a rod of iron, kept their wage dockets a secret from their wives, gave them house-keeping money and required them to wait on the 'head of the house'. Times have changed. For an increasing number, marriage is a partnership and husbands take their share by helping with domestic chores, and it is by no means exceptional for the father of the family to do the shopping, take the baby out, help with the cooking and the washing up. This has contributed immensely to raising the status of women, and from the Christian standpoint this new equality of status between man and woman is a valuable expression of *agape*.

The relationship between parents and children is one which alters as children grow up. When children are young they regard their parents as infallible, but when they reach adulthood their parents are simply the best friends they have in the world. There is at least one fundamental principle, however, which always applies, namely that the tone of the family is set by the parents. Young children grow in knowledge and behaviour by the imitation of those they love and admire, and thus courtesy and good manners are best taught not by being imposed, but by being absorbed from the example of their parents. Where parents treat each other and their children with kindness and consideration, children will learn to do the same. 'The things that are taken for granted at home make a deeper impression on children than what they are told.' (T S Eliot). Paul points out that while children have a duty to their parents, the obligation is reciprocal:

> You fathers, again, must not goad your children to resentment, but give them the instruction and the correction which belong to a Christian upbringing. (Ephesians 6:4)

To treat children without respect and consideration is to train them to treat other people in the same way. Parents also have to be careful not to do violence to their children's freedom by indoctrinating them with their prejudices.

Duty to one's parents

The Bible lays great emphasis on the duty of children to their parents and the first of the Ten Commandments, after the four dealing with duty to God, is that of the duty of honouring one's parents.

Jesus himself paid deference to his parents. On his return to Nazareth after his boyhood visit to the Temple he 'continued to be under their authority' (Luke 2:51). One of his last concerns during the agony of crucifixion was that his mother should be provided for. He commended her to the care of his young disciple John. He said to her 'Mother, there is your son.' and to the disciple, 'There is your mother' (John 19:26).

Jesus strongly condemned the practice of Corban by which a Jew could evade his responsibility to support his parents by making a gift of money to the Temple.

23.4 Divorce

There can be no question that the ideal of Christian marriage as laid down by Jesus is the union of one man and one woman for life. His teaching on this subject, however, does present difficulties.

The Law of Moses stipulates that a man may divorce his wife 'if she does not win his favour because he finds something shameful in her' (Deuteronomy 24:1). In the days of Jesus there was a dispute about the definition of 'something shameful', Rabbi Shammai taking it to mean some grave offence like adultery while Rabbi Hillel took it to mean any trivial fault. When asked whether divorce was lawful, Jesus replied that:

1 in this matter the law of Moses was a concession to the hardness of men's hearts;

2 marriage was an indissoluble union, as stated in Genesis 2:23–24;

3 remarriage after divorce involved the sin of adultery and was not permissible.

Note that in Jewish Law a woman could not divorce her husband, and in Mark 10:2–12, Jesus places both husband and wife on a level, in that neither has the right to divorce the other.

Matthew 5:31–32 introduces an exception to the rule, that of the unchastity of the wife. Some scholars have grave doubts about whether this is an original word of Jesus, because if it is, Jesus is merely taking sides with Shammai against Hillel. We note that neither Mark nor Paul mentions it, and their writings are much nearer to the event than Matthew, who has clearly inserted the clause into a verse he has taken from Mark. The insertion may not be attributable to the writer of Matthew's Gospel, but may have been added later by Christians who found the teaching of Jesus on this matter too rigorous.

It is important to note that this word of Jesus can only justly apply in Christian marriages. It would be morally wrong for the Church to impose its own laws on those who owe it no allegiance. Within the Church, however, it is important that the divine ideal is upheld and at the same time compassion is not withheld from those who fall short of its high standards.

The position adopted by the different churches varies a great deal. The **Roman Catholic Church** admits no exceptions to the rule that marriage is for life, and divorcees cannot receive the blessing of the Church on a second marriage while their first spouse is still alive, whatever the reason for the breakdown of the marriage. This ruling is consistent with the Christian principle that there can be no limit placed on the duty to forgive (Matthew 18:21). Those who remarry after divorce close the door on the possibility of forgiving the offending party. The Church is ready to admit that it would be wrong to expect some marriage partners to remain together, but the answer to this predicament is separation, not divorce. Even a divorce can be regarded as a form of separation, and the divorcee may be admitted to communion, but remarriage to another person involves exclusion from the sacrament of the altar.

This rigorous line can be softened somewhat by the remote possibility of having the first marriage 'annulled', that is, declared to be deficient in some important respect and therefore not a marriage at all. Grounds for annulment include lack of consent and the existence of a previous valid marriage; marriages which take place under duress or marriage between people who are under the legal age are also not valid, and neither is marriage between the insane, though this is not easy to define.

It seems that in the past a great deal of ingenuity has been exercised in granting annulments, but the process is prolonged and therefore expensive. However unjustified the charge may be, the Church's practice of the annulment of marriages does lay itself open to the claim of cynics that the rich and powerful have an advantage denied to others.

The position adopted by the **Church of England** is similar in important respects particularly with regard to the indissolubility of the marriage bond. In what some regard as a typically English attempt at compromise, the Church of England modifies its official view of divorce by admitting to communion, under certain circumstances and with the Bishop's permission, those who have been divorced but have entered into a second marriage. No clergyman can be compelled to solemnize the marriage of a divorced person in his church, or to permit such a marriage there, but it is regarded as a practical solution to the problem to advise the parties to marry in a registry office and then to proceed to the church for prayers afterwards. This practice bears a superficial resemblance only to the practice in France, of having the legal part of the marriage in the town hall and the blessing afterwards in the church; it is, however, in fact, very different.

It has been pointed out that it is taking a peculiar attitude to say:

1 marriage after divorce is permitted as long as it does not take place in church;
or

2 that it is not permissible but it may be forgiven if, unfortunately, it takes place.

If a Christian does wrong he knows he will be forgiven if he is sorry and intends never to do that wrong again, but how can he be forgiven if he has done wrong and has every intention of continuing to do so? From the point of view of logic, this places the Church of England in a difficult position, and it can only be said that compassion for those involved in 'impossible' marriages has been allowed to over-ride theory. The heart is sometimes permitted to rule the head. The latest proposal is that divorcees may be married in church under certain stringent conditions, but not as a right.

The Free Churches take a different line. While upholding the ideal of marriage as a life-long union, the Church is confronted with the fact that some marriages fail irretrievably, and it seems inhuman to expect the victims of such marriages to live for the rest of their lives bereft of the joys of a happy married life. It is pointed out that it would be odd if this moral precept of Jesus, alone of all those in the Bible, should have to be taken literally, and as absolute without any possibility of interpretation or adaptation to differing circumstances: Jesus himself upheld the spirit of the Law of Moses when it was in conflict with its letter. It is argued that it was Jesus' anger with the male superiority of his day, and his obvious respect for womanhood, that made him reject a law by which a man could put away his wife.

The concept of the 'innocent party' to a divorce has been largely abandoned: neither party can possibly be completely blameless, and even in seemingly clear cases of adultery or desertion it is possible for the 'guilty party' to have been driven to desperation by the private behaviour of the one who, in the eyes of those outside the home, is the innocent one.

The Free Churches are usually prepared to marry, in church, those whose first marriages have failed, and many Christians of other denominations are grateful to be able to take advantage of this generous attitude.

The most important matter in the breakdown of a marriage is the future of the children. It must not be forgotten that whereas parents chose each other, children do not choose their parents. They are not responsible for the situation and it is a serious matter of conscience that their welfare should be safeguarded.

The Divorce Laws of 1969 enacted that the only ground for divorce is the complete breakdown of a marriage. Either party may apply for a divorce if they have not lived together as man and wife for two years, (and have been married for at least three years) provided that both parties agree to this. In the event of one party refusing consent, the period is five years.

23.5 Family Planning

The older and more popular phrase for this is birth control, but this is now considered an inaccurate title in that it stresses the negative side, the prevention of conception, whereas there has now developed a positive side: the treatment of infertility, by which families which would otherwise be childless are enabled to have children. Incidentally, this has given rise to another moral problem – artificial insemination and test-tube babies.

Most Christians agree that it is undesirable and even immoral that any woman should have to submit to a yearly pregnancy during her child-bearing years. To bring into the world so many children that they damage the health of the mother and put such a strain on the family resources that none of the children has a reasonable chance in life, is irresponsible. Disagreement occurs however, over the means by which the size of the family may be limited.

The attitudes of the various Christian bodies stem from their view of the purpose of sexual intercourse in marriage. The **Roman Catholic Church** thus teaches that its primary purpose is the begetting of children, and that intercourse is a sin unless the procreation of a child is intended or at least, not hindered, because children are a gift from God and His Will must not be interfered with. The use of any form of contraception, except total abstinence or the use of the so-called 'safe period' is therefore forbidden.

The **Church of England** and the **Free Churches** take the view that the children are a by-product of a marriage rather than its purpose, and that the union of the hearts and minds of the married partners, in which the sex act plays an important part, is the prime purpose of marriage. In this view the act of sex has a value of its own distinct from that of the means of producing children. The number of children desired by a married couple is a matter for their own consciences as Christians and is the result of a positive choice on their part, before God.

To sum up: various points of view stem from different interpretations of the meaning and purpose of the sex act.

1 It is merely an animal appetite like hunger and thirst which it is normally legitimate to gratify at any time. (If so it is morally neutral.)

2 Its sole purpose is the procreation of children.

3 It is the supreme expression of the deepest possible relationship between man and woman – not merely the gratification of an appetite, nor only the means of producing children, but also the means of expressing and deepening the love that exists between two people who are utterly committed to each other. It has, in fact, the nature of a sacrament – the physical means of expressing and conveying a spiritual value.

The use or non-use of contraceptives is a matter of personal conscience for everyone. Sexual intercourse is not a crime (except in the case of rape, and in South Africa under the Race Laws) and there is a presumption that we are all free to interpret the meaning and purpose of intercourse in our own way.

Family planning methods

1 The Rhythm Method – the only method allowed by the Roman Catholic Church. It has grave weaknesses:

(a) Only relatively intelligent people can work it out and it demands considerable skill and patience.

(b) It assumes that ovulation is regular, but in fact the delicate balance can be upset by worry, anxiety or illness.

(c) It demands considerable self-control and the price paid in strain and irritability is high.

(d) It destroys the spontaneity essential to lovemaking, since there is a strong element of calculation.

(e) It is notoriously unreliable. The very reason that it is permitted is that conception is possible.

2 Coitus Interruptus or Withdrawal. This can have serious physical and psychological effects and is very unreliable.

3 The Pill. This is very reliable (99·7 per cent) but it needs careful monitoring by doctors, since there can be side effects. It works by imitating the state of pregnancy. Its indefinite use is unwise.

4 The Coil. This probably works by making it impossible for the fertilized egg to embed itself in the wall of the uterus (no one knows exactly how it works, it just does). Roman Catholics believe that human life is present at the moment of conception; it could therefore be argued on this view that the coil is a method of early abortion.

5 Sterilization. It must be noted that this is largely irreversible and carries the risk that if a person loses his/her partner and marries again, the second partner may want children.

6 The Sheath. This is efficient, especially when used in conjunction with chemical spermicides, and it is relatively cheap. The main objections are aesthetic.

7 The Cap This must be used together with spermicides and, with careful use, is as effective as the coil. It works by preventing sperm reaching the cervical canal.

Principles of family planning

1 Whatever the method used, it should be by mutual agreement between husband and wife.

2 Ideally, the method should be reliable, easy to use and cheap.

3 The method used should normally be chosen by the woman, since it is she who bears the main consequences of malfunction.

Should contraceptive advice be freely available to all?

This is a problem which bristles with difficulties. Should a doctor give contraceptive advice to a girl who is under age, without the knowledge of her parents? This raises the spectre of encouraging promiscuity in the young, but it must be faced that the alternatives are not chastity *or* contraceptives, but contraceptives or sexual intercourse without them, with the attendant risk of pregnancy. In most cases the girl's sexual activity has begun before, perhaps in desperation, she consults the family planning doctor. The sympathy we may feel for both parent and doctor is of little practical help.

Despite the availability of abortion or adoption, very many young unmarried mothers keep their babies, and in many cases by doing so become part of the social problem attending illegitimacy. While it is not possible to generalize, there is still a disquieting relationship between underprivilege and illegitimacy.

Sexual intercourse outside marriage

Christians generally uphold the rule of chastity, that is, no sex before the total commitment of marriage, but many others who certainly do not believe in promiscuity allow sex between people who genuinely love each other and wish to share each other's lives by setting up a permanent relationship (while love lasts) 'without benefit of clergy'. Whatever our views, we have to take account of the fact that sex is concerned with human relationships at the deepest possible level, and that wherever there are intimate personal relationships, people can be badly hurt. The problem is the more acute because sex differs from other instincts in that it vitally affects the life of another person. In any personal relationship people are liable to get hurt, but in this case consequences can be serious even if pregnancy does not occur, and disastrous if it does. Few indiscretions carry so disproportionate a penalty.

It could be argued that a girl under the age of consent is especially vulnerable, having the body of a woman but the emotions and judgment of a child. This is also an area of gross inequality between the sexes in which double standards popularly prevail. Thus the girl who is readily available is a 'slag' while the boy is 'a bit of a lad'.

The moral weakness of casual sex is that it conveys false information in that the intimacy of the act implies a depth of feeling which is not present; certainly it implies more than the mere gratification of an animal appetite. The boy who blackmails a girl into sexual intercourse by saying that she cannot love him if she resists him, could not be more wrong: she may resist his advances because she truly loves him and sees the act as an expression of deep commitment and care which as yet does not exist. She may be quite unable to express her feelings in this way, but almost by instinct it is what she deeply feels. The intention which the act expresses is all important, and it is in the exchange of vows in marriage which is the critical test of the intentions of both parties. It is the public promise of life-long commitment in the marriage ceremony which creates the complete intimacy which it is the purpose of sexual intercourse to express.

On these grounds 'trial marriage' is an impossibility, because the total commitment which is essential to marriage is absent. Whatever kind of trial it may be, it cannot be a trial marriage, for in marriage the couple take each other 'for better, for worse, for richer, for poorer' until death parts them; they do not take each other on approval.

24 A Christian View of Work

24.1 Introduction

All work, except that which damages the structure or functioning of society, is a contribution to the welfare of the community at large. In a primitive society this is easy to see: communal work on the land and in the hunting field produces the food and clothing which the community needs, and those who cannot work, like infants, the sick and aged, are given their due share. **Work is an expression of responsibility to the community**. Modern industrial society is infinitely more complex, but the same principle applies: the vast machinery of production, manufacture and distribution is for the purpose of providing the necessities of life and all honest jobs contribute to it. To accept this is to acknowledge the dignity of labour.

In addition to the basic needs, there are other aspects of man's life which need to be provided for: his health and his intellectual and spiritual needs, for 'man doth not live by bread alone'. The doctor, the sewage worker, the dustman, are each in their way, complementary, while the actor, artist, poet, musician and priest all contribute to man's mental, aesthetic and spiritual needs. This 'unproductive' work is vital to man's personal fulfilment.

God, as the Creator and sustainer of the universe, is a worker, and since man is made in the image of God, he too, is by nature, a worker and creator: his work ought to be such that he finds fulfilment in it, what is called 'job satisfaction'. For most people, however, it is probably true to say that the main object of working is to secure an adequate income, and that when this has been achieved, and there is reasonable security of employment, the next important element in work is that it gives the worker a status, a sense of individual value and belonging in the community. To put this simply:

1 We work to live, to support ourselves so as not to become parasites on the rest of society.

2 We work to be useful. The drop-out and the playboy run away from their responsibilities towards their fellow men. Paul said, 'The thief must give up stealing, and instead work hard and honestly with his own hands, so that he may have something to share with the needy' (Ephesians 4:28). We cannot give unless we have something to give, either in service or in kind. We work in order to contribute to the welfare of others.

3 We work to achieve something, to find satisfaction in the use of our special talents.

The world of work

The transition from school life to work is often a difficult time for young people, hence the importance of 'work experience' before they leave school. Work is so often uninteresting, monotonous and repetitive, and in some cases long hours have to be put in. This is a result of the creation of assembly-line production which 'refines' division of labour down to the simplest repetitive process like fitting a single small unit of a complicated product. To sit or stand all day pulling levers, pushing buttons or watching gauges, reduces to a minimum any satisfaction a worker might obtain from his job, and does nothing to suggest that his job is worthwhile. Karl Marx predicted that the revolution would take place in the most highly developed industrial societies (Britain was the favourite) as a reaction to the loss of job satisfaction among the workers, but in the event, the workers have largely transferred the grounds of satisfaction from the job to the amenities which the wages can provide – housing, cars and all the improvements in lifestyle which good wages can make possible. Oddly enough the revolution took place in Russia, a country which was very undeveloped industrially.

The point being made is that, for many, work could be described as 'the way we earn our leisure'. This is in complete contradiction to the Christian view, that work should be a vocation, a calling. This is beautifully expressed in the Prayer Book Catechism: my duty to my neighbour includes 'to learn and labour truly to get my own living, and to do my duty in that state of life unto which it shall please God to call me'. There are jobs which are easily recognized as callings, like nursing or the priesthood (if your job is a 'calling' in this sense, it appears to entitle you to less pay), but for a Christian every honest job is a means of serving God and neighbour.

24.2 Leisure

If work is regarded as the means by which we pay for our leisure, the implication is that work is only necessary if we haven't otherwise enough money to enjoy ourselves; so the ideal existence is a round of pleasurable activity: this was the attitude of the young workman who, when asked why he came to work on only four days in a week, replied, 'Because I can't quite manage on three.'

As we have seen, the Christian attitude towards work is quite different. Work represents the contribution of each individual to the welfare of the community at large and is a basic means by which the individual fulfils his own nature, as one made in the image of God. Man, however, is not a machine, and he needs a break from work from time to time, because work itself, except in rare cases, does not fulfil all his needs.

It is impossible to estimate the benefit mankind has derived from the fourth commandment, 'You have six days to labour and do all your work. But the seventh day is a sabbath of the Lord your God; that day you shall not do any work' (Exodus 20:8). The text goes on to include in this benefit, children, servants and even the farm animals. The reason given for the sabbath rest is that God Himself took a rest after the six days' work of creation. In the Jewish religion the day is set apart for the worship of God and for rest and recreation, and is observed very strictly by orthodox Jews.

The observance of Sunday replaced the sabbath in the very early days of the Church. Gentile Christians soon outnumbered the Jewish, and it became more appropriate to keep the first day of the week in commemoration of the Resurrection of Jesus, rather than the seventh, Saturday. Already in New Testament times Paul and the Christians of Troas met on the first day of the week 'to break bread' (Acts 20:7) and Sunday is called 'the Lord's Day' in Revelation 1:10. Sunday observance began to be regulated in the fourth century, and in 321 AD the Emperor Constantine forbade townspeople to work on Sunday, though farm labour was permitted. The regulations later became more strict.

Sabbatarianism, or excessive strictness in the observance of 'the divinely ordained day of rest', took a rigorous form in England and Scotland during the Reformation, though not on the continent. It was relaxed during the Restoration but again revived at the end of the 18th century as a result of the Evangelical Revival. The Lord's Day Observance Act of 1781 forbade the opening on Sunday of any entertainment to which admission was gained by payment. In recent years, the rules have been progressively relaxed, and some Christians are very distressed by this. Two things, however, need to be remembered:

1 Sunday is the first day of the week not the seventh, a day Christians observe as one of joyful commemoration of the Resurrection of Jesus, an event without which there would be no Gospel. It is a day of joy on which the priority for Christians is the thankful worship of God, and next, for recreation of body and mind as well as spirit. Sunday is not the sabbath.

2 It would be indefensible for Christians, or any other minority, to force their ideas and observances on an unwilling majority.

The observance of Sunday as a day set aside for worship and leisure is a matter for the conscience of every Christian.

One very great advantage of having a set day (two in the case of the five-day working week) is that most people are free at the same time so that families and friends can share in each other's leisure activities. The fact that people generally value their 'Sunday off' is reflected in the fact that employers have to offer 'double time' as an inducement for Sunday work.

The needs of each person cannot be met by the same leisure activities: there are a vast number of ways of relaxing, many of which may seem quite extraordinary to those who do not take part in them, but there are certain general underlying principles:

1 Leisure can rest and relax our minds and our bodies.

2 Unless our work is physically demanding, our bodies need physical activities. Exercise is essential to health and physical well-being.

3 Work may be mechanical and monotonous and yet need mental concentration. We need time 'to stand and stare', to get to know ourselves.

4 Our daily work limits the number of people we really get to know, and we may be in the unfortunate position of having around us people whose way of life is not ours. We need time to get about to meet other people and in this way to achieve a balance.

5 Work can set up tensions and frustrations and we need leisure time to 'unwind'.

6 We need life to present us with a challenge. (The difference between a rut and a grave lies only in the width and depth.) Our work alone may leave us in a rut and we then need to find something challenging, for which we need leisure.

7 There are many ways in which we can serve those in need. The Welfare State achieves a great deal, but there is no substitute for personal contact, and a caring attitude. Leisure time is needed to enable us to offer practical help.

We see the care and concern of Jesus for his overworked disciples, and his recognition of their need for rest and refreshment, in his invitation, 'Come with me, by yourselves, to some lonely place where you can rest quietly' (for they had no leisure even to eat, so many were coming and going. Mark 6:31).

24.3 Unemployment

The development of steam power made it possible to construct machines which needed far fewer men to operate, and did a great deal more work than machines operated by manpower alone. The result was a vast increase in production on the one hand and increasing unemployment on the other: more goods but less work. One of the slogans of the 1930s was 'Starvation in a world of plenty'. Men were 'thrown on the industrial scrap-heap' with no work and no prospect of it, and in some parts of Britain such as Durham and South Wales there was grinding poverty.

It is not surprising that in the early days of the Industrial Revolution men attacked and wrecked the machines (the Chartist Movement), a process now known as industrial 'sabotage', from the French *sabot*, a wooden clog, because one of the early forms of machine wrecking consisted of workers throwing their clogs into the machinery.

In recent years the Welfare State has mitigated the financial plight of the unemployed, but only work can solve the personal problem. 'The first thing that matters about work is to have it.' It was G B Shaw who said that it is easy to say that we believe in the dignity of labour if we have never done it. Of course this is true, but nevertheless the unemployed do suffer a loss of dignity, a feeling of being useless, inadequate, unwanted, a drain on the resources of the community, which is dehumanizing in its effect. The situation in the home where a lower or declining standard of living soon becomes obvious, only reinforces the personal predicament of the unemployed, and leads to friction and unhappiness. (See unit 7.3 the Parable of the Labourers in the Vineyard (Matthew 20:1–16). Note the farmer's compassion for those who were unemployed through no fault of their own.)

The increasing use of automation has, in recent years, greatly accelerated the problem. We live in a world in which the computer and the silicon-chip have greatly reduced the need for man-power, and where competing industries and national economies are geared to producing more and better products at lower cost. Machines could be used not only to take over heavy, monotonous and uncongenial work, but also to reduce working hours and thus to enable the more equitable distribution of jobs. To do this, however, would require world wide economic agreement, which at present seems to be a remote possibility. The Communist saying, 'From each according to his ability; to each according to his need', is surely in harmony with the mind of Christ.

24.4 Trade Unions

Trade unions came into being in the atmosphere of class warfare but they still have a vitally important function even when these tensions have diminished.

It must be admitted that a great deal of the industrial unrest of present days is rooted in the social injustices of the not too distant past when the employers had the whip-hand and used it ruthlessly. The memory of years of privation and of exploitation by employers who took advantage of the 'pool' of unemployed to subject their employees to sweated labour, often under appalling conditions, has bedevilled industrial relations ever since. The trade unions provided the only defence of the workers during these bad times, and they continue to represent and defend the interests of working people today.

The largest single cost of production is wages, and it is therefore the business of management, in its task of co-ordinating the means of production, labour, raw materials and capital, to take decisions about wages and working conditions as a vitally important factor in the price which may be charged for the end product. However, even for the good manager, the welfare of his workers can be only one consideration among the many to be borne in mind; (it is certainly not in the interest of the worker if the end-product is priced out of the market and he becomes 'redundant'). In striking a balance, it is of vital importance that there should be a means by which the interests of the worker can be pressed effectively; that is what trade unions are for. Where wrong or unacceptable decisions are made by management, the individual is powerless, but as a member of a trade union, his cause has seriously to be considered. In this sense, the trade union is a bastion of personal freedom.

We may sympathize with the theory that there should be no conflict between labour and management, (David Steel, the Liberal leader, is on record as having said that 'We talk about both sides of industry when there should be only one side in industry; we talk about concessions, defeats, victories for one side or the other, we even talk of peace negotiations') but as industrial society is at present organized in the West, conflict between management and unions is a fact of life. If conflict must exist, it is better that its expression should formally be controlled in a disciplined fashion in an organization which has the authority to represent the workers.

The ultimate weapon of a trade union is a strike, which involves a trial of strength between the two parties. Every worker has the right to withdraw his labour and this is the only basic sanction he has, but it is a very powerful one and needs to be used with discretion. There is little point in striking for higher wages if, at the end, the firm is bankrupt and there is no job to go to.

Britain's competitors abroad refer to our strikes as 'the British disease' as though our record in this respect is much worse than that of other countries. Generally speaking, statistics show that this is not so; the record could be better, but there are worse. It is certainly unfair to say that trade unions are to blame for all strikes; undoubtedly they are to blame for some, but probably no more than can be laid at the door of poor management. In the defence of management it must be said that businesses can make unreasonable demands on managers as well as on workers.

24.5 The Closed Shop

One of the most controversial rights of the union is to prevent non-unionists from being employed in their trade.

1 If it is a craft union they do not wish to see their skills diluted by labour which is not properly qualified. A man who has served a craft apprenticeship requiring perhaps five years of training, cannot be happy about the employment of less skilled workers.

2 People who work in a trade enjoy benefits in wages and conditions won for them by the union and it is only fair that they should pay the union subscription. We never hear of non-unionists refusing to accept a pay-rise negotiated by the union to which they do not subscribe.

3 The genuine conscientious objector who refuses to join a union because of its affiliation to a political party can arrange to have the political levy withdrawn from his subscription. However, his vote at the Trade Union Conference probably remains at the disposal of the executive. Objections on other conscientious grounds ought to be able to be accommodated by the payment of an amount equal to the union subscription to an agreed charity.

The unions have certain privileges in law not shared by any other organization, and this too, is a very controversial subject.

It is not possible to whitewash the Trade Union Movement, nor would the movement desire it, but it can safely be said that whatever their faults, the trade unions have made a unique contribution to the quality of life in this country. Some of the most glaring faults are due to the apathy shown by the membership in local trade union meetings, where often a mere handful of people decide policy and action for the whole branch. Christians should accept the obligations of their trade union membership as a serious responsibility: they should attend meetings, express their views and exercise their vote. There are occasions when it will be a Christian's duty to be at his trade union meeting on Sunday, rather than in church.

Many people believe that the unions are too powerful and attempts have been made by both Labour and Conservative Governments to curb this power by law, The public image of the unions has suffered in recent years for a number of reasons:

1 The Trades Dispute Act of 1906 established that unions could not be sued for damages if a company sustained losses because of strike action. Some feel that this places the unions above the law, but union leaders argue that the only protection this affords is freedom from legal action in a situation which would otherwise make strikes impossible. The law was enacted in order to protect the right of men to withdraw their labour.

2 The unions provide the greater part of the money required to run the Labour Party, which, in 1900 was formed by the TUC. Judging by the large number of trade unionists who vote Conservative, this seems unfair. On the other hand, the main source of income for the Conservatives is subscriptions from companies.

3 The influence of the unions in government. Some Labour MPs are sponsored by unions, and the unions have great influence in the selection of the leader and deputy leader of the party. On the other hand it can be shown that on major issues a Labour Government can reject the advice of the unions.

4 The unions are accused of using their power for political ends. When irresponsible leaders boast of having brought down the Conservative Government of 1974, the public can be forgiven for getting the wrong impression, and feeling that democracy is under threat. In fact, direct political strikes aimed at bringing down governments are unknown in this country. The miners' strike began with a wages claim which clashed with government policy, and when the government 'went to the country', it lost.

5 Union leaders are described as being 'bosses' rather than servants, but in fact the idea that a union leader can dictate to his members is less true than people think. The miners' leader, backed by his executive, recently called for a mine-workers' strike, but this was rejected by the membership. A leader is elected to lead, but the members have no obligation to follow.

6 The 'closed shop' (See above).

7 Trade union officials at all levels complain of the biased and distorted reporting of the media. They claim that the press and television concentrate on the hardship which strikes inflict on the public and ignore the hardship inflicted on the workers by management. The media reply that when the unions do things that are unpopular they find it easier to blame the media than themselves.

25 Money

25.1 Jesus' Teachings

'Money is the root of all evil' proclaims the old song, quoting ostensibly the New Testament, but in fact, very inaccurately, for Paul actually says, 'The love of money is the root of all evil things' (I Timothy 6:10), a very different statement.

A number of stories in the Gospels illustrate the teaching of Jesus about wealth and we shall examine some of them.

1 The Rich Young Man (Mark 10:17–31; Luke 18:18–30; Matthew 19:16–30)

He wanted to know how he might win eternal life and Jesus told him to keep the commandments. 'But Master,' he replied, 'I have kept all these since I was a boy.' Jesus looked straight at him, 'his heart warmed to him, and he said, "One thing you lack; go, sell everything you have, and give to the poor, and you will have riches in heaven; and come, follow me."' It was a call to total commitment and the young man could not offer it; love of his wealth held him back.

We have here a story of a young man whose besetting sin was the love of his wealth, and Jesus saw that in his case there was no alternative to complete renunciation. It is useless to advise an alcoholic to control his drinking habits; for him there is no other 'cure' than complete abstinence. Jesus saw that for this young man complete renunciation of wealth was the only way to deeper spiritual life.

2 Zacchaeus, the Tax-collector (Luke 19:1–10)

In this instance, complete renunciation is not demanded. Zacchaeus said, 'Here and now, sir, I give half my possessions to charity; and if I have cheated anyone, I am ready to repay him four times over.' Jesus replied, 'Salvation has come to this house today.'

One man is asked to give up everything, while another is approved for an act of extravagant generosity only. What seems obvious, therefore, is that our attitude towards our money and possessions is the deciding factor in what we are called by God to do. Money can become an idol, something to which we give our allegiance instead of to God. To possess money is not wrong until it comes to possess us; it then becomes our first love, our 'ultimate concern' or what Jesus would call 'our treasure'. When what we treasure has become a substitute for God, we put our trust in what we have, rather than what we are, children of God. Jesus tells us, 'Do not store up for yourselves treasure on earth, where it grows rusty and moth eaten, and thieves break in to steal it. Store up treasure in heaven where there is no moth and no rust to spoil it' (Matthew 6:19).

3 The Rich Fool (Luke 12:20–21)

Not all rich people are possessed by their riches and not all poor people are lovers of God, but Jesus saw that the dangers were greater for the rich. He tells us about a rich man whose land yielded such heavy crops that he decided to pull down his barns and build bigger ones, so that he could lay up a store of good things to last many years, enabling him to take life easy, eat, drink and enjoy himself. 'But God said to him, "You fool, this very night you must surrender your life; you have made your money–who will get it now?" That is how it is with the man who amasses wealth for himself and remains a pauper in the sight of God' (Luke 12:20–21). The things in which the Rich Fool put his trust were not part of his essential being and when his life was demanded, had to be left behind. 'Even when a man has more than enough, his wealth does not give him life.'

4 The Rich Man and Lazarus (Luke 16:19–31)

The sin of the rich man was not that he was rich, but that he lacked any sense of responsibility towards the crippled beggar who lay at his very door. In the after-life there was 'a great chasm fixed between' them but it already lay between them in this life, and it had been created by the rich man who took it for granted that one could expect to have beggars at the gate, for whom there was no responsibility or obligation. He had no twinge of conscience, no compassion, only an assumption of natural superiority which entitled him to maintain the social barriers. Even in hell he feels entitled to ask that Lazarus should be sent, not asked to go, with a message to his socially exclusive family.

It is this social snobbery and arrogance which is always a danger to the rich, and it is attacked explicitly in the Epistle of James 2:1–8. Rich and poor alike must be given equal honour in the Church.

What Jesus himself did is as important as what he said

1 He paid his taxes–the incident of the coin in the fish's mouth (Matthew 17:27).

2 He approved the payment of the tribute-money (Mark 12:13–17).

3 He applauded the giving of alms.

4 He and his disciples held a reserve of money to meet their needs. 'Judas was in charge of the common purse' (John 13:29).

25.2 Stewardship

Christians believe that since they owe to God their very existence and all the blessings of His Creation, they are directly responsible to Him for the use they make of all His gifts. In God's sight a man can own nothing; he merely has the use of what he calls his own and he is accountable to God for his 'stewardship'. Money, time, skills, intelligence are held in trust from God.

No true Christian can avoid a stirring of conscience when he asks himself such questions as:

1 What proportion of my earnings do I spend on myself, such as entertainment, recreation, holidays?

2 How much do I give away to help needy people or worthy causes?

3 How much do I save?

4 How much do I waste?

The Church itself needs to ask similar questions about its priorities. Is it an organization concerned mainly or even only, with its own survival, with paying its way, rather than serving the world? Is it more concerned with beautifying its buildings than helping with such things as the relief of world hunger and disease? Such questions are searching indeed, and the Church needs to remember that 'it is the only organization that exists for the benefit of those who are not its members', and also that Jesus said, 'If then you have not proved trustworthy with the wealth of this world, who will trust you with the wealth that is real?' (Luke 16:11–12).

25.3 Gambling

Gambling has been defined in many ways and here are two examples:

1 A gamble is an artificial risk created solely for the purpose of gain.

2 Gambling is a method of redistributing wealth by an appeal to chance.

People indulge in gambling for a number of reasons:

1 Some people when very short of money stake part or all of it in the hope that they will make more. There is no other way in which they can increase their share in the necessities of life. It gives them a hope, however unrealistic it may be, that one day everything will change for the better.

2 People who live dull, monotonous, colourless, lives find in gambling something which brings excitement into their drab existence.

3 Some people cannot sufficiently enjoy sport for its own sake; a gamble provides what they feel to be the necessary stimulation. Rich people are often in this category.

4 Some people look on gambling as an interesting form of entertainment, or of passing the time.

5 Extraordinary as it may seem, the person severely 'hooked' on gambling sometimes does so not to win, but to express his contempt for himself when he loses, rather like the alcoholic who drinks so that in the misery of a hangover, he may express his self contempt. This happens in the most severe cases of addiction, and could be classed as perverse.

Some people look on gambling as merely a harmless 'flutter' which, since they can afford it, does no harm; others look upon it as a dangerous habit and a serious sin.

The argument against gambling

1 Our money represents our claim on our share of the necessities of life which are provided by the community at large. It provides us with food, shelter, clothing and recreation. All these come from God through man's co-operation with nature; they are therefore gifts that God gives to meet man's need, and that being so, the only fair basis for distributing these resources is people's needs. 'From each according to his ability, to each according to his needs', is a Marxist slogan, but it is thoroughly Christian in its practical application. If by gambling a man acquires more than he needs someone else goes short. God's gifts are in any case badly distributed because of man's refusal to accept his obligation to meet the claims of those in need. This is no reason why we should make matters worse. Gambling is wrong because it redistributes the necessities of life by an appeal to chance instead of need.

4 Gambling ministers to 'covetousness', to greed of gain. For people 'on the make' everything has a price and sometimes the assumption is made that people too can be bought. It is an attitude that strikes at the heart of trust of other people, and seriously distorts the sense of value. Thus, 'a cynic is one who knows the price of everything, and the value of nothing'. It encourages the selfish pursuit of one's own ends.

J B Priestley wrote ' . . . the trouble about gambling, habitual gambling, is that it drains a man's interest in the more rewarding things of life . . . it tends to create a world of trumpery or rotten values. It makes "spivs" and not good citizens. It encourages people to see life in terms of easy money and silly spending instead of in terms of useful work and zestful absorbing play.'

It introduces a thoroughly alien element into sport. When money is at stake, it is the winning that matters rather than the game itself, and this introduces all kinds of gamesmanship, cheating and foul play which often besmirch the good name of healthy sport.

Gambling is very big business of a kind which is often associated with crime. The Gaming Act was an attempt to effect a much needed reform, but in the event it led to London becoming one of the great gambling centres of the world. There was a great influx of the Mafia and international criminals and it led to the growth of protection rackets in which criminals, by threats, intimidation, and even murder, grew rich on the proceeds of gambling.

3 It discourages the habit of generous giving. Raffles and sweepstakes in aid of good causes are easy ways of raising large sums of money, but appeals made to direct giving, with no strings, and no possibility of reward, are nothing like as popular or effective. People give because there is always a chance that they may win something valuable–they feel generous when they buy the ticket, but they give themselves a false sense of generosity, for the hidden motive is largely selfish.

4 Gambling is one of the most addictive of habits; there is now an organization called 'Gamblers Anonymous' which is based on the older 'Alcoholics Anonymous'. The habit is associated with wrecked homes, spoiled lives, prison sentences, poverty, despair, and even suicide. At Monte Carlo there is a cemetery for suicides.

5 There are only three justifiable reasons for taking money from anyone:
(a) in payment for goods;
(b) in payment for a service rendered;
(c) taking a present.
None of these justifications applies to gambling.

Is life a gamble?

Human life involves risk, and it would be a poor existence if it did not. Most people would rather not know exactly how their lives are going to work out, and every day involves making decisions, some trivial and some important, which will influence the future. We are daily faced with 'the changes and chances of this fleeting world', as the prayer puts it, but we would be foolish indeed to make our decisions on the toss of a coin rather than by reason and conscience. The man who takes a risk by putting to sea in a small boat when a gale is forecast is a fool, and because he does so, the lifeboat crew may have to take unavoidable risks to bring him to safety. Because life is uncertain, we insure against possibilities; we insure our property against fire, flood, theft, and any other possible disaster. Motorists are compelled by law to insure against 'third party risks', designed to provide compensation for any injured party. Gambling exploits chance, but insurance seeks to minimize it, and to protect us against it.

In spite of the strong case that can be made against it, most people believe that there is little harm in gambling provided it is indulged in with 'moderation', difficult as it is to define such a term. Many Christians regard it as permissible while acknowledging that it is open to grave abuse. A rough guide to the regulation of its use is:

1 the amount staked must not be excessive;

2 the subject of the gamble must not be unlawful;

3 the motive must not be pure greed, but for such purposes as recreation.

26 The Christian and the Community

26.1 Yesterday and Today

It is a basic Christian principle that the state exists for the sake of the individual, and not, as in totalitarian regimes, the individual for the state. In his parable of the Great Assize (Matthew 25:31–46) Jesus tells us that our love for God can be tested in a very practical way, by feeding the hungry, clothing the naked, housing the homeless, visiting the sick and those in prison, 'I tell you this: anything you did for one of my brothers here, however humble, you did for me.' (Matthew 25:40).

Whereas years ago, the relief of human need was left entirely to private charity and the Church, it is now accepted by every British Government, whatever its political stance, that the relief of basic human needs must be the responsibility of the whole community. The aim is the abolition of all unjustifiable privilege and the establishment of equal opportunity for all. It will take a long time to achieve, but a great deal has been accomplished by the Welfare State in constructing a 'safety net' through which the underprivileged should not be allowed to fall.

Some feel that there is a danger of encouraging lack of self-reliance and independence, which can damage rather than assist people who are prepared to allow the state to take over responsibilities which properly belong to themselves, to lay greater stress on their rights than on their duties. Undoubtedly harm can be done by and to the professional 'skiver', and examples of this kind lead from time to time to a tightening of procedures, which does not help those in genuine need. It is easy, however, for those who have never known a day of real hardship all their lives to criticize the Welfare State. While it is true that we are confronted with a new set of problems, they are infinitely preferable to the want and misery which existed under the old system.

26.2 The Individual

Any social system, however perfect it may be on paper, can only be as good as the people who work it, and there is also a limit to the services the state can provide. There is a mass of human need which cannot be covered by the state organization, and which can only be met by regular personal contact in the home. This is especially true of service to the weak, the old and the lonely. A regular visit (young people are often among the best at this) can be a great service–reading aloud, writing letters, helping to fill in forms, doing an odd job or a minor repair which would be costly to have done professionally–these and many such activities would be of practical assistance, and more importantly, make the recipients feel cared for.

26.3 The Voluntary Organizations

These are most valuable for the specialized contribution they make to the welfare of those in need. There are so many in number that it would be beyond the scope of this book to deal with them all, and unfair to deal exhaustively with a selection, for they are all equally worthy and almost all rely on advertisement to bring in the necessary funds for their work. They range from organizations caring for the old, the young, the sick, the mentally ill, the homeless, orphans, prisoners, alcoholics, drug addicts, to those who care for people in immediate and desperate need requiring a 'life-line'. Inclusion of a fair-sized stamped addressed envelope should normally procure basic information from the following:

Action Research for the Crippled Child
Vincent House
North Parade
Horsham
Sussex RH12 2DA

Alcoholics Anonymous
See Local Telephone Directory

Association for Spina Bifida and Hydrocephalus
Tavistock House North
Tavistock Square
London WC1 9HJ

Dr Barnado's
Tanners Lane
Barkingside
Ilford
Essex IG6 1QG

Christian Aid
240 Ferndale Road
London SW9 8BH

The Church Army
Independents Road,
Blackheath
London SE3 9LG

The Guide Dogs for the Blind Association
Alexandra House
Park Street
Windsor SL4 1JR

Help the Aged
32 Dover Street
London W1A 2AP

The Imperial Cancer Research Fund
P.O. Box 123
Lincoln's Inn Fields
London WC2 APX

The Iris Fund for the Prevention of Blindness
York House (Fifth Floor)
199 Westminster Bridge Road
London SE1 7UT

The Leukaemia Research Fund
43 Great Ormond Street
London WC1N 3JJ

The Multiple Sclerosis Society of Great Britain & Northern Ireland
286 Munster Road
London SW6

The Muscular Dystrophy Group of Great Britain
Nattrass House
35 Macaulay Road
London SW4 0QP

The National Kidney Research Fund
184b Station Road
Harrow
Middlesex HA1 2RE

The National Society for Cancer Relief
30 Dorset Square
London NW1 6QL

The National Society for the Deaf, Blind and Rubella Handicapped
164 Cromwell Lane
Coventry CV4 8AP

The National Society for the Prevention of Cruelty to Children
1 Riding House Street
London W1P 8AA

Oxfam
274 Banbury Road
Oxford

Royal Commonwealth Society for the Blind
Commonwealth House
Heath Road
Haywards Heath
West Sussex RH16 3AZ

Royal National Institution for the Blind
224-8 Great Portland Street
London W1

Royal National Lifeboat Institution
West Quay Road
Poole
Dorset BH15 1HZ

The Salvation Army
Press Office & Information Services
101 Queen Victoria Street
London EC4P 4EP

The Samaritans
See Local Telephone Directory

Save the Children Fund
Mary Datchelor House
Camberwell Grove
London SE5

Shelter National Campaign for the Homeless Ltd
157 Waterloo Road
London SE1

The Spastics Society
12 Park Crescent
London W1N 4EQ

Stoke Mandeville Hospital
Aylesbury
Bucks

War on Want
467 Caledonian Road
London N7 9BE

27 Discrimination

27.1 Racial Discrimination

The House of Lords has recently drawn up a check-list against which one could establish whether a particular group is a racial group. Essential features are:

1 a long and shared history, and
2 a cultural tradition of its own.

In addition, some of the following are likely to be present:

3 either a common geographic origin, or descent from a small number of common ancestors;
4 a common language;
5 a common literature;
6 a common religion;
7 being either a minority or a majority within a larger community.

The problem is not new and it is strange that the Jews, who have suffered from discrimination of this kind more severely and for longer than any other race, themselves practised it in early days. In about 397 BC, many years after the Jews had returned from exile in Babylon, Ezra, a descendant of the Jews who had remained in that city, returned to Jerusalem to find that many of the Jews there had married foreigners. He was outraged at this and persuaded them to divorce their wives and renounce their children (Ezra 9:1–5 and 10–15). The incident must be judged against the background of Israel's history. The books of Joshua and Judges are records of how the pure religion of Yahweh, the desert God, was corrupted by the contact of the Hebrews with the immoral Canaanite religion, largely through inter-marriage. Solomon's wives introduced the worship of strange gods, and the supreme example was King Ahab's wife Jezebel, a devoted worshipper of Baal-Melkart, who almost succeeded in rooting out the worship of Yahweh in Israel. The history of mixed marriages in Israel was dismal, a long record of national disaster.

However, the book of Ruth contains one of the most beautiful stories of the Bible and tells of events which led up to the marriage of Ruth (a Moabite) to Boaz (an Israelite). The writer makes the point that David, the great hero of the nation, was a descendant of this mixed marriage.

Several incidents recorded in the Gospel indicate the attitude of Jesus towards race:

1 The Healing of the Daughter of the Woman of Canaan (Matthew 15:21–31) (See Unit 8.2)

Though she was not a Jewess, her faith released the healing power of Jesus.

2 The Centurion's servant (Luke 7:1–10) (See Unit 8.2)

A foreigner received the most fulsome praise Jesus ever uttered, 'I tell you, nowhere, even in Israel, have I found faith like this.'

3 The Samaritan Woman (John 4:1–30) (See Unit 7.4)

Jesus asked a Samaritan woman for a drink of water. She was astonished and said 'What! You, a Jew, ask a drink of me, a Samaritan woman?' Then the account goes on, 'Jews and Samaritans, it should be noted, do not use vessels in common' (John 4:8–9). When the disciples returned to the well they were astonished to find him talking to this woman.

4 The Ten Lepers (Luke 17:11–19) (See Unit 8:4)

Jesus makes the point that of the ten who were cleansed the only one who returned to give thanks to God was a hated Samaritan. Jesus refers to him as a 'foreigner'. These incidents show clearly that Jesus did not distinguish between people on the ground of race.

His parable of the Good Samaritan (Luke 10:25–37) makes the point even more emphatically (see Unit 7.4). A hated Samaritan fulfils the law of love, while the official custodians of the Law pass by.

Several incidents in the Acts of the Apostles illustrate the attitude of the early Church:

1 Philip and the Ethiopian (Acts 8:26–40)

This incident is of special interest in that it records the conversion to Christianity of a black man. When the Ethiopian said to Philip, 'Here is water: what is there to prevent my being baptized?' (Acts 8:36), the colour of his skin was not a consideration.

2 The Vision of Peter (Acts 10:9–23) and the **Conversion of Cornelius** (Acts 10:24–48)

The message of both of these passages is summed up in one verse. Peter said, 'I now see how true it is that God has no favourites, but that in every nation the man who is God-fearing and does what is right is acceptable to him' (Acts 10:34–35).

3 The Christian Church at Antioch (Acts 11:19–26)

The first Christian missionaries were Jews and carried their message to Jews only, but quite early on, some of them preached to pagans at Antioch, with great success. Barnabas and Paul ministered to this Church and lived in fellowship with the mixed congregation of Jews and gentiles for a year; 'It was in Antioch that the disciples first got the name of Christians' (Acts 11:26).

4 Paul's Sermon in Athens (Acts 17:22–33)

Preaching before the Court of Areopagus he declared that God 'created every race of men of one stock, to inhabit the whole earth's surface'. The Authorized Version puts it more dramatically: God 'hath made of one blood all nations of men for to dwell on the face of the earth' (Acts 17:26). What makes us brothers and sisters is blood, not skin.

In Paul's letter to the Colossians, we find the often quoted words, 'There is no question here of Greek and Jew, circumcised and uncircumcised, barbarian, Scythian, freeman, slave: but Christ is all, and is in all (Colossians 3:11). Discrimination on grounds of race or class is a denial of the Gospel itself, which is God's good news for all men everywhere.

The Race Relations Act of 1976

The existence of the act is irrefutable evidence that racial discrimination exists, and that it is a serious problem. It defines racial discrimination as 'treating a person less favourably than another person on grounds of colour, race or ethnic or national origins'. The act applies especially to opportunities for training and employment, and to housing; it forbids discrimination in selling or letting houses, businesses or land, and in the treatment of tenants. Complaints are dealt with by the Commission for Racial Equality which may refer any unresolved cases to the appropriate court, or Industrial Tribunal which has power to serve injunctions and award damages.

Some reasons why racial discrimination exists

They are many and varied but what is common to each is their lack of rationality: prejudice of this kind is emotional rather than reasonable.

They are many and varied but what is common to each is their lack of rationality: prejudice of this kind is emotional rather than reasonable.

1 People tend to generalize about groups other than their own. Thus the Germans are described as militarist, the French are volatile, eat snails and frogs legs, the Irish are charmingly zany, the Welsh sing and play rugby, the Scots are thrifty and the English are stand-offish. When we meet real people we find them very different from the stereotyped picture; one potent source of prejudice is ignorance.

2 When we have a low opinion of people, and show it, they tend to react by living up to our opinion or becoming angry and thus confirming our prejudices.

3 We tend to find security in our own group, among people who have the same outlook, the same standards, the same politics and religion. People belonging to other groups pose a threat to our security; they frighten us.

4 We tend to despise what we do not understand. (Modern art is treated with contempt by those who do not understand it.) We tend to dismiss people whose way of life we do not understand.

5 In the old cowboy films the hero wore a white hat and often rode a white horse, while the villain was easily identifiable by his black hat and horse. This was of course symbolic: many people associate black with dirt and evil, and white with purity and the good. It is of course a nonsense to apply this symbolism to the colour of someone's skin but it can be quite unconsciously the source of prejudice.

6 White people sometimes falsely assume that coloured people are mentally inferior. This often arises because account is not taken of the language difficulties of those for whom English is a foreign language. Many of the original immigrants came from societies very different from our own and in which educational opportunities were not as easily available as in this country. The new opportunities have often been eagerly grasped and coloured people can justly point to great achievement. Parental support is often excellent.

7 Different cultural expectations about what constitutes good housing can lead to misunderstandings. For example, Chinese from Hong Kong have a very different view of what constitutes overcrowding. For Asians, the importance of the extended family far outweighs inconvenient accommodation. British subjects of African, Asian and Chinese descent frequently find it difficult to acquire appropriate accommodation and they are sometimes the victims of racketeering as well as prejudice. These problems point to a major social issue, the question of appropriate housing policies at governmental level.

Racial prejudice is irrational. It nurses hatred and other strong emotions which usually remain secretive and grudging, appearing from time to time as insensitivity and awkwardness, but occasionally finding violent expression. All people look for acceptance, and a society which requires any of its individual members to lose their identity in order to be acceptable is acting unjustly and contrary to the teaching and example of Jesus. The goal of all people is to enjoy the fundamental freedoms set out in the United Nations Declaration of Human Rights.

27.2 Social Discrimination–Class Distinction

Class is a distinction which sociologists find impossible accurately to define, but there are many indications.

1 Money The rich often consider themselves to be superior to the poor and the poor mostly wish to be rich, but money alone is no indication of class. The newly rich can be very working class.

2 Birth Each county has its 'county families' mostly drawn from land-owning gentry. They tend to marry into their own class. 'Gentleman' can be a formal title as well as a comment on a man's character.

3 Education This is a ladder by which it is possible to climb in the social scale. The introduction of comprehensive schools was an attempt to provide more equality of educational opportunity, but there still lingers the prejudice that those who go to grammar schools and even more, to public schools (in Scotland, private schools), are superior.

4 Where you live Housing areas are graded from very good to bad. The good areas have larger, better spaced, more expensive houses, and are much more select and prestigious.

5 Speech Regional accents tend to indicate lower class. Upper class accents are sometimes expensive to acquire.

For a Christian, social prejudice and snobbery are indefensible. Jesus himself was at home with all sections of society and earned himself the title of 'a friend of publicans and sinners'. Two incidents in his life serve to illustrate this:

1 Jesus and the Prostitute (Luke 7:36–50). The Pharisees would have no contact whatever with 'A woman who was living an immoral life in the town' (Luke 7:37) but for Jesus she was a woman in need of forgiveness.

2 Zacchaeus (Luke 19:1–9). The respectable Pharisees would have no dealings with a man whom they regarded as a traitor, in the pay of Rome. Jesus accepted his hospitality without a trace of patronage. The Letter of James 2:1–7 is quite explicit in its condemnation of snobbery. Paul says 'There is no such thing as Jew and Greek, slave and freeman, male and female, for you are all one person in Jesus Christ' (Galatians 3:38). This means that within the Church there can be no distinctions of race, class, or sex.

27.3 Religious Discrimination

1 Among Christians This is one of the saddest results of the divisions within the Church, the Body of Christ. Paul emphasizes again and again the essential unity of the Church:

> Spare no effort to make fast with bonds of peace the unity which the Spirit gives. There is one body and one Spirit . . . one Lord, one faith, one baptism; one God and Father of all. (Ephesians 4:3–6)

The prayer of Jesus was 'May they all be one' (John 17:21) and the divisions within his Church are a denial of his will. Thankfully the Church is becoming more aware of the scandal of its unhappy divisions, and more penitent for them. Recent years have witnessed increasing friendliness and co-operation among the separated Christian bodies. It is increasingly being realized that unity is not the same as uniformity: each separated part of the Body has its own unique contribution to make to the well-being of the whole. The Ecumenical Movement has already achieved great things and much is expected of it.

2 Among world religions Christians have been as uncompromising about the superiority of their religion as anyone else, and their past record of intolerance and persecution is an unenviable one. In particular their attitude towards the Jews has been an acknowledged disgrace, and there is much reason for penitence. Happily those days have passed for ever, but there is still religious discrimination in some parts of the world.

But for the accident of birth we might be Muslim, Sikh, Hindu or Buddhist. Christianity may not be the most holy path of all for others, and we need to remember the words of Max Warren:

> Our first task in approaching another people, another culture, another religion, is to take off our shoes, for the place we are approaching is holy. Else we may find ourselves treading on men's dreams.

Each religion, in theory, prizes humility and tolerance, and it is important that theory should be translated into practice.

28 Crime and Punishment

28.1 Introduction

The break-down of law and order in our society has led many people to advocate a tightening up of the law and increased severity of punishment of offenders. Those who believe in human freedom argue that any restriction placed on freedom has to be justified by those who wish to impose it. A distinguished British soldier once said, 'When a boy takes out his bicycle to go for a ride, that is freedom: when he rides off on the left hand side of the road, that is discipline.' The rule of the road is, however, more than a matter of discipline, for its observance is necessary to our survival. It would be impossible to run even a pack of Brownies without at least some rules, either stated or taken for granted. Any group of people, from Hell's Angels to democratic governments, need to work according to rules, and must have means of enforcing those rules if there is not to be anarchy.

The law of the land exists for the protection of society from those who threaten the freedom of the individual, and the most useless of laws is the one that cannot be enforced. Punishment of one kind or another is a means of enforcement, but since it involves the deliberate infliction of pain (restriction of freedom, fines, and corporal punishment are included in this word) normally against the wishes of the recipient, those who wish to impose it must be able to offer a moral justification for it.

28.2 Justification for Punishment

1 **Retribution** This may be defined as the infliction of pain by an appropriate authority on a person because he is guilty of an offence.

2 **Deterrence** This means the infliction of pain on a person to deter him from repeating the offence, or to deter others from doing the same. The concept of deterrence is sometimes subdivided. It is 'preventive' in that it is intended to discourage the miscreant himself from repeating the offence and 'deterrent' in that its intention is to deter others from imitating him. Its aim is not reform, for if the offender does not repeat the offence it is from fear of punishment, not because he now realizes that what he did was wrong.

3 **Reform** Its aim is to help a person to will to do right where formerly he willed to do wrong. Thus the young thug who finds his violent crime to be entertaining and amusing, learns through punishment that society does not agree. Punishment expresses society's abhorrence of an offence, and this may induce a change of mind and heart on the part of the offender. The best way of protecting society is to reform the criminal.

Retribution is morally the most justifiable. Laws have to be enforced and if the penalty is known, there is a sense in which the offender chooses to be punished by committing the offence. (Naturally we exclude here those who are not responsible for what they do.) The theory of retribution associates punishment with what is deserved and therefore with justice, for only if a punishment is deserved can it be just. Justice gives lawful authority the right to punish up to a certain limit, according to the offence, but there is no obligation to punish to the limit; justice need not conflict with mercy. The law may choose to award a lighter penalty than it is entitled to, with the object of providing an opportunity for reform, and there is also no reason why retribution should be incompatible with deterrence. It is certainly true that in some cases a succession of sentences of ever increasing severity can persuade some criminals to feel that crime is no longer a paying proposition, though it must be agreed that many, if not most, habitual criminals seem undeterred by imprisonment. They hate it, but it is looked on as an 'overhead cost'; many are convinced that if they have not yet 'hit the jack-pot', inevitably they will.

Retribution is not revenge, for revenge is private, personal, knows no law, and has no limit. We see this clearly in the vendetta where private revenge can condemn to death a whole family for the offence of one if its members. It needs to be stressed that the Biblical law of 'an eye for an eye and a tooth for a tooth' is a legal principle which controls rather than supports revenge, setting an upper limit to what may be exacted. As members of the state we surrender our right to obtain justice to the proper authorities appointed by the state.

Retribution is thought by some to be vindictive and revengeful, but only the injured party can be vindictive or revengeful, and what the law expresses is rather indignation. While vindictiveness is immoral, moral indignation is not. The use of 'exemplary' sentences with the object of deterring others who commit, or are likely to commit, a particular crime, is difficult to defend on moral grounds. We can only justly punish an offender for his own crimes, and if a sentence is awarded which is more severe than the particular crime warrants, in order to deter others, serious injustice is done: to punish an offender for the crimes and possible crimes of others, is not justice but victimization.

The policy of the British Prison System has been for many years based equally on reform and deterrence, and the problem in practice has been to achieve a balance, a most difficult task; the prison authorities deserve a great deal more sympathy and understanding than they are ever likely to be given. Judges may when sentencing have motives of reform in mind, but if for some reason the desired reform is not effected, this does not mean that the punishment was unjustified. Prison authorities may try to help prisoners to reform, but it is not in their power to ensure this, and if they fail, the sentence is not the less justifiable.

We may observe the following:

1 Law is a necessity to civilized existence.

2 Laws must be obeyed by all, and if they are broken there must be penalties. A law which is not, or cannot be enforced, is useless.

3 Punishment is a 'blunt instrument' and is at best a necessary evil. It may be a means of reform by helping an offender to appreciate the seriousness of his offence, or on the other hand it may embitter him, turning him into an anti-social being with a chip on his shoulder, bent on getting his revenge on society by indulging in further crime. In this case a main object of punishment, the protection of society, is denied. The result is never predictable.

4 The law generally achieves its end without punishment. It does not choose to punish and does not wish to. The offender brings punishment upon himself.

28.3 Personal Responsibility

In all moral argument it has to be assumed that we are normally responsible for what we do: it would not be just to punish anyone for conduct beyond his power to control. The second of the Ten Commandments says, 'I punish the children for the sins of the fathers to the third and fourth generations' (Exodus 20:5), but the prophet, Ezekiel, corrected this idea. He saw that the people were not prepared to acknowledge their own sins, but placed the blame for the national disasters on the sins of their ancestors. The prophet referred to the popular proverb, 'The fathers have eaten sour grapes, and the children's teeth are set on edge' (Ezekiel 18:2), and told the people that God had forbidden the use of it ever again. Each individual must bear the responsibility for his own sins alone and 'the soul that sins shall die' (Ezekiel 18:4). This was an enormous advance on the second of the commandments.

The result of research in the Social Sciences, Biology and Psychology suggests that no one is 100 per cent responsible for everything they do, on the grounds that there are factors which quite unconsciously influence our choice of action.

1 Sociological factors The effect of social pressures on our choice of action is seen most clearly in extreme situations like juvenile delinquency. It is not possible to be precise, but it is not surprising that young people brought up in an environment of heavy drinking, sexual promiscuity, broken homes, slum conditions, in a section of society in which police are regarded as 'pigs' or 'the filth', should have very different standards of accepted behaviour, from those in more affluent areas. Social conditioning can unconsciously limit responsibility.

2 Biological factors Young people who suffer from nervous and physical handicaps, sometimes more than one, can fail to cope with stresses which could leave a more normal person unaffected. Family rows, for example, which many children accept as almost normal behaviour, can leave such children maladjusted and delinquent. There is an association between the crime rate and very large families living in crowded and inadequate accommodation.

3 Psychological factors Psychologists believe that much of our behaviour is influenced by past events which we have long forgotten but which motivate our present actions.

These factors do not totally explain our behaviour, and though we may be aware of them we still feel that we are responsible for what we do, and that it is fair that we can be held to be accountable.

28.4 The Christian Idea of Responsibility

1 We are responsible for our actions. The Creation stories of Genesis tell us that God made man free to obey or disobey, for he could not respond to love if he were not free. God created people, not robots.

2 Jesus said that we are not to judge others. Commonsense, however, suggests that we are often compelled to judge others, but when we do so our judgment is not to be a condemnation, but rather in the nature of a diagnosis. A doctor makes a diagnosis on the evidence of the symptoms, and is then able to prescribe effective treatment. He does not blame a patient for being ill; all he wants to do is help. In like manner a Christian needs to make a judgment not in order to condemn, but to understand and to offer effective help.

3 Jesus teaches in the Sermon on the Mount the duty of renouncing revenge. 'Love your enemies, and pray for your persecutors' (Matthew 5:43–48). Paul in the Letter to the Romans expands on this, 'Never pay back evil for evil . . . do not seek revenge, but leave a place for divine retribution . . . "Justice is mine," says the Lord, "I will repay"' (Romans 12:17–19).

At the heart of a Christian's attitude to others there must be love and forgiveness, but this does not mean that rights and duties, rules and obligations can be dismissed. Love is not a matter of feeling but of attitude and action; it involves a great deal more than kindly feelings. Our 'on the spot' reaction to a human need, prompted by love, may turn out on reflection to be misguided, and even to have been harmful. Love needs to be guided by truth so that action is based on discretion. Love is not soft sentimentality, it is not to be seen in the attitude of the mother who so swamps her child with affection that he is prevented from growing up into a responsible and independent adult. It has the ultimate good of the loved one at heart and that can mean taking a very strong line. Love in action is responsible action.

The most important person in any British court of law is not the judge, the barrister, or any other official, but the accused. The criminal can be the kind of person for whom it is difficult to feel any sympathy, but he can be in greater need of help than the person he has in some way injured. He may need help more than punishment and some of the most appalling crimes are committed by the most pathetic people. One thinks of the mother convicted of gross neglect of her children who presents a picture of such unbelievable incompetence and helplessness that she is to be pitied.

28.5 Corporal Punishment in Schools

Discipline is an essential ingredient of school life, for without it an 'educational situation' would not be possible. In the last resort discipline, if it is not present by consent, has to be enforced.

The case for corporal punishment

1 The 'short sharp shock' is sometimes effective and does not need to be repeated if it fails. No form of punishment can always be relied on to be effective.

2 A punishment is the better for being inflicted close to the occasion of the offence, and over and done with quickly. Long delay or drawn out punishments like detention on the next four Fridays, leave the impression that the punisher is being vindictive and unforgiving.

3 If used sparingly and for serious offences only, it serves to emphasize the school's abhorrence of certain gross violations of discipline.

4 In societies where violence is the accepted instrument of authority (mum clouts them, the peer group enforces discipline by violence) disruptive children will ignore anyone who in the last resort will not use force.

5 A teacher has a duty to each member of the class as well as to a disruptive pupil. Without corporal punishment a great deal of valuable teaching time can be wasted going through the alternatives, before the final sanction of suspension from school becomes necessary.

The case against corporal punishment

1 The past unhappy experience of schools in which each teacher had a cane (hidden behind the blackboard) resulted in a permanent atmosphere of intimidation, which seriously inhibited learning. Severe corporal punishment was common well into the 1930s.

2 It cannot reform in the sense of making the offender will to do right where he formerly willed to do wrong. It can only teach prudence–better not to 'risk it'.

3 It is demeaning to those who suffer it and to those who inflict it.

4 It involves a conflict of principles when pain is inflicted on a bully for inflicting pain on others.

5 It is often ineffective. The regular caning of some pupils is strong evidence of this.

6 If given the option many would prefer to be caned than placed on detention. What entitles them to choose their own punishment?

7 It is a blunt instrument in that while in some cases it may arguably do good, in others it does harm, causing resentment and an anti-social reaction. It is impossible to know the result before-hand.

8 It gives status to some of the 'hard boys'. Education does not need heroes of that kind.

9 Corporal punishment is no longer awarded by the courts. If no longer appropriate for criminals, why should it be for school children?

10 We train a dog, but we educate a child. Schools are for education and it is inappropriate to train a pupil, by the use of the cane, so as to condition him against repeating behaviour which authority regards as being against the rules.

11 Sex discrimination. It is very unusual for girls (who can give at least as much trouble as boys) to be subjected to corporal punishment.

Judgment should be based on the weight, rather than the number, of the points for and against.

28.6 Capital Punishment

The case for

1 It is justifiable in the case of terrorists, who kill for minority political ends, employing the bullet rather than the ballot. They often kill indiscriminately, and when sentenced their companions kill to set them free. Examples include the Bader-Meinhof gang, the PLO and IRA.

2 It deters potential murderers. There has been a marked increase in the use of fire-arms in the pursuit of crime; criminals would hesitate to use such weapons if they were likely to be executed.

3 Capital punishment would protect our unarmed police, and especially prison officers who are in charge of people with little to lose by killing them.

4 It expresses, as does no other sentence, society's abhorrence of the crime of murder.

5 In the case of particularly brutal or callous murders people may be tempted to take the law into their own hands if the punishment of imprisonment is not regarded as severe enough.

6 Spies or traitors may have much more information than interrogation brings to light. Escape is always possible, and in the event, the lives of many people would be at risk.

The case against

1 It makes reform of the criminal impossible.

2 There is no firm evidence that it deters; if it did, it could be argued that the murderer owes it to society that his life should be forfeit. The murder rate, in fact, is fairly static.

3 It coarsens the moral sensitivities of the public. The sensationalism engendered in the media ministers to a debased and degenerate side of human nature.

4 There is a certain 'ghoulish' attraction about it for some abnormal people who are tempted to imitate the murder and suffer the punishment. Some would prefer to be a notorious murderer than a nobody.

5 The law condemns murder and then proceeds to murder in the name of the law.

6 It is inhumane. There is the torture of the long wait in the condemned cell of up to six weeks. The murderer dies a thousand deaths.

7 Society hires a killer. There is no shortage of volunteers but the executioner's job dehumanizes him.

8 It turns criminals into notabilities and terrorists into martyrs.

9 The case of Timothy Evans demonstrates that a miscarriage of justice is possible and cannot be rectified. Evans was condemned on the evidence of Christie, a man who at the time, unbeknown to the police, was a multiple murderer. No jury on earth would have convicted Evans had this been known. He received a posthumous pardon.

10 Human life is sacred. God alone has all the evidence, and He is the final judge. The Christian cannot be unconcerned about the good of the criminal as well as the protection of society.

28.7 Violence and War

Many Christians believe that any form of violence, and especially war, is contrary to the teaching of Jesus, and there are many equally devout Christians who believe that in certain circumstances resort to violence and war is not wrong. We live in a violent world in which people seek to exploit others by any means at their disposal, and the question is this, 'Are people who are suffering from gross injustices, with no possibility of redress, to accept their situation, or are they justified in using violence if all else fails?'.

Christians naturally look to the teaching and example of Jesus: he too lived in a violent world, in a country dominated by a ruthless foreign invader. The Christian pacifist takes as his key text 'Do not set yourself against the man who wrongs you. If someone slaps you on the right cheek, turn and offer him your left. If a man wants to sue you for your shirt, let him have your coat as well. If a man in authority makes you go one mile, go with him two' (Matthew 5:39–41). Jesus said, 'Love your enemies and pray for your persecutors' and this in a very provocative situation indeed. His enemies eventually hounded him to death.

The possibility of using force to further his cause was probably in his mind at the Temptation. 'The devil took him to a very high mountain, and showed him all the kingdoms of the world in their glory. "All these, he said, I will give you if you will only fall down and do me homage"' (Matthew 4:8–9). The known world was under the rule of Caesar, by right of conquest, and held down by force of arms; by adopting the methods of Caesar he could ensure a hearing for his message, but he rejected the temptation. God's purpose cannot be achieved by the use of the devil's methods.

In the Garden of Gethsemane one of his disciples tried to resist the arrest of Jesus by attacking the servant of the High Priest with a sword. Jesus said,

> Put up your sword. All who take the sword die by the sword. Do you suppose that I cannot appeal to my Father, who would at once send to my aid more than twelve legions of angels? (Matthew 26:52–54)

At his trial before Pilate he rejected the use of force by his followers

> My kingdom does not belong to this world. If it did, my followers would be fighting to save me from arrest by the Jews. (John 18:36)

It is quite mistaken to regard this as a gospel of non-resistance: it preaches non-violence and non-retaliation, but these are potent weapons of resistance. It does not involve giving in to evil, but rather fighting it with spiritual and not physical weapons. Gandhi realized this, and so did Martin Luther King: both offered non-violent resistance, and both proved it to be a powerful weapon indeed. Reaction was so strong that, like Jesus, both suffered violent deaths. Jesus made it clear that he was engaged in a war, a spiritual war against the forces of evil, and the imagery he uses is of a fight. 'When a strong man fully armed is on guard over his castle his possessions are safe. But when someone stronger comes upon him, he carries off the arms and armour on which the man had relied and divides the plunder' (Luke 11:21–22).

Paul saw that the Christian life was a battle against evil 'against cosmic powers, against the authorities and potentates of the dark world' (Ephesians 6:12) and he exhorts the Ephesians to take up God's armour, the belt of truth, the coat of mail of integrity and the shield of faith.

> Take salvation for helmet, for sword take that which the Spirit gives you–the words that come from God. (Ephesians 6:17–18)

It is all 'fighting talk', but it must be observed that non-violent resistance can still leave opponents as enemies, whereas the intention of Jesus is reconciliation. Christians are to love and forgive: they are to be peace-makers, but as Martin Luther King said, 'Peace is not the absence of tension, but the presence of justice.'

28.8 War

This is an extreme expression of physical violence but its principles apply in some respects to all forms of physical violence.

There are situations when without question it is proper for physical force to be used, indeed when there is a duty to apply it: when a mental patient becomes violent it may be in his own interest as well as that of others that he should be restrained. There are occasions too when there is no time to explain to someone that they stand in imminent physical danger, and that immediate action is called for. A violent push may be the only resort if a life is to be saved. In such cases it is the intention which justifies the violence. Few would argue that we do not need a police force, and a majority would feel that it is asking too much to expect the police to go unarmed into confrontation with armed and known killers. There is a presumption in British Law that everyone has the right to defend himself and even to kill in self-defence if there is good reason to believe that there is danger to life. However, it seems that if a Christian takes the words of Jesus literally, he must be prepared to allow himself to be killed without resistance, presumably in the interests of the killer, whose conscience may be touched by the refusal of his victim to resist. This is not, however, the experience of many thousands of old people who each year are mugged by hooligans, or of women who are raped.

Moral problems seldom present themselves in terms of black and white; there are usually 'grey' areas where there is no clear-cut answer, and even where the choice to be made is not between the good and the evil, but between the lesser of two evils. I may feel that it is my Christian duty, in the interest of an assailant, to allow him to kill me, but I should not forget my duty to my sorrowing wife and fatherless children. My assailant may not be influenced in any way by my action, but my wife and children most certainly will be. My decision has to be made between a possibility and an absolute certainty. It is this predicament which is presented in the case of war. Christians are bound by the law of love, but its application can sometimes involve agonies of conscience when there is no clear cut answer to how best it can be applied.

War is an obscenity, and it can safely be said that no Christian should ever engage in it without the conviction that all other alternatives have been explored and that it is the last possible resort. The reluctance of Britain to oppose Hitler by force was in the event exploited by him in his programme of conquest; the helplessness of nearly six million Jews did not save them from the gas-chambers, and it would have been unforgiveable to offer non-resistance as a weapon 'to the last Jew'. No Christian should be in favour of war as a way of settling disputes between nations, and it can only be engaged in as a very last resort in the face of blatant and unjustified aggression. Joel 3:1–12 presents such a case. In the face of gross injustice the prophet proclaims:

> 'Declare a holy war, call the troops to arms! Beat your mattocks into swords and your pruning hooks into spears' (Joel 3:9).

The 'Just War'

In medieval days the principles of a just war were laid down by the Church. The conditions may be summarized as follows:

1 the cause must be just;

2 there is no other way of achieving it, all other legitimate means having been exhausted;

3 the sufferings involved in war will be less than those which would result from submission to tyranny;

4 there must be a reasonable chance of success;

5 war is to be pursued as humanely as possible.

On the grounds of conditions 3 and 5 no Christian could approve of atomic war. The effect would be so horrendously destructive of human life that Christians cannot regard it as an option. Micah 4:1–6 says that there is a hope for the future: 'They shall beat their swords into mattocks . . . nation shall not lift up sword against nation.'

28.9 Violence in Society

For some years there has been an increase in violent behaviour, from the violence of football supporters, demonstrators, teenage gangs, muggers, violent criminals, to terrorist bombings, and there has been a predictable demand from the public for stronger measures to combat it. Very often the victims of attack are innocent of any involvement with the attackers.

Far more effective than harsh counter measures would be to remove the causes of violence, but it is a complex subject and there appears to be no simple explanation. Experiments with higher animals suggest that gross overcrowding is one cause of violence. When their living space is limited, they resort to defending their territory, and a rigid hierarchy of rank is established by brute force: females and young animals suffer most. Behaviour of this kind is almost unknown in the wild where there is unlimited space.

Whether or not it is valid to apply the lessons of animal behaviour to the human situation, it does seem feasible that poor, overcrowded housing conditions should be a factor. A society whose educational system lays such emphasis on achievement, increases the frustration of young people who because of the disadvantageous conditions of life, or their lack of ability, are unable to 'get on'. They work out their disgruntlement by violence and vandalism, a conscious or unconscious protest against a society which makes the rules which place them in a permanent state of disadvantage. It is requiring a great deal of young people to expect them to conform to the moral rules of a society which provides them with the worst housing and schooling, and offers them the least attractive and worst paid jobs–if any. Another factor may be the violence displayed in the media, especially in TV and films, which seems to invite imitation. So too when violence is tolerated on the field of play it is not surprising if it finds an echo on the terraces.

The weakening of the influence of the family is another possible cause of delinquency. In former days a family might live in the same community for generations and close relatives lived very near each other, with grandma often the custodian of the family tradition of respectability, a situation which was supportive of the individual and effective in its influence on behaviour. With changing patterns in industry came the necessity to find work further afield, and eventually to move house away from the old community, resulting in a weakening of the family support. More and more people live anonymously in flats or housing estates.

Human nature does not change, and horrified as we may rightly be by the violence of society, we need to ask if some of it at least is not generated by the failings of society itself.

29 Drug Abuse

29.1 The Drugs

This is a subject which arouses strong emotional reaction from older people and it is very necessary to examine the problem in perspective to get a balanced view. It must be appreciated that almost everyone takes drugs, for tea, coffee, cocoa, and chocolate all contain the drug caffeine, while smokers take nicotine in tobacco. Most people have taken aspirin and its derivatives from time to time, and also various cough medicines, all of which contain drugs. One of the most dangerous drugs is alcohol, and this is insufficiently taken account of, for it is socially acceptable and condemned by most people only when it is grossly misused. Its moderate use is regarded as respectable, though it is more addictive than some drugs which are forbidden by law. We have every reason to be grateful for the use of drugs as pain removers and anaesthetics. Their use in the treatment of depression and mental illness is especially beneficial.

There are drugs which are legal only when given under medical supervision:
(a) the opiates (heroin, morphine, opium) derived from the opium poppy; (b) cocaine (originally derived from coca plant but now by synthesis); (c) cannabis (marijuana, hashish); (d) amphetamines; (e) barbiturates (pheno-barbitone); and (f) LSD. Opium and marijuana have been known and used for nearly 5000 years.

In addition there are substances which are in everyday use and are usually safe, but dangerous when used as drugs! Examples are modelling glue, and methylated spirit. It is impossible accurately to estimate the number of people who take the illegal drugs. The only figure we have is of the number of registered addicts receiving medical treatment each year, and this represents the tip of a very large iceberg. In 1960 it was about 200, and in 1975 about 2000; by 1982 the number had more than doubled to over 4300. These figures take no account of unregistered addicts, or of users of marijuana or LSD. The law provides severe punishment for possessors of the 'hard' drugs and is even more severe on 'pushers', those who sell and distribute them; sentences for possession of the 'soft' drugs such as marijuana are lighter, but still quite stiff. It is now recognized however that the increased use of alcohol by young people presents a greater social menace than all the rest of the drugs put together.

29.2 The Dangers

The greatest danger of drug taking is eventual dependence. Amphetamines are especially likely to produce dependence because their use leads to depression, which is relieved by further doses of the drug. Tobacco and marijuana (pot) may be difficult to give up, but with will-power it can be done. Alcohol, barbiturates, and hard drugs can be seriously addictive because tolerance builds up, and more and more of the drug is required for the same effect. The symptoms of withdrawal (DT's in the case of alcohol and 'cold turkey' with heroin) are extremely dramatic and severe.

People who are really addicted (hooked) on heroin, barbiturates, cocaine and alcohol eventually deteriorate. The majority become so obsessed with their drug that what money they have is spent on it. They do not feed themselves properly or bother to keep themselves and their living quarters clean. Heroin addicts seldom sterilize their syringes and often end up with blood poisoning and related conditions. Cocaine sniffers usually lose the partition between their nostrils: they are also prone to mental illness. Alcoholics lose their jobs, their families, their friends, their health, as well as their self-respect. Amphetamines can produce mental illness and so can LSD.

It is not certain that the use of 'soft' drugs leads on to 'hard' drugs, but it is likely that once someone is involved in the drug scene, and mixing with people who are using hard drugs, the opportunity is presented.

A very great danger with drugs is of not knowing the strength of what is being taken. They are all very easy to dilute by mixing with harmless ingredients of similar appearance, and 'pushers' often do this; a person buying drugs can never be sure of what he is buying and the effect can be disastrous.

All drugs have the effect of removing inhibitions. Some people say they drive a car better when they have had a drink, but the evidence shows that the effect of alcohol is to make them think they have better control, when in fact their reactions are much slower. The drug which most seriously affects control is LSD. It can cause hallucinations which make the users think that they can fly, with tragic results. There is no guarantee that the hallucinations will be beautiful, for they can equally be so horrific that there are permanent bad effects.

29.3 Why People take Drugs

With all the warnings about the dangers it is amazing that so many people become drug users.

1 Laws against drug taking are severe so those who take them have to be very secretive. They congregate in groups with very close ties, and adopt a superior attitude to non-drug users. This select and closed society presents attractions to people who want to be members of a close, and in their minds, prestigious group, which expresses its contempt for the conventions of society.

2 People who feel hopelessly inadequate to face the problems of life because of their immaturity and psychological instability.

3 Those who find it impossible to build up relationships with other people. They feel rejected and in their loneliness and misery find solace in drugs.

In general it is people who suffer from severe personality disabilities who become addicts. In a sense it is like a suicide attempt, a cry for help. The drug may seem to help for a while, but the only permanent solution is to treat the personality problem, and that is much more difficult.

29.4 The 'Socially Acceptable' Drugs

These vary according to the different types of society. In Islamic countries alcohol is not socially acceptable, though it is in most other parts of the world. We shall consider the two most dangerous drugs of this kind, tobacco and alcohol.

1 Tobacco This has not always been socially acceptable: for example King James I wrote a famous anti-tobacco pamphlet in 1604 entitled 'A counter-blast to Tobacco', and in recent years it is becoming less and less acceptable as a result of anti-smoking propaganda. Pupils are warned at school about its dangers, and advertisements on hoardings and newspapers have to carry a government warning, though sponsorship of sporting events provides tobacco firms with valuable publicity. Smoking is restricted in public transport to designated areas, and in shops and public places smoking is increasingly being prohibited. What only a few years ago was regarded as an expensive but relatively harmless habit is now seen to present a serious danger to public health.

It is now generally accepted that smoking adds considerably to the risk of death from lung cancer and heart disease, and to the incidence of chronic bronchitis, a very debilitating disease. Perhaps the most persuasive argument against smoking is the fact that the great majority of doctors do not now smoke.

In November 1983 the Royal College of Physicians published its fourth authoritative warning about the dangers of smoking, entitled, 'Smoking or Health?' The Report states that 100 000 people per year in Britain are dying prematurely from the effects of smoking. Doctors believe that non-smokers are also at risk from inhaling other people's cigarette smoke. Women smokers who take 'the pill' are ten times more at risk of suffering a heart attack, stroke or other heart disease.

Pregnant women who smoke harm their unborn children and impair their future development. Children of smokers are more susceptible to chest infections, are shorter than their primary school friends whose parents do not smoke, and may lag behind them by six or seven months in intellectual ability at the age of eleven.

The figure of not less than 100 000 deaths per year 'completely dwarfs the number of deaths that can be reliably attributed to any other known external factor such as alcohol, road accidents or suicide.'

The problem is described in the report as 'comparable with that of the devastating epidemics of infectious diseases of the past' such as polio, diphtheria and tuberculosis.

It is true that it is not certain that a smoker will suffer from any of these serious diseases, but the chances of it happening are greatly increased. The person who says 'I've smoked all my life but no harm has come to me.' has just been very lucky, and has also taken no account of the effect the habit may have had on other people: not only those who smoke are at risk but also those who have to live or work in a smoke-filled atmosphere.

Smoking is a difficult habit to give up because it is very addictive, and the withdrawal symptoms can be distressing. The best course is never to begin to smoke and it is very sad that most people begin the habit at school, in spite of the active discouragement of teachers.

2 **Alcohol** Serious as the drug scene has become, many informed people believe that alcohol presents by far the more serious problem. Contrary to popular opinion, it is not a stimulant, but a depressive, slowing down the process of thinking and the action of the muscles. It affects decision-making and also judgments like those required in driving a car. It removes normal restraints on behaviour, 'taking the brake off' and making it easier to do silly or dangerous things. It induces a feeling of well-being; indeed one euphemism for being slightly under the influence is being 'happy'.

In most societies alcoholic drinks have an important place in celebrations of any kind, and appreciation of good food and wine is looked upon as evidence of a civilized approach to life. Drink contributes to an atmosphere of warmth at parties, and in moderation it can be socially useful. However, alcohol can be dangerous in many ways:

1 Great discretion is advisable in its use, and young people are especially at risk. The law is not easy to enforce and young people under the legal age, when drinking with their friends, are very liable to drink to excess. Tragic accidents are too often the result.

2 Even in moderation it reduces self control and it is a major factor in violence, vandalism and crime.

3 Motor accidents have a strong association with drinking, and the toll of death and injury on the roads is appalling. If we had lost a division of men (about 15 000) killed and wounded in a military operation such as the Falklands, no government would survive, but road casualties take on something of this scale every year.

4 Alcohol is readily available in many homes and is often a great danger to young children wishing to experiment. It also proves a temptation to lonely wives who have not enough to occupy themselves, and who become dependent and eventually alcoholic, through secret drinking.

5 The incalculable misery caused to individuals and families by excessive drinking.

6 The problem of alcoholism. There is a state of such complete addiction that the victims are obsessed by the drug and live in a continual state of intoxication. Whatever money is available is spent on drink and in the end they lose their jobs, family, friends and self respect. They often live among groups of vagrants in the slum areas of our great cities in conditions of filth and squalor. It

is a life of degradation, and there is little hope for any who do not eventually find their 'own personal gutter'. No one can tell at the beginning, who will become an alcoholic.

7 Vast sums of money are spent by the drink trade to persuade people that alcohol is associated with manliness, sexual potency, career success, the fashionable trend, and health. These associations have been created by the media and have little relation to fact. Alcohol is a poison, not a food.

Because governments derive so much money from taxes on both tobacco and alcohol, they are reluctant to ban either of them, but in fairness it must be said that it would be impracticable to make their use illegal. The government of the USA tried to enforce 'Prohibition' of alcohol from 1920–1933, but it was a disastrous failure leading to a vast increase in crime. Gangsters like Al Capone and Legs Diamond made huge fortunes in smuggling and distributing alcohol, murdering and generally using strong arm methods to preserve their 'empires'. No one seems keen to revive the experiment. In Britain some kind of control is imposed through the licensing laws which regulate the opening hours of public houses.

Alcoholics Anonymous

This is an organization which has branches in all big cities. Its members all have a drink problem, must have acknowledged themselves to be alcoholics (this is one of the most difficult things for an alcoholic to do), and be resolved to be 'cured'. The members meet regularly to discuss their problems, and to support and encourage each other: it is group therapy in which each works for all. It is acknowledged that there is no cure for alcoholism–once an alcoholic always an alcoholic, and a single mouthful of drink can put a 'dried out' addict back where he started, on the bottle. Every telephone book lists Alcoholics Anonymous, and a call can put a victim in touch with someone who will offer help.

Biblical references

The Bible is not against drink, but is very much against drunkenness.

1 Psalm 104:15. For the psalmist, one of the gifts of God is 'wine to gladden men's hearts'.

2 Amos 2:12. 'You made the Nazarites drink wine.' The Nazarites were people who protested against the corruption of the pure religion of Israel's years in the wilderness, by the influence of the immoral religions of the people of Canaan. They did this by abstaining from any product of the vine, for the vine is a plant of cultivated land, not of the desert. Their abstention was a symbol of their desire that Israel should return to the pure faith.

3 (a) Matthew 11:19. 'Look at him! a glutton and a drinker.'
(b) John 2:1–10. The wedding at Cana in Galilee.
The first reference is to a jibe made by his enemies because Jesus ate and drank with sinners. John's reference, if taken literally, indicates that Jesus could not have disapproved of the drinking of wine.

4 Luke 22:17–19. The Institution of the Lord's Supper. Wine was used, not grape juice, as some would plead. The grape harvest in Palestine is in October, and it would not be possible to keep grape juice for six months without fermentation. The lesson is that what Jesus used for the highest sacramental purpose, some use to make beasts of themselves.

5 1 Corinthians 8:13. 'If food be the downfall of my brother, I will never eat meat any more.' The reference is to the eating of meat sacrificed to idols, but some Christians argue that it applies equally to alcohol. If drink is the cause of my brother's downfall I will abstain so as to strengthen him by my example.

6 Ephesians 5:18. 'Do not give way to drunkenness and the dissipation that goes with it.'

7 1 Timothy 5:23. 'Stop drinking nothing but water; take a little wine for your digestion, for your frequent ailments.'

Paul commends the use of wine as a medicine.

30 Medical Ethics

30.1 Abortion

The intention of the Abortion Act of 1967 was to regulate a serious social evil. It was calculated that up to a quarter of a million illegal back street abortions were being carried out in Britain every year, many of which resulted in serious injury or death. The evil existed, and simply would not go away.

The new act gives doctors the right to perform an abortion under certain conditions, provided that two doctors agree that the patient qualifies, under the terms of the Act. The conditions are:

A 'That the continuance of the pregnancy would involve risk to life of the pregnant woman, or of injury to the physical or mental health of the pregnant woman, or any existing children of her family, greater than if the pregnancy were terminated;
or
B that there is substantial risk that if the child were born it would suffer from such physical or mental abnormalities as to be seriously handicapped.' It adds that the actual or foreseeable environment of the woman should be taken into account.

There is a conscience clause protecting doctors who do not wish to take part in abortions. The alternatives were seen to be, either:

1 back street abortions with the appalling attendant dangers and suffering;
or
2 legal abortions under controlled conditions, in hospitals, with the minimum of risk involved.

There can be no doubt of the good intentions behind the Act, but its application has led to a great deal of disquiet even among some of those who strongly supported it in Parliament. One effect has been to make London the abortion centre of Europe: in 1982, 20 000 Spanish women travelled to London for abortions and the BBC has recently documented the existence of abortion traffic from France. Every weekend bus-loads of continental women travel to London for a quick weekend abortion, and since abortion is illegal in Eire, many Irish girls travel to London for the same purpose. There is even an accusation that taxi drivers tout for business at airports. None of this was intended by most of those who voted the Act through Parliament.

Some people feel that the more serious problems arise because it is possible to interpret the Act with a fair amount of freedom. It leaves a great deal of discretion to the doctors, and the inclusion in the conditions of such factors as the well-being of the existing family, is capable of a loose interpretation. It is felt that whereas the Act does not legalize abortion on demand, in practice it very nearly does so.

The moral issue

Behind the problem of abortion lies the difficulty of definition. There is no single point in the process of conception, pregnancy and birth at which it is universally agreed that we are dealing with a foetus rather than a human being. Crucial to the argument is the point at which the foetus becomes a human being. Medical opinion is very much divided on the question of when independent life may be said to exist, and the reason why this is so important is that to abort when independent life exists or could exist with proper medical care, is to kill a human being. The Roman Catholic Church takes the firm line that the unborn child is a human being from the moment of conception, and therefore forbids abortion. This attitude has the advantage of offering the only firm point of definition, but others feel that it is unrealistic, taking the view that abortion is morally acceptable up to the point at which the foetus is capable of independent existence. This has the grave disadvantage of being incapable of definition because it is not possible to know with sufficient accuracy how far gestation has advanced. In 1983 there was a celebrated case in which a surgeon was prosecuted after aborting a baby which survived, and has since been adopted.

The case against abortion

1 The life and well-being of the mother and the existing family take precedence over the life of the unborn child. However, the foetus is a life which already exists, and though incomplete, it has some rights, difficult as it may be to establish what. There is a moral presumption that a foetus has a right to life.

2 The evidence of surgeons and nurses shows that abortion is very much a matter of killing a living creature, and this evidence is reinforced by illustrations in many colour magazines which have brought this fact to the notice of the public. Improvements in medical techniques have made possible the survival of even younger premature babies, and the point is being reached at which the distinction between abortion and infanticide is blurred.

3 Abortion involves the acceptance of a dangerous principle, that life is expendable under certain conditions. In the violent world in which we live, we need rather to stress the sanctity of human life; to fail in this is to 'devalue' human life to the level of an intelligent animal.

4 The clause in the Act legalizing the abortion of a foetus which is at risk of being seriously deformed or handicapped, presents particular problems. The Humanist point of view is that life is valuable but not sacred, and that there must be a certain quality of life before it can be worth living: it must be possible to enjoy the world through the the senses (how many of the senses are needed is not clear). However, many physically and mentally handicapped people live happy lives while some people who have all their faculties are very unhappy indeed. The Humanist view that somewhere a line has to be drawn below which life is not worth living, still begs the question of where to draw the line, and who is to draw it.

5 Abortion is not the easy way out that some suppose. There are physical risks, and also psychological risks, which are insufficiently taken into account. Bearing a child is so much part of the physical and mental make-up of women that an abortion can lead to considerable psychological disturbance, particularly a deep sense of guilt, as well as physical risks.

The case for abortion

1 It has been practised for thousands of years and is certain to continue whatever laws are passed prohibiting it.

2 Women are driven to resort to back-street abortions in sheer desperation. They will risk injury and death to resolve their predicament. The alternative to legal abortion is therefore back-street abortion with all its attendant dangers. Before the Act a great deal of money, time and medical skill had to be devoted to saving the lives of women badly damaged in illegal abortions.

3 No one but the mother knows what her motives are, the stress which she is under, or her ability to cope with her situation. Only she can decide whether she can cope with another child, or with a child who may be deformed, and of course the same applies to an unmarried mother. It has to be a matter for her conscience alone.

4 A strong case can be made for abortion in the following circumstances:

(a) Where there can be shown to be serious risk to the life or mental stability of the mother.

(b) In a case of criminal rape. There was a celebrated case in 1938 when a doctor terminated the seven week-pregnancy of a raped 14-year-old girl, and then invited prosecution: he was acquitted.

(c) When the girl is mentally deficient–on the grounds of lack of consent.

(d) Arguably when the girl is under age–on the grounds of lack of consent.

(e) Arguably when there is grave risk of serious physical or mental deformity. Only parents with grossly handicapped children know the agony of their situation, not forgetting any other children of the marriage.

30.2 Euthanasia (Mercy killing: the right to death)

'I would not treat a dog the way some old and sick people are treated.' This may bring to mind memories of a very dear old family dog who had reached the end of his life. Blind, arthritic, incontinent, in pain. Was it kind to allow him to suffer when the only prospect was further deterioration and suffering? However much we grieved over it we had him painlessly put down. If we consider that to be an act of love towards an old dog, how can we refuse such relief to an old person dying painfully from an incurable disease? Why prolong the agony when life cannot be worthwhile? We all have the right to die in dignity. It is in these terms that the case for euthanasia is often expressed.

There is a strong case against it:

1 If a patient is in unbearable pain, the doctors are to blame. There is a wide choice of pain-killing drugs and they can be varied to avoid the build-up of tolerance.

2 The patient has the right to life and should therefore be the one to take the decision; he might not however be capable of decision, and if so who is to decide? The patient may change his mind but be incapable of indicating this.

3 We cannot be sure that the doctor has made the correct diagnosis. Many people have recovered after being 'written off' by doctors.

4 Old people often feel themselves to be a nuisance to their younger relatives and may feel that they owe it to their loved ones to accept the offer of euthanasia if it is available. Even if they are in a geriatric ward and being well cared for by devoted staff, visits by relatives can be costly in terms of time and money and they may feel they ought to be willing to be 'put to sleep' though they would prefer to go on living. Old people have as much right to life as anyone else, and it would be wrong to put such pressures on them.

5 For the Humanist, if life is no longer 'worth-while' it is immoral to prolong it. For the Christian, however, life is sacred: it is a gift from God, and only God can take it.

6 Euthanasia devalues life by making it disposable. Many feel it to be the first step on to a slippery slope: once the principle is breached there is no knowing what the consequences will be. Hitler began with the mentally deficient, then tramps, gipsies, communists, Jews, Lutherans, Catholics, and anyone else who opposed him.

7 It would be wrong to expect doctors to kill patients. The Hippocratic Oath which all doctors take, expressly forbids it. The relationship of trust which exists between doctor and patient is of vital importance to successful medical practice, and it is based on the patient's conviction that the doctor can be relied upon to do all that is humanly possible to preserve life. This relationship would be eroded or destroyed if doctors were empowered in certain cases to take life. Drugs which alleviate pain can kill, and a doctor who is using all his skills to enable his patient to die without unbearable pain, may be on a knife edge. It could be difficult to know in such cases whether the disease or the drug was the final cause of death, and what counts in such cases is the intention of the doctor. If it is to kill it is morally wrong, if to relieve pain, it is not.

'Exit'

This is a pressure group which advocates that people should be given the means of taking their own lives in an efficient and painless manner if they become 'terminally ill'. It is against the law at present to encourage or assist anyone to commit suicide. Popular opinion is probably still strongly influenced by the fact that until fairly recently, suicide was a crime punishable by law. There was the ridiculous situation that only those whose attempt at suicide failed, could be punished, for those who successfully committed the 'crime' were unpunishable.

The old law derived from the teaching of the Church that to take one's own life was the ultimate sin, because it involved throwing back into God's face His gift of life. Though it is no longer punishable by law, suicide is still under normal circumstances, a sin (not in the case of those whose mental balance is disturbed). There is a distinction to be made between a crime and a sin: a crime is an offence against the state, and in this country is almost always at the same time a sin, but a sin is an offence against God. 'Exit' then involves the approval of a sin. It also makes old or terminally ill people subject to subtle but powerful pressures from relatives.

Biblical references

1 2 Samuel 1:6–16. The death of Saul. He was mortally wounded and the enemy was closing in on him. He asked an Amalakite soldier to dispatch him. When this soldier brought the news to David, expecting to be rewarded, David had him put to death for killing 'the Lord's anointed'. Rather harsh under the circumstances we may think.

2 I Kings 19:1–9. Elijah was forced to flee from the wrath of Jezebel after the massacre of the prophets of Baal. He came to the wilderness and sat down under a bush, praying for death. 'It is enough' he said, 'now Lord take my life, for I am no better than my fathers before me.' An angel of the Lord came and provided him with food which sustained him until he reached Horeb, the mount of God. When the prophet felt that his life was over, God still had important work for him to do.

3 Matthew 27:3–10. The suicide of Judas. Peter and the rest had denied Jesus though they had sworn allegiance until death. There is no reason why Judas also should not have obtained the forgiveness of Jesus and joined with Peter and the rest in the mission of the Church.

4 The Ten Commandments (Exodus 20:1–17) (See Unit 13.10).

5 The Sermon on the Mount (Matthew Chapters 5–7) (See Unit 6.5).

31 The Population Explosion

31.1 Introduction

Between 1900 and 1970 the population of the world doubled and it is expected to double again by the year 2000 by which time it will reach the estimated total of over 7300 million, of which 5500 million will be living in the Third World. The population explosion as it is called is due to the increased ability of medicine to control disease. Fewer babies die in their early years and the expectancy of life for adults has become much higher. In the wild, animal populations are controlled by natural forces, including the toll of life taken by predators, overgrazing and disease, and in the past human population has also been controlled by natural wastage of the same kind. It would say little for man's intelligence or compassion if, in the 20th century, he relied on famine, disease and war to control the increase in population.

Every country is aware of the danger of having too few resources and too little food to provide for an increasing population, and it is recognized that the problem has to be tackled on two fronts: (a) by growing more food and finding means of sharing it more equitably and (b) by controlling the population explosion.

31.2 To Grow More Food is not Always Easy

1 The Third World countries are mostly tropical or semi-tropical and a large part of the land is either mountainous, jungle, or desert. In other words there are geographical problems. Agriculture requires a regular supply of water and in many of these countries rain is seasonal and if it fails to arrive, there is drought and famine. In countries like India the river systems are subject to severe flooding. The desert is spreading every year, mainly because the cutting down of trees and overgrazing leads to erosion of the soil.

2 In many countries farmers cultivate very small plots of land which make it impossible to use anything more than primitive tools. The land provides enough food for the farmer alone in a good year, but there is no surplus to sell in order to buy fertilizers and better tools. In a bad year he has to borrow and is then saddled with the repayment for many years. This is called subsistence farming. In some countries dried cow manure is the principal fuel for cooking, and therefore the land is starved of a valuable fertilizer which, incidentally, is also the only one available. Increasing infertility is the result.

3 Local traditions and religious customs can also delay improvements in farming methods. The Masai of Central Africa measure their wealth by the number of cattle they own rather than their quality and this results in overgrazing of land which in fact would be better used for growing crops. In India millions of sacred cows wander the countryside, eating the herbage but producing almost nothing.

4 To produce more food by increasing the size of farms and employing modern machinery and farming methods, could make the problem worse by throwing thousands out of work.

5 The economy of many Third World countries depend mainly on a single crop–coffee in Uganda, tea in Kenya, cotton in Sudan, cocoa in West Africa–and as a result much food has to be imported. Fluctuations in world prices can lead to great hardship.

6 In spite of the breeding of new strains of corn and rice which have greatly increased yields, the population of the Third World is rising faster than the supply of food can cater for.

7 Vast sums of money, far beyond the resources of the poorer countries, are needed in order to control the effects of climatic conditions, conserve natural resources, educate the farmers, provide them with better tools, and assist them in building a balanced economy.

31.3 Population Control

1 There is a sense in which poverty is a major cause of the population explosion rather than its effect. In countries where it has been the normal expectation that many children will die in infancy, large families are necessary in order that some may survive to support the parents in old age. There is no other provision for the aged.

2 Since many more babies survive as a result of improved medical attention, and old people for the same reason live longer, the working man finds himself with a large number of children to support, in addition to his aged parents. This is often a crippling burden, reinforcing an already poverty-stricken situation. Poverty is a vicious circle. Insufficient food leads to sickness and reduced energy which lead to less work done, which leads to poorer crops and less money, which lead to less, and so on.

3 Illiteracy increases the problem because vast numbers of people are unable to read the information about family planning which has therefore to be communicated through personal contact, a laborious process. In some countries very few women can read.

4 Religious problems. In Muslim countries, and others, where a man may have several wives this can lead to large families. The Roman Catholic Church forbids the use of artificial methods of family planning and some of the highest birth rates in the Third World are to be found in Roman Catholic countries.

5 Improved standards of living could contribute to the solving of the problem of over-population, for it can be seen that in richer countries most people do not desire large families. However, improvements could only be financed from outside the Third World as a result of the richer nations resolving to share the resources of the world more fairly. The United States has six per cent of the world's population but consumes forty per cent of the raw materials produced in the world each year, and most of the Western nations could show similarly embarrassing figures. The rich countries of the world in the United Nations have agreed to give one per cent of their income to help poorer countries, but the amount given has in most cases fallen short of this very modest target.

6 The best assistance is direct grants of money to pay for such things as dams, irrigation systems, roads, machinery such as ploughs and tractors, medical programmes, education, salaries for engineers, agricultural experts, and teachers. Loans are helpful but are a second best because the need for repayment adds to the burden. Private investment by richer countries is also a help but there is a danger that private companies can be more concerned with profit than with helping the poor country. The stimulation of trade could greatly contribute, but in view of fluctuation in the price of raw materials, the richer countries would need to agree to the stabilizing of prices at a fair level. Help given by charities such as Christian Aid, Save the Children, the Red Cross, Roman Catholic organizations and Oxfam, though relatively small, is of great practical assistance.

Biblical references

1 Amos 6:1–6. For his message see Unit 17.2. In this passage the prophet condemns those who live in luxury 'who loll on beds inlaid with ivory . . . feasting on lambs from the flock and fatted calves . . . who drink wine by the bowlful,' while the peasants live in poverty.

2 Amos 8:4–7. They have all the appearance of being religious but they cannot wait to plunder the poor, giving short weight and selling inferior grain. They are able to bribe the judges for the price of a pair of shoes. In both of these references the rich are condemned for exploiting the needy.

3 Matthew 25:81–46. The Great Assize. Men are judged by their willingness to meet any human need.

4 Acts 6:1–6. The early Church accepted responsibility for the support of widows.

5 Acts 11:27–30. The Christians in Antioch sent relief to those Christians in Jerusalem who were suffering as a result of a famine.

6 James 2:14–17. Christians have a duty to supply those who have insufficient food or clothing.

7 I John 3:17. Love of God needs to be expressed in deed not words.

Charitable organizations

Christian Aid, P.O. Box 1, 2 Sloane Gardens, London, SW1
Oxfam, 274, Banbury Road, Oxford

Part IV World Religions

32 Judaism

32.1 Background

The central belief of Judaism is expressed in what Jews call the **Shema** (Deuteronomy 6:4–9; 11:13–21 and Numbers 15:37–41), the first words of which are, 'Hear, O Israel, the Lord our God is one Lord, and you must love the Lord your God with all your heart and soul and strength.' The Hebrew word 'shema' means 'hear'.

This is the most important Jewish prayer, repeated by every devout Jew every morning and evening. It is the first prayer he learns as a child and the last he is required to say at death. It is both the foundation stone and key-stone of the Jewish religion.

1 The Shema states emphatically the concept of monotheism, which means that God is not dependent for His existence or His nature upon any reality other than Himself: He just is, as He said to Moses at the Burning Bush, 'I am; that is who I am' (Exodus 3:14). It follows that God is without beginning or end, for if He had a beginning there would have to be a reality capable of bringing Him into existence, and for Him to have an end, there would have to be a reality capable of ending that existence.

2 God is the creator of everything that is: He did not create the universe out of already existing material as a builder builds a home, but out of nothing. If anything existed before God, it would take priority over Him, and Jewish belief is that there is a complete distinction between God and everything He has created. If the whole universe should perish, God would still exist; the universe, including man, depends upon God for its continued existence from day to day. Man has existence not by any natural right, but as a gift from God.

3 The Bible makes it clear that God is personal, which is not to say that He is a person, for that would be to limit His being, suggesting that He is merely an enormously powerful man. It means that He is at least personal, for since He has created the human personality, He cannot Himself be less than personal. The Bible speaks of God in personal terms as 'the God of Abraham, the God of Jacob, and the God of Isaac' (Exodus 3:6).

4 The prophets of the Hebrew Bible, especially Second or Deutero-Isaiah (See Unit 21), proclaimed a God of righteousness and love. Jews believe that God created the universe for a purpose; it is not an illusion, as some religions teach, nor is it an endless cycle, but rather a progression towards the fulfilment of a promise–the coming of the 'Kingdom of God', which will be built on earth by men, under the guidance of God who works through all things including individuals and nations. It will be inaugurated by the Messiah, a leader sent by God: it is believed that Israel will be restored to the promised land, and peace, justice and freedom will be established all over the world. Some Jews believe in a Messianic Age rather than a personal Messiah, but this alternative is not regarded as unacceptable.

5 Life is not a burden to be endured but the gift of God which is to be received thankfully. This delight in and gratitude for God's gifts is expressed in the opening words of the typical Jewish prayer 'Blessed art thou, O Lord, our God, King of the universe, who . . . ' and it goes on to specify some reason for grateful thanks which might be anything–a sight of the sea, a tree in blossom, beautiful music, new clothes, good food or drink, a peaceful night–anything at all, for which thanks are felt to be due, There is no distinction in the Jewish Law between religious and secular, spiritual and material; everything is made by God so everything in life is religious and as a result the life of the devout Jew is a round of blessings.

6 While the later books of the Hebrew Bible look to an after-life, the emphasis in Jewish practice is not so much on preparing for the next world as guidance through this one. Judaism is a highly moral religion. Man is a partner of God in working out the divine purpose, and this is to be achieved by the 'imitation of God', the practice of mercy and righteousness as laid down in the Torah, the Law, a heavy responsibility for every Jew. Men are not perfect and sometimes fail, but true repentance brings forgiveness from a loving God.

7 The Bible says that the Jews are the 'chosen people' of God. This is not taken to mean that God has favourites nor that Jews are in many ways superior to other people, but it does mean that they were chosen for the special task of revealing the nature of God to mankind; this is not disputed by either Christians or Jews.

8 It is believed that the wicked go to 'gehenna' after death, a place of fire, and after a short time they go to 'Gan Eden', where, with the good, they enjoy the Shechinah, the presence of God. This is not a static state of being, for progress in goodness can be made. The non-Jew who lives a good life may also achieve this heavenly existence. Some Jews believe that since the Scriptures do not give much detail about the after-life, it is best to treat what is given as symbolic and trust in the mercy of God.

32.2 The Scriptures

The most important section of the Hebrew Bible is the **Torah**, sometimes called the Law, but it could more accurately be translated the 'teaching' or 'instruction'. It consists of the first five books of the Bible, Genesis to Deuteronomy (the Pentateuch), and tradition attributes its authorship to Moses.

The second section, **Nev'im**, consists of the books of the prophets, which includes the books of Joshua, Judges, Samuel, Kings as well as Isaiah, Jeremiah, Ezekiel and the ten 'minor' prophets. The word 'minor' does not refer to their value, for some, such as Amos, Micah and Hosea are very important, but rather, to the size of the books.

The third section, **Ketuvim**, or writings, contains eleven books which are miscellaneous in character, consisting of Chronicles, Psalms, Job, Song of Songs, Ruth, Lamentations, Ecclesiastes, Esther, Daniel, Ezra and Nehemiah.

The canon of Jewish Scripture, that is, the books accepted as authoritative in Judaism, was not complete until the second century CE. (Note that since Jews do not accept that Jesus was the Messiah, they prefer not to use the abbreviations BC and AD when quoting historical dates: instead they use CE–the Common Era, and BCE–Before the Common Era.)

Moses is the chief of the Jewish prophets, and **Isaiah** is held in particular regard. **Elijah** has a prominent place because he is to return to earth to announce the Messiah and the subsequent Resurrection. The Patriarchs or founding fathers of Judaism, Abraham, Isaac and Jacob, are constantly remembered.

Jews believe that Moses was given two Torahs, the written Torah, the Bible, and the oral Torah, handed down from generation to generation. This latter was eventually collected and written down in about 200 CE by the great rabbi, **Judah, 'the Prince'** (135–217 CE) in the interest of its accurate preservation; it is known as the **Mishnah**, meaning 'teaching by repeating'. It contains rules and regulations relevant to every aspect of Jewish life down to the most minute detail. This huge work is divided into six sections, the most widely known of which is the **Pirke Avot**, which deals with behaviour, and contains the sayings of many important rabbis over a period of 500 years.

After the death of Rabbi Judah, the Mishnah became the subject of much study and discussion; the teachings which resulted, called **Gemara**, were added to the Mishnah and the complete work was called the **Talmud**. It is said to contain about 3 million words and about 2000 authors contributed to it. It is not, however, the last word. Rabbi Moses Maimonides (1135–1204) made an extensive contribution to it and rulings are still being made today. One modern problem, for example, was how to keep the sabbath when flying over the international date line.

The **Midrash** is another important writing consisting of the exegesis, or explanations and comments on Biblical passages, from the sermons of early rabbis.

There are **two forms of the Talmud, the Palestinian and the Babylonian**, originating in the two main centres in which Rabbi Judah's work was studied. The Talmud ranks second only to the Bible in importance in the Jewish religion.

Jews believe that God revealed Himself completely on Mount Sinai when He gave the Torah to Moses, and nothing needs to be added to it, for it contains everything which is necessary for man to know. The duty of the Jew is quite clear, namely that he should carry out the Law in his daily life. Moses is the only man who ever talked with God face to face, 'There has never yet risen a prophet in Israel, a prophet like Moses, whom the Lord knew face to face' (Deuteronomy 34:10). His revelation of God is unique.

32.3 The Synagogue

In 586 BCE Jerusalem was captured by the Babylonians, the Temple was destroyed and the people were led away into captivity. Up to this time the Temple was the only place where the worship of God could take place, so its destruction, as well as the exile of the people, threatened to bring the worship of God to an end. The people, however, gathered on the sabbath 'by the rivers of Babylon' to read and meditate upon the Scriptures and to pray, and it is likely that this was the origin of synagogue worship. It needed no Temple or priest and on the return from exile the practice was preserved. Certainly by the time of Jesus of Nazareth there were synagogues in every major city in the Roman world as well as in every village in Palestine. In Jerusalem, as well as the Temple, there were several hundred synagogues belonging to the many foreign communities of Jews who visited the city on pilgrimage. (For the synagogue in New Testament times, see Unit 1.5.)

The last Temple was that of King Herod, which was destroyed by the Romans in 70 CE during the capture of the city: many thousands of Jews were killed, and the disaster would probably have put an end to the Jewish religion had it not been for the synagogues. No priest was required at synagogue services as at the Temple because there were no sacrifices, and prayer and the reading of the Scripture became the sole form of worship. There was only one Temple but hundreds of synagogues, and this enabled the faith to survive the destruction of the central shrine.

Important as the synagogue is, it is not the centre of Jewish life; as we shall see, the home is much more important. The synagogue building is not itself holy, it is the people who are holy, in the sense that they have been chosen and set apart for the service of God.

The word 'synagogue' comes from a Greek word meaning a meeting place, and this accurately describes the function of the building. Most have rooms and a hall attached in which all kinds of social activities take place such as Youth Clubs, Young Wives, organizations for older people, but the main purpose of the synagogue is prayer and study.

32.4 Architecture and Furnishings

There is only one set architectural feature in any synagogue, namely that the interior is so planned that the congregation faces Jerusalem. The exterior can be in any style but frequently Jewish symbols such as the star of David and the menorah (the seven-branched candlestick) appear as decorations.

Furniture

Whatever their style or size, there are certain features common to all synagogues.

1 The Holy Ark, **Aron Hakodesh**, a cupboard set in the wall which faces Jerusalem, in which are kept one or more copies of the Pentateuch, the Five Books of Moses (Sefer Torah). These are handwritten in Hebrew on parchment with a quill pen and specially prepared ink. It takes more than a year for a skilled scribe (a Sofer) to complete a scroll, so exacting is the standard required. The ark is a reminder of the Ark of the Covenant in the Temple which contained the two 'Tablets of Stone' which Moses brought down from Mount Sinai, and on which were written the Ten Commandments. It is the focal point of the synagogue.

2 The Ark was placed in the innermost shrine of the Temple and the entrance to this windowless room was covered by a veil. The ark in the synagogue, as a reminder of this, is covered by an embroidered curtain called a Parochet.

3 Above the ark is a representation of the two tablets of stone on which were written the Ten Commandments. The opening words of each of the commandments is inscribed in Hebrew on the tablets, five on each.

4 In front of the ark is the **Ner Tamid**, the perpetual light, as a remembrance of the light kept burning day and night in the Temple before the Ark of the Covenant.

5 In the centre of the synagogue is a raised platform called a **bimah** on which is a reading desk from which long passages from the Torah are read on sabbaths and festivals throughout the year.

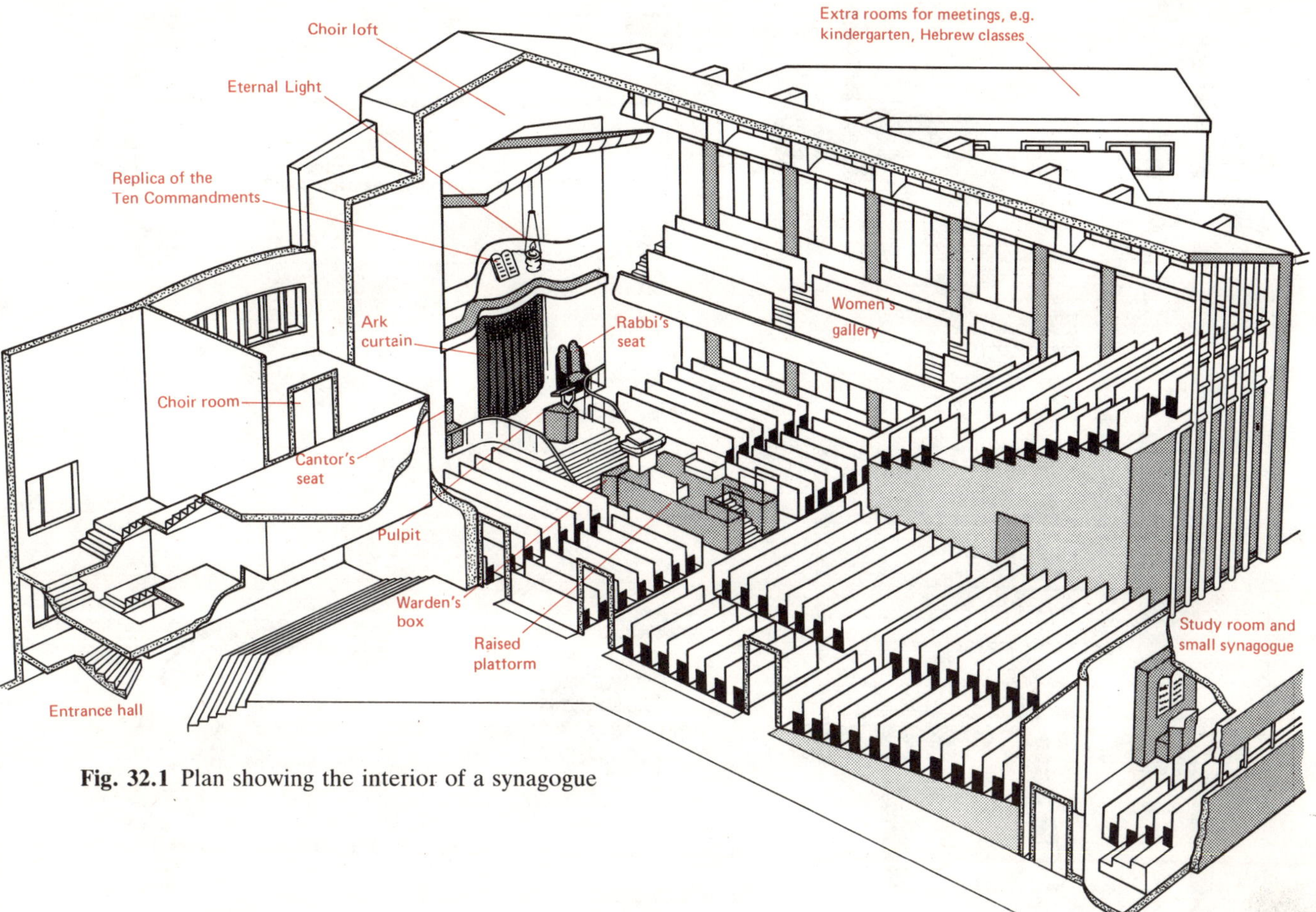

Fig. 32.1 Plan showing the interior of a synagogue

This is a reminder that Ezra on his return from Babylon read the book of the Law to the assembled people and 'stood on a wooden platform made for the purpose' (Nehemiah 8:4). Each of the five books is read through in a year and Jews all over the world read the same portions each week. The services may also be read from a small lectern standing before the ark.

6 The **scrolls** are on wooden rollers and are covered by an embroidered mantle of velvet or silk. On top of the rollers is placed a crown signifying the majesty of God's Law, and small bells are attached which tinkle when the scrolls are carried in procession. This is a reminder of the golden bells worn on the garment of the High Priest when he entered the holy Place in the Temple (Exodus 28:34). On the mantle hangs a silver breast-plate, recalling that worn by the High Priest when ministering in the Temple.

7 When the Law is to be read, the attachments are removed and the scroll is opened. A finger-shaped pointer called a **yad** is used to indicate the passage so as not to soil the sacred scroll.

8 The **menorah**. In the Temple there was an oil-burning candelabrum with seven branches which was kept burning day and night. Each synagogue has one and the menorah has become a distinctive symbol of Israel, as well as its most ancient.

32.5 Worship

In orthodox synagogues men and women are seated separately, recalling the Court of the Women in the Temple, by which women were segregated. Men wear hats or a skull cap known as **kippa** or **capple** and also the **tallith** or prayer shawl, which has fringes at each of the four corners to serve as a reminder to obey all the commandments of God (Numbers 15:39–40).

The phylacteries or tefillin are little black leather boxes worn by men only, containing four Biblical passages relating to the deliverance of Israel from bondage in Egypt and the giving of the Law on Mount Sinai. It is decreed in each case, 'You shall have the record of it as a sign upon your hand, and upon your forehead as a phylactery.' (Exodus 13:16 and Deuteronomy 6:8). One phylactery is placed on the inside of the left arm close to the heart, attached by a strap which is then wound down the arm and around two fingers. The other is attached to the centre of the

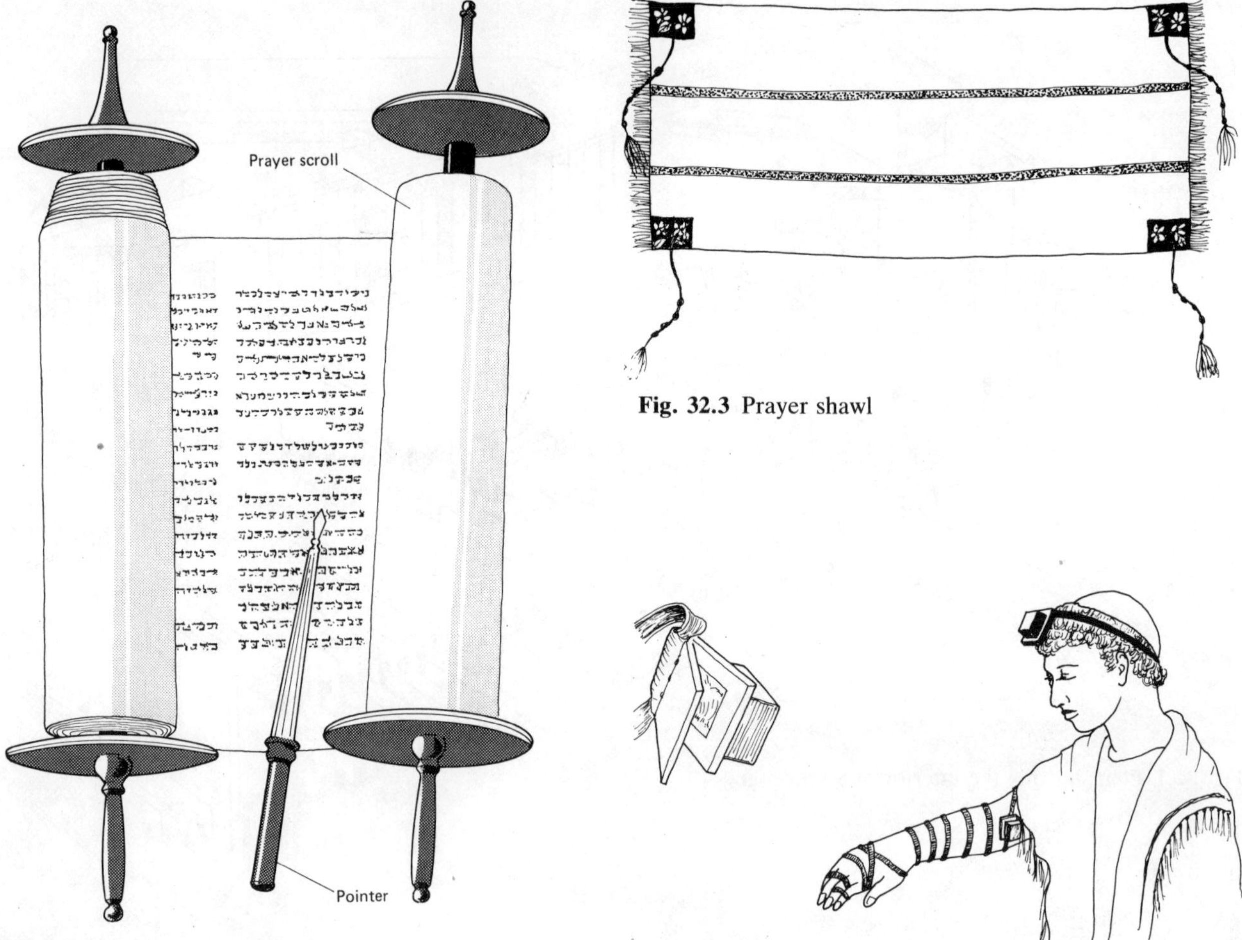

Fig. 32.2 Prayer scroll with pointer

Fig. 32.3 Prayer shawl

Fig. 32.4 Phylacteries or tefillin

forehead by straps which are knotted at the back. They serve to remind Jews of their duty to devote hand and heart and mind to the service of God. They are worn at the weekly morning service whether at home or at the synagogue. They are not worn at sabbath or festival morning services.

There are three fixed periods of prayer at home or in the synagogue, morning, afternoon and evening, corresponding with the hours of sacrifice in the Temple. Some of the prayers are very ancient, probably about 2500 years old.

The climax of weekly worship is on the **sabbath** (Saturday) morning when the ark is solemnly opened and a scroll is lifted out and carried around the synagogue, the bells on the crown tinkling as it proceeds. There is a reading from it in the original Hebrew, normally chanted in an ancient mode. Members of the congregation are called up to recite blessings before and after the reading. After the reading the scroll is again carried around the synagogue, and members may touch it with the fringes of their talliths; they then kiss the fringe.

Passages from the prophets which have some link with the reading from the scroll are read, and there is usually a sermon delivered by a **rabbi.** The word 'rabbi' means a teacher, and his original function was to interpret the Torah and teach how it was to be observed, but since the early nineteenth century rabbis have taken on duties which are similar to the pastoral duties of Christian ministers, for example visiting the sick, as well as officiating at marriages and funerals.

Any male Jew over the age of thirteen may lead services provided that he has the consent of those present, but congregations try to have a prayer leader who is both experienced and gifted with a good singing voice, because prayers are largely sung to traditional modes. He is known as the **chazan** or cantor. The singing is unaccompanied, possibly because it is thought wrong to play music on the sabbath. Larger synagogues have a rabbi and a chazan, but the rabbi plays relatively little part in the service itself. He gives the sermon and recites the prayers for the Royal Family and the State of Israel. Visitors are often surprised at the informality of synagogue services. Many prayers are offered in silence but at other times people come and go, talk to each other, and children sometimes play among the worshippers. This is an expression of the fact that they are 'at home' in the synagogue and are unlikely to be silent at home.

Worship is regarded as public when a quorum (minyam) of ten men over the age of thirteen are present.

The traditional greeting on the sabbath is '**Shabbat Shalom**', a peaceful sabbath. No work of any kind may be done except when a life is endangered, no writing, no handling of money or operating a machine. At service no collection plate is carried around and it would be considered improper to associate money with worship. Orthodox Jews live within walking distance of the synagogue since they must not drive or use public transport on the sabbath. This tends to concentrate the Jewish population around the synagogue so that Jewish districts grow up.

The five-day week has made it much easier for Jews to observe the sabbath, and may account for the present trend towards orthodox observance. The prohibitions give the impression that the sabbath is a very dull day, but this is not so. It is a joyful day devoted to retreat from the world, to the worship of God and enjoyment of family life. The sabbath is welcomed as a great blessing: it is a memorial of the Creation and is a reminder of man's utter dependence on God for his very life and all the good things he enjoys.

32.6 Education

Children begin their religious education at the age of five at the religion schools after normal school hours and on Sunday mornings. Because Hebrew is the language of the Torah and of synagogue worship, children are taught to read and understand it.

At the age of thirteen a Jewish boy becomes **Bar-Mitzvah**, meaning 'son of the Commandment'. He will have completed an intensive course of instruction and will have learned to translate the passages of Scripture from the scroll of the Law, and the writings of the prophets which he will read in public at the Bar-Mitzvah ceremony by which he is made an adult member of the Israel of God. He receives his tallith (prayer shawl) on this day and also his **siddur** (prayer book).

In ancient times girls received their education from their mothers but increasingly girls attend religious classes and are 'confirmed' at a service at which they become Bat-Mitzvah, or 'daughter of the Commandment' when they are twelve, not thirteen years of age. It is usual to have a party at home to celebrate these ceremonies.

32.7 Rites of Passage

1 Birth

This is an occasion of great rejoicing. Jewish tradition teaches that there are three agencies in the creation of a child, the father, the mother and God. If the baby is a girl the father is '**called up**' to the Torah on the next sabbath and the child is given a name, and if a boy, he is **circumcised** on the eighth day by a **mohel**, a specially qualified man. Circumcision was the sign of the covenant between God and Abraham, the founding father of the nation. 'Every male among you in every generation, shall be circumcised on the eighth day' (Genesis 17:12). At circumcision the boy is given a Hebrew name which will be used on all future religious occasions.

2 Bar-Mitzvah (See Unit 32.6)

3 Marriage

The unmarried life is not commended in Judaism and there is no tradition of celibacy. English Law as well as Jewish Law does not permit the marriage of a non-Jew in a synagogue, so there can be no synagogue ceremony for a mixed marriage. The wedding ceremony is conducted under a canopy (**chuppah**) supported by four poles, representing the future home. The rabbi gives an address, and a goblet of wine is blessed by the cantor and is sipped by the bride and groom. A plain wedding ring is then placed on the forefinger of the bride's right hand while the bridegroom says, 'Behold thou art consecrated unto me by this ring according to the Law of Moses and Israel.' The ring is then placed on the usual finger of the left hand. The marriage certificate is presented (the **ketubah**) after being signed by two witnesses. The bride and groom again drink from the goblet of wine and the final part of the ceremony is the breaking underfoot of a small wine glass by the bridegroom. The meaning of this ritual has been forgotten, but some think it is symbolic of the fact that marriage can bring sadness as well as joy; others say that it is a reminder that the Temple has been destroyed.

4 Funerals

Larger synagogues have their own cemeteries and smaller ones have a section of the public cemetery allocated to them. Jewish funerals are by tradition simple ceremonies, for the Talmud decrees that the dead should be buried in simple coffins and in a linen shroud so that the poor should not be embarrassed by a display of affluence. Flowers are not encouraged for the same reason. At death rich and poor are equal and Jews believe that this should be seen to be so.

There are many **mourning customs**, including the tearing of a lapel or the cutting of a tie. In homes, the mirrors may be covered and a special candle is lit in memory of the departed. There is a week of mourning when relatives and friends visit the bereaved, and many Jews do not take part in any entertainment for eleven months. Death is not believed to be the end: the body dies, but the spirit returns to God who created it. Embalming and cremation are not permitted.

32.8 Dietary Laws

Jews are forbidden to eat certain foods. They may eat any animal that chews the cud and has cloven hooves but none which has one only of these features. Thus the cow and sheep may be eaten but not the pig or hare. Any fish which has scales may be eaten, but eels, octopus and whales are forbidden. All birds may be eaten except birds of prey, swans, storks or partridges. The regulations are to be found in Leviticus 11 and Deuteronomy 14.

Even meat which may be eaten has to be killed ritually by a person called **a shochet** who slaughters in a way which causes the animal the minimum of pain. (Jews have always stressed the humane treatment of animals.) The blood is carefully drained away and the meat is soaked in salt water for at least an hour to remove as much blood as possible to comply with the law against consuming the blood (Leviticus 17:12).

Meat and milk dishes must not be mixed so that a beef sandwich spread with butter is not permitted and tea or coffee after a meat meal must be without milk. Many Jews use separate dishes for meat and milk meals and some even have separate dishwashers for each set of dishes.

The foods Jews may eat are called **kosher** and those they cannot eat are called **teraifa**.

Not all Jews keep the dietary laws and some do so simply because the Torah commands it, and to abide by the Law is looked on as an act of obedience.

32.9 The Jewish Home

Judaism is first and foremost a religion of the home. A **mezuzah** identifies a Jewish home at the very door, a hand-written parchment scroll on which the Shema is written, contained in a small case and attached to the right-hand door post just above eye level. Many Jews fix them on other door posts in the house. A devout Jew will, on leaving the house, touch the mezuzah and then kiss his fingers, thus expressing love for the Law of God. The use of the mezuzah is prescribed in Deuteronomy 6:9 which states that the commandments are to be written on the door-posts of the houses.

The Jewish wife holds a central position in the home as the one who organizes the religious ceremonies.

32.10 The Sabbath

To the Jew the sabbath is not only a holy, but also a joyful day. It begins on Friday evening, traditionally at sundown, which means that in winter it could begin as early as 3.30 p.m. and in summer as late as 10.30 p.m. In practice, however, it does not begin later than 8.00 p.m. All Jewish festivals begin on the evening before the day because Genesis 1 says of each day of the Creation, 'So evening came, and morning came . . . '.

Preparations for the day take place on Friday, and Orthodox Jews, since they keep strictly to the traditional time, have to ask that their children be released early from school during winter months. The sabbath is welcomed into the home by the wife lighting two candles and reciting a blessing.

When the father returns from the synagogue he recites a blessing (**Kiddush**) over a cup of wine which is then passed around. He then blesses the **two challots**, twisted loaves of bread, and everyone takes a portion. Two loaves are used as a reminder of the manna which fed the children of Israel during their wanderings in the desert–a double portion was provided on the day before the sabbath so that they would not profane the holy day by gathering food (Exodus 16:5). The father then reads the passage from Proverbs 31:10–31 in praise of the capable wife, thus emphasizing the importance of the wife in a Jewish home. 'Many a woman shows how capable

Fig. 32.5 Mezuzah

she is; but you excel them all' (Proverbs 31:29). The meal follows and hymns in honour of the sabbath are sung between the courses in many homes.

There is no breakfast on the sabbath day but after service at the synagogue a meal similar to that of the previous evening is eaten

32.11 Festivals

1 Rosh Hashanah (Jewish New Year)

The Jewish New Year begins in September/October and Jews calculate their years from the creation of the world, 4004 BCE. Rosh Hashanah is celebrated as the birthday of the world, and it is also the first of the ten days of penitence leading up to the most solemn day in the Jewish calendar, Yom Kippur. It is not a holiday, and the emphasis is on the fact that the whole world stands under the judgment of God: God's people are called upon to account for their use of the Creation which He has placed in their care. God is thought of as keeping a **Book of Life**, an account of the good and bad deeds of everyone, and Jews are called upon to express sorrow for all they have done amiss during the year. On this day Jews greet each other with a hope that their record in the Book of Life may be a good one. A portion of apple or bread dipped in honey is eaten at the meal on the eve of the day as a symbol of the hope that the coming year will be a sweet and pleasant one.

In the synagogue the robes of the ministers, the curtain of the ark and the mantles of the

Torah are all in white, the colour of purity. Morning service in the synagogue is introduced by the blowing of the **shofar**, a ram's horn, a very dramatic and meaningful part of the service because of its rich associations. It is a reminder of Abraham's sacrifice of the ram caught in a thicket by its horns, in place of his son, Isaac. It is also a reminder of the covenant given on Mount Sinai, for God came down upon the mountain in fire 'and the sound of the trumpet grew ever louder' (Exodus 19:19). On the day when Israel shall triumph over her enemies, 'the Lord God shall blow a blast upon the horn' (Zechariah 9:14). The blowing of the shofar proclaims God as the King of the universe and calls men to repentance. It also sounds at the end of the Rosh Hashanah morning service.

2 **Yom Kippur** (Day of Atonement)

Devout Jews meet at synagogue for each day of the ten day penitential period ending in Yom Kippur, the **holiest of all days** in the Jewish year. It is kept as a strict fast for the twenty-five hours, and only pregnant women, the sick and children under the age of thirteen are exempt. Many Jews give the money saved on this day to the hungry of the world. The whole day is spent at the synagogue in prayer. Atonement means at-one-ment, being at one again with God, restoring with Him the relationship which has been strained because of sin.

The opening prayer on the eve of the fast is for forgiveness for the breaking of promises made to God by not listening to, or not keeping, commandments; so before Yom Kippur Jews seek forgiveness from anyone they may have wronged. The constant theme of Yom Kippur is the duty to return to God in confession, repentance, and resolve to amend. Devout men wear the **kittel**, a long white garment in which they will eventually be buried.

The highlight of the day is the **Additional Prayer** which is in the context of a remembrance of the celebration of the Day of Atonement in the Temple. Once in each year, the High Priest entered the Holy of Holies and offered the blood of the sacrifice as an atonement for the people. Tradition describes how the priests and people standing outside the Temple courts prostrated themselves when the High Priest uttered the name of God, and when the passage of Scripture which records the events of the Day of Atonement is read (Leviticus 16:1–34) all present in the synagogue prostrate themselves at each mention of the name of God.

The afternoon service is inspired by the **Book of Jonah**, which describes how the city of Nineveh, having been warned by Jonah, the messenger of God, that it would be destroyed because of its wickedness, was spared because the king and his people heeded the warning and turned to God in prayer and fasting. The service emphasizes that atonement is available for all people.

The day ends at sunset with the **Neilah**, the Closing Service which refers to the closing of the Temple gates at sunset. The prayer beginning, 'O keep the gate open for us at the time of shutting the gate' is repeated, among others, expressing the desire of the people to be received by God. The service ends with the congregation repeating seven times, 'The Lord, he is God', the words with which the people acclaimed the triumph of the prophet Elijah over the prophets of Baal on Mount Carmel (I Kings 18:39). One long blast of the shofar in front of the open ark marks the end of the day.

3 **Succot** (Tabernacles)

Five days after Yom Kippur comes the joyful feast of Succot. For a few days before, some Jews build in the garden a **succah** (tabernacle), a small hut or shelter. It has no proper roof and it must be possible to see the stars through the branches that cover it. Its purpose is to remind Jews of the days when Israel wandered in the desert of Sinai for forty years with no permanent dwellings. The journey from captivity in Egypt to the promised land has always been an important source of inspiration to Jews, because this was the time when their religion was at its purest. In warmer climates Jews live in the tabernacles during the festival, taking meals and sleeping there, and it is usual in this country, weather permitting, to take meals there.

The tabernacle is decorated with fruit and foliage; Jewish pictures, especially scenes from Israel, decorate the walls, for Succot is also a harvest festival, the time when in Israel the harvest is gathered from the fields.

Succot is held in obedience to the command of God, 'You shall live in arbours for seven days . . . so that your descendants may be reminded how I made the Israelites live in arbours when I brought them out of Egypt' (Leviticus 23:42–43). Leviticus also says, 'On the first day you shall take the fruit of citrus trees, palm fronds and leafy branches, and willows from the riverside, and you shall rejoice before the Lord your God for seven days' (Leviticus 23:40). This is observed in the worship of the synagogue. The citrus fruit used is the **etrog**, a relative of the lemon, but larger and more expensive. These offerings are carried in procession around the scroll of the Torah while psalms and hymns are sung. They are waved in all directions to show that God is everywhere.

4 Simchat Torah (Rejoicing of the Law)

On the last day of Succot comes what is, in effect, a distinct festival which is of recent origin by Jewish standards (several hundred years old). All Torah scrolls are taken out of the ark and carried in procession around the synagogue seven times. At the completion of each circuit those carrying the scrolls dance, while the rest of the congregation sings and claps. Children carry flags and miniature scrolls in the procession and everyone is 'called up' while the Torah is being read. In most places in Israel the scrolls are taken out into the street where old and young sing and dance around them. This happy festival is an expression of gratitude for the Torah, without which there could be no Judaism. The end of the festival marks the end of the cycle of the reading of the Torah in the synagogue, and a return to the beginning.

5 Chanukah (The Dedication of the Temple)

In 168 BCE a Greek king called Antiochus Epiphanes conquered Palestine and tried to make the Jews conform to Greek customs by rooting out the Jewish religion. He was foolish enough to set up a statue of Zeus, the chief of the Greek gods, in the Temple, and sacrificed pigs on the altar. These studied insults to God could not be tolerated, and a priest named Mattathias rebelled, escaping to the desert mountains with his sons. When he died his son **Yehudah (Judas)**, who earned the title **Maccabee** (the Hammer), carried on the struggle. After a successful guerrilla campaign during which many young Jews flocked to his side, he was able to defeat the forces of Antiochus in open battle.

In 164 BCE when the enemy had been driven out, the Temple was cleansed and re-dedicated, that is, made holy again. When they came to light the seven-branched menorah, the lamp that stood before the Ark of the Covenant, there was oil for only one day's burning, but miraculously it lasted for eight. That is why the Jews celebrate this event every year with an **eight-branched candlestick**. One candle is lit on the first day of the festival, another is added on the second and so on until on the last day all eight are alight, standing straight and strong like the brave Maccabee soldiers.

Jews derive great inspiration from these heroes of their nation, especially at this time when Israel is under constant threat from its neighbours. In Israel there is a national holiday, and an eight-branched candlestick (**chanukiyah**) burns in every window. Every year at the festival, a torch is lit at the graves of the old Maccabees in the desert and relays of runners carry it to Tel Aviv where it is used to light the great chanukiyah in the city centre.

Food cooked in oil is a speciality during the festival and children enjoy the game of 'Spin the dreidle'; this is a spinning top with four sides, and a player wins or loses according to the side which is uppermost when the top falls. Parties for the children are often provided and the children themselves do the entertaining in the form of pantomime.

The early rabbis celebrated Chanukah not as a military victory but as a spiritual one, for it established the right of the nation to follow its own religion.

6 Purim (Casting of Lots)

This is another celebration of a Jewish deliverance, recorded in the Book of Esther. The Persian King Ahasuerus chose a queen who, unbeknown to him, was Jewish. The king's chief minister, Haman, hated the Jews and planned to kill them. To select the day for the massacre he cast lots (Purim). Esther, aided by her cousin Mordecai, was able to intervene on behalf of her people and they were saved from destruction. Haman was executed.

Whenever the villain Haman's name is mentioned in the reading of the Book of Esther, the children hiss and stamp to blot out the sound of his name. In some countries effigies of Haman are burned. According to the Talmud it is permissible to drink rather too much on this day. Parties are held and children put on fancy dress. A special biscuit filled with poppy seed is eaten, called a **hamantashen**; it is three-cornered in shape. Some say it represents Haman's ears, while others say it represents his hat.

7 Pesach (Passover)

This week-long celebration commemorates the **deliverance of Israel from bondage** in Egypt (See Unit 13). For the previous eight days, no leavened bread is allowed in the house and ceremonial searches are made to make sure that not a crumb remains. 'You shall rid your house of leaven.' (Exodus 12:15). Special dishes and crockery are used. The festival begins with the celebration of **Seder**, the service in the home.

On the table will be the special Seder plate on which is placed the various items of food required–a roasted shank bone of lamb, roasted egg, maror (bitter herbs, usually horse-radish), green herbs (parsley, watercress or lettuce), charoset (a mixture of finely chopped nuts, apples, cinnamon and wine), unleavened bread (matzoth) over which a cover is spread; a jug of salted water stands beside the plate. One large wine goblet, wine glasses for all present and two unlighted candles are also on the table.

The head of the family begins the service with the lighting of the candles and the filling of the glasses with wine. The large one is for the prophet Elijah and remains on the table throughout the meal in anticipation of his return to inaugurate the Messianic Age. It is poured away at the end of the meal. The door is left open to make it easy for him to come in. Each item of food has deep symbolic meaning: the shank bone is a symbol of the Passover Lamb whose blood was sprinkled on the door posts of the Hebrews; the roasted egg symbolizes the free-will offering presented by the worshipper at the Temple when the Paschal Lamb (Passover Lamb) was sacrificed; the maror, or bitter herbs, symbolize the bitterness of slavery; the charoset stands for the mud from which the Hebrews made bricks for Pharaoh's buildings; the green herbs are symbolic of the spring festival; the matzoth is a reminder of the haste with which they left Egypt–there was no time to allow the dough to be leavened. The salt water represents the tears of the suffering slaves.

As the service proper begins, three pieces of matzoth lie in front of the head of the house, representing the priests, the levites and the people of Israel. The middle one is broken and part is eaten at the end of the meal. The head of the house then points to the matzoth and says, 'This is the bread of affliction that our fathers ate in the land of Egypt. Let all who are hungry come and eat. Let all who are in want come and celebrate the Passover with us.' At this point, the youngest child present, boy or girl, asks, 'Why is this night different from all other nights?' In response to which, and to three other questions, the story of Israel's captivity in Egypt and how they were delivered by God at the hand of Moses is dramatically told. (Moses is not mentioned because though he was God's agent, the praise is due to God alone.)

Four cups of wine are drunk at various points to recall the promises made by God to Israel in Exodus 6:6–8:

(a) 'I will release you from your labours in Egypt.'

(b) 'I will redeem you with an outstretched arm.'

(c) 'I will adopt you as my people.'

(d) 'I will lead you to the land which I swore with uplifted hand to give to Abraham, to Isaac and to Jacob.'

The service ends with the psalms of Hallel (praise), Psalms 113–118, the final cup of wine and a blessing.

The service is much more than a mere remembering of past events: it is a bringing of the past into the present where it is relived. The devout Jew does not remember the events of the Passover, he is there, with Moses, experiencing it. Most Jews hold another Seder on the second night, perhaps in the house of a friend; or there might be one in the hall of the synagogue, where large numbers might attend.

8 Shavuot (Pentecost or the Feast of Weeks)

The Bible tells us that the long journey from Egypt to Mount Sinai took seven weeks. The Ten Commandments were given on Mount Sinai and this is celebrated at Shavuot. It is also the festival of the 'First Fruits'. People brought baskets of the first fruits of their land to the Temple, and the synagogues on this day are decorated with flowers as an expression of gratitude to God.

32.12 Pilgrimages

In the days before the Temple was destroyed there were three great pilgrim festivals during which people came in many thousands to the Temple; these were the **Feast of Tabernacles**, the **Feast of Weeks** and the **Passover**. The Temple having been destroyed, the actual pilgrimages cannot now take place, but the feasts are still kept as Pilgrim Feasts in the home and in the synagogue. There are two principal places of pilgrimage in Israel, the **Wailing Wall** and the **Masada.**

The Wailing Wall (Western Wall)

This consists of several courses of huge hewn stones which some Jews believe is all that remains of the Temple of Solomon. Others believe it more likely to be part of the platform constructed to accommodate the enormous temple built by Herod the Great, the building destroyed by the Romans in 70 CE. Jews from all over the world visit it to pray.

The Masada

This is an immensely strong fortress on the western side of the Dead Sea, in shape rather like a ship and with precipitous cliffs on all sides. The Zealots who remained alive after the Fall of Jerusalem in 70 CE occupied it with their families, determined to carry on the struggle. They refused to surrender even when the Romans built a huge ramp of earth and stone up to the foot of the wall. They fought the Romans every inch of the way until the day of the final assault when the enemy, to their amazement, found themselves unopposed: inside the fortress they discovered the dead bodies of almost a thousand men, women and children. Each father had killed the members of his family, and a few soldiers, selected by lot, killed their comrades, finally killing each other. They did this as a final act of defiance against those who occupied their country.

The site has recently been excavated by skilled archaeologists and many important discoveries have been made. The Masada is now a national memorial, and the heroism of its defenders is an inspiration to every Jew.

32.13 Revered People

Among these are Moses, the Patriarchs (Abraham, Isaac and Jacob), Samuel, David, Solomon, Elijah (See Units 12–16). For Judas Maccabeus, see Unit 32.11 **Chanukah**, the Dedication of the Temple.

32.14 Reform

Some Jews believe that some of the ancient rituals and laws are no longer relevant and need to be interpreted in the light of modern knowledge. They do not accept the view of the Orthodox Jews that the Bible and Talmud contain the literal word of God, and they believe that God reveals Himself to every generation, so that His revelation of Himself is not static but progressive. They therefore do not observe laws and rituals which they feel are no longer suitable to the modern world. For example, many do not observe the food laws very strictly, and they discard some of the old prayers, substituting new ones. The majority of Jews in this country are Orthodox, and what has been written here is about Orthodox practice.

33 Islam

33.1 Historical Background

Muhammad was born in 570 CE in the town of **Mecca**. He belonged to a tribe called the Quraish, who were the custodians of the **Kaaba**, the sacred shrine at Mecca believed to have been founded by Abraham and associated also with his wife Hagar and son Ishmael (Genesis 21:1–21). Muhammad's father died before he was born and his mother when he was six; he was brought up by his uncle **Abu Talib**.

He spent his youth working for his uncle and travelling with the caravans to Syria, but when he was twenty-five, a rich widow named **Khadijah**, impressed by his reputation for high moral conduct, asked him to take charge of her caravan on a journey to Syria. He served her so well that she offered herself to him in marriage; they had three sons, each of whom died in infancy, and four daughters who survived. It was a very happy marriage though Khadijah was fifteen years older than her husband, and he remained devoted to her for twenty-five years until she died.

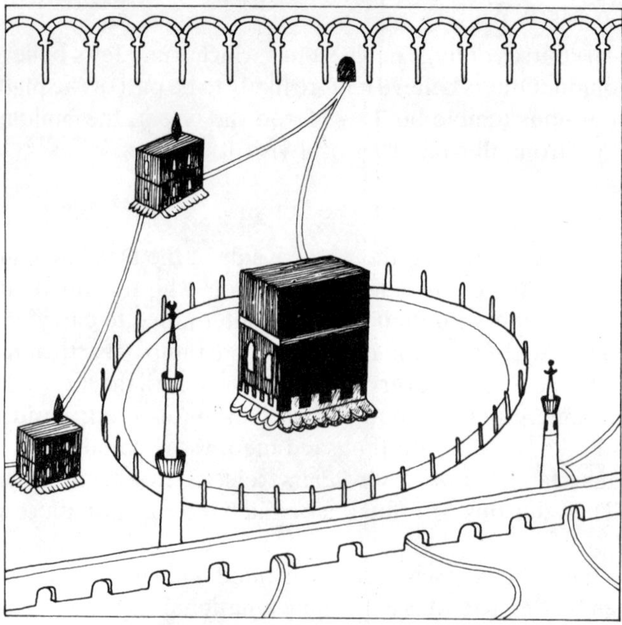

Fig. 33.1 The Kaaba at Mecca

The religion of Mecca caused Muhammad much disquiet, for it was given over to idol-worship: there were 360 images, one for each day of the Arab year. When he was thirty-five the Quraish were obliged to repair the **Kaaba** when it was damaged by flood, and a dispute arose as to who should have the honour of restoring to its place the sacred black stone which had been displaced by the flood. This black stone is believed to have fallen from heaven (it is probably a meteorite) and is held in special reverence. They agreed that the first man to pass by should settle the dispute, and Muhammad happened to be that man. He told them to take a sheet of cloth, place it under the stone, and then each man should then share in carrying and replacing the stone.

After this he began to spend periods of time alone in meditation in a cave on Mount Hira, which was near the city. It was there, when he was about forty, that he had the first of the great visions which changed his life. He saw first the eyes and then the face of a supernatural being whom he later identified as the angel Gabriel. He was ordered to read a scroll and when Muhammad confessed that he could not read, the command was repeated three times until he was able, miraculously, to read and recite by heart what is now the sura 96 of the Qur'an ('sura' means 'chapter'), 'Recite in the name of your Lord the creator, who created men from clots of blood, recite! Your Lord is the most bounteous one.' Muhammad was terrified and fled home, where he was comforted by his wife. Over a period of twenty-three years he received similar messages in visions and these were written down in the **Qur'an**. Immediately before these visions he was usually in an anguished state of mind and complained of throbbing in his head like the sound of muffled bells. (This is why bells are not used to call Muslims to worship.) This has given rise to considerable speculation by those hostile to Islam, but it must be said that ecstatic visions are phenomena well authenticated in all religions including Judaism and Christianity.

Muhammad began to preach publicly that there was only one God, **Allah**, and that he, Muhammad, was his prophet, the Apostle of God. His condemnation of idolatry soon alienated him from his tribe, the Quraish, and he and his followers were persecuted by every means open to his enemies. They tried to persuade his uncle to withdraw his protection but he refused to do this when Muhammad declared that only death could stop him from preaching the message of God. Eventually he was outlawed and had to flee for his life from Mecca.

Fortunately, by this time he had some followers in Yathrib, later known as **Medina**, a town some 430 kilometres north of Mecca; he was invited to be their governor, but he still had the problem of how to escape from Mecca without being murdered. He did so by slipping out of the town by night, travelling south, the opposite direction expected by his enemies, and after hiding in a cave for three days he later journeyed to Medina by a circuitous route. He was received with joy and there were many offers of hospitality, so in order not to offend anyone, he mounted his camel and said he would build his house on the spot where the animal halted. He did this, and his house was later converted into a mosque where his body lies buried. The **flight to Mecca** is called the **Hijrah** and took place in the year 622 CE which is now year 1 in the Muslim calendar.

His rule at Medina was a success. He established law and order on the lines of his religious principles, equality and freedom. He attacked a Meccan caravan and defeated a force sent out to protect it and this made relations with the people of Mecca worse than ever. Soon there was open war and the prophet was fortunate, at the **Battle of Uhud**, to escape with the loss of two teeth, rather than his life. On one famous occasion he foiled a superior Meccan force by digging a deep trench on their line of advance. A storm destroyed the enemy camp. About this time Muhammad suspected that the Jews with whom he had made a pact, were guilty of treachery and he drove them out of Medina with considerable violence.

From then on, Muslims were taught to say their prayers in the direction of Mecca instead of Jerusalem, and Friday became the Muslim holy day instead of Saturday.

Muhammad now gave his attention to converting neighbouring tribes, with considerable success. Recourse to war was permitted in the spreading of the faith and in attacking the enemies of God: this was called **Jihad, a holy war**.

As his power and influence increased, he was able, in 630 CE, to lead an army against Mecca which, taken by surprise, fell easily into his hands. He destroyed all the idols in the city and proclaimed the worship of Allah, the one true God. He returned to Medina after the conquest of Mecca and busied himself in sending letters and delegations to other countries, telling them about Islam.

In 632 CE he returned to Mecca on a pilgrimage, and preached his 'farewell sermon' on Mount Arafat. On his return to Medina he fell ill and died in the arms of his favourite wife Ayesha.

After his death, **Islam spread rapidly** into Syria, Persia and Egypt and along the north coast of Africa into Portugal and Spain. Charles Martel at the Battle of Tours in 732 saved France and Western Christendom from subjugation. Constantinople preserved Eastern Europe until 1453 when it fell to the Turks, and the banner of Islam was brought almost to the gates of Vienna. It was a remarkable movement by any standards. In the East, the faith spread to India and Java.

The conquests in the West were assisted by rivalries among Christians and also by the teaching that those who fall in a Holy War are immediately transported to paradise.

33.2 Fundamental and Distinctive Beliefs

Muslims believe that the Qur'an, the words spoken to Muhammad by the angel Gabriel, are the very words of Allah, so that the Qur'an is literally the word of God. The Qur'an teaches that there are five major principles which Muslims must believe and obey. They are called the **Five Pillars of Islam.**

1 Shahadah (The Creed)

(a) This is contained in one simple sentence, 'There is no God but Allah, and Muhammad is the prophet of Allah.' Allah is given ninety-nine names, such as the 'Merciful, Compassionate, the Guide, the Giver-of-Life; to help Muslims repeat them there are strings of beads called 'misbaha' which are used in a way similar to the Christian use of a rosary. Every sura of the Qur'an, except one, begins with 'In the name of Allah, the compassionate, the merciful . . . ' and these words are repeated many times a day by devout Muslims, at prayer and before every kind of activity.

(b) The second phrase of the creed emphasizes that Muhammad is the last and greatest of the prophets. There have been many prophets in the world from Adam to Jesus including Noah, Abraham, Ishmael and Moses, but after their deaths their messages were either destroyed or corrupted, so a new messenger from God was needed to give a final and complete revelation. There are many prophets but only one message, and Muhammad alone declares it in its fullness. He is 'the Seal of the Prophets', the one whose work and person sums up all that previous prophets have taught. Though he is held in great reverence, he is not worshipped.

(c) God's message was conveyed to Muhammad by the angel Gabriel and Muslims believe in angels who are given special duties by Allah. They also guard the true believer, help him in battle and pray for him in the day of judgment. There are also evil spirits.

(d) There will be a day of judgment after death in which men will be rewarded or punished according to their deeds. Heaven and hell are described in material terms, heaven being a place of sensual delight, and hell a horrifying place of torment.

(e) Muslims believe in predestination. Man has been given the ability to choose for himself what he is to do but God has pre-knowledge of what man's choice will be. Man is called upon not only to believe, but to submit to the will of God.

2 Salat (Prayer)

The Five Daily Prayers, said at dawn, midday, during the afternoon, after sunset and after dark. A Muslim can pray at any time of day and face in any direction for ordinary prayer, but prayer at the five compulsory times is governed by three conditions:

(a) A clear conscience–a pure intention.

(b) Ritual purity. This involves a complicated ritual of washing, beginning with the hands and wrists three times, rinsing the mouth three times, the nostrils three times, washing the arms up to the elbows three times, the head, the ears inside and out with thumbs and forefingers, the right foot to the ankle three times, and finally the ablution is completed with the washing of the left foot to the ankle, three times.

(c) A clean place. A Muslim may pray anywhere but it must be a clean place. To ensure this, it is usual when not in a mosque (which is richly carpeted) to use a **prayer rug**. This must be in simple colours, usually black, red, blue and white, and bear the design of an arch which is pointed in the direction of Mecca. It is called the **Prophets Window**. In the mosque a niche (**mihrab**) in the wall indicates the direction (**quiblah**) of Mecca and all worshippers face this. By all facing the same direction at prayer, each Muslim feels himself to be a member of a world-wide fellowship engaged in a common activity.

The Friday midday prayer in the mosque is of special importance and is generally preceded by a sermon. Prayer commences with the words of the first sura which begins, 'In the name of God, most gracious, most merciful', and afterwards private prayer may be offered.

During the prayers Muslims stand and then prostrate themselves to indicate submission to the will of Allah. A definite sequence of movement is required. The Muslim stands then bows deeply, stands upright again, then prostrates himself kneeling forward with his forehead on the ground and then kneeling back on his knees. This set of movements is called a **raka** and in some prayers it can be performed four times. Great emphasis is laid on the correct movements; thus during the prostration the feet must be upright and the toes must be flat on the ground with the tips pointing towards Mecca. The other details of the posture are equally minute.

The prayers are said in Arabic and usually the sermon is given in the language normally spoken by the congregation.

Fig. 33.2 Prayer movements in Islam

3 Zakat (Charity)

There is an obligation on all Muslims to pay 2.5 per cent of their income and of the value of certain kinds of property to charity. In the early days its collection was well organized but nowadays the payment is left to the conscience of the individual. The Qur'an lays down that the proceeds are to be given to the poor and needy and also 'in the way of God'. This means that if it is not required to meet individual human need, it can be used for education, hospitals and the propagation of the faith. It is usual in Britain for the money to be sent to the mosque for distribution. Zakat is looked upon as an act of worship.

Zakat does not absolve the Muslim from giving private charity to anyone in need, but this is called **Sadaqa** and is distinct from the obligatory Zakat.

4 Siyam (Fasting)

The only fast which all Muslims are obliged to observe takes place in **Ramadan, the ninth month of the Muslim year**, and it is very rigorous. No food or drink may be taken from sunrise to sunset and there must be complete abstention from sexual intercourse. Since Muslims observe the lunar calendar, each month beginning with the new moon and ending with the appearance of the next new moon, Ramadan falls on a different day in each solar year. Thus in 1980 Ramadan began on 13 July while in 1986 it will fall on 9 May. This means that when the fast falls in the long days of the hot season it is very demanding. The aged and infirm, pregnant women and young children are excused, but quite often young boys who wish to prove their devotion and manhood voluntarily observe the fast. **Ramadan is obligatory**, but the Prophet commended voluntary fasting at other times.

The purpose of fasting, according to the Qur'an, is the practice of self-discipline and restraint. Ramadan provides a valuable act of collective discipline, accepted equally by rich and poor. Every night there are special prayers in the mosque. Ramadan ends with the great **feast of Eid-ul-Fitr.**

5 Hajj (Pilgrimage)

Every Muslim is obliged once in his lifetime to make a pilgrimage to Mecca, if he is able to do so: this qualification exempts those unable to travel for health or financial reasons.

The pilgrimage represents a willingness to give up all that life holds dear, in obedience to God. Though the dangers involved are not as great as in earlier days it is still possible in the heat and crush of a huge crowd to die on pilgrimage. Parting from the family is compared with the parting of Abraham from Ishmael.

The Hajj takes place ten weeks after the feast of Eid-ul-Fitr, which ends the Ramadan fast, and since the fast is held at a different time each year according to the solar calendar, the Hajj follows suit. It is held in Dhu l' Hijja, the twelfth month.

(a) On arrival at a point near Mecca male pilgrims remove their clothes and put on two white sheets, one to cover the lower half of the body and the other thrown over one shoulder to cover the upper half. Women put on a long-sleeved plain white dress reaching to the ankles.

(b) On arrival at the Kaaba the pilgrim goes in procession seven times around it anticlockwise, kissing the sacred stone or touching it and kissing his fingers if he can, though the vast crowd may make this impossible.

(c) From Mecca the pilgrim proceeds to the two hills of Safa and Marwa; he runs between them and drinks from the well Zamzam. The Bible, (Genesis 21:1–21) tells the story of how Hagar, the wife of Abraham, having been forced to leave her husband because of the jealousy of Abraham's first wife, Sarah, took her young son Ishmael and wandered in the wilderness until her water skin was empty. An angel of God heard her bitter weeping and led her to a well full of water. The Qur'an tells how Hagar ran between the two hills of Safa and Marwa, calling on God for help, and how her prayers were answered when she found the well Zamzam.

(d) On the ninth day the pilgrims go to Mount Arafat, about twenty-four kilometres from Mecca where they spend the day in prayer and meditation. It is called the **Wuguf**, meaning 'standing

Fig. 33.3 Plain of Arafat surrounding walled Kaaba at Mecca

before God'. This is the place where Muhammad preached his 'Farewell Sermon' and on the Hajj an Imam preaches a sermon and leads the prayers. The night is spent at Muzdalifah, a valley between Arafat and Mina.

(e) Before dawn on the tenth day the pilgrims arrive at Mina where there are three pillars of stone. Muslims believe that it was Ishmael, not Isaac, whom Abraham intended to sacrifice, and the place was Mina. Ishmael was tempted by the devil to resist his father but he overcame the temptation by throwing stones at the tempter. Pilgrims re-enact the event, each of them throwing seven stones at the pillars. After the stoning of the first pillar, the Eid-ul-Adha (Festival of Sacrifice) takes place, with the sacrifice of a lamb, recalling the sacrifice of Abraham. (See Unit 33.3)

When the Hajj is completed, pilgrims often go on to Medina where the prophet is buried, to pay their respects, though this is not a part of the official Hajj.

The value of the Hajj

1 It emphasizes the importance of complete submission and surrender to the will of God. It represents a willingness to give up all life holds dear.

2 All distinctions of race, social status, wealth, education, are done away with. The simple uniform dress of each pilgrim emphasizes that all men are equal in the sight of God.

3 It reminds the pilgrims in a most dramatic way of the roots of their religion.

Another form of pilgrimage is the **Umra** which is to Mecca alone, and can be made at any time.

Jerusalem is also a Muslim holy place. At a time when the Meccans were persecuting those who followed him, Muhammad was passing through a difficult time, and matters were made worse by the death of his uncle and protector, Abu Talib and his beloved wife, Kadijah: he needed reassurance, and there came a sign from God–the Journey to the Seven Heavens. The angel Gabriel put him on **al-Burag**, a winged animal which took him from Mecca to Jerusalem and then to the Seven Heavens. He met other prophets including Abraham, Moses and Jesus. During his stay in the heavens he received instructions about the daily prayers. Since that time Jerusalem has been revered by Muslims, and the **Dome of the Rock**, a mosque built over the traditional place of Abraham's sacrifice, and the site of Solomon's Temple, is the **third most important holy place of Islam**.

33.3 Feasts

There are only two feasts, the **Eid-ul-Fitr** and the **Eid-ul-Adha.**

Eid-ul-Fitr takes place at the end of the fast of Ramadan, and everyone takes part. Children are given gifts and money and the family goes to the mosque in the morning in their best clothes for the dawn prayers and to pay Zakat-ul-Fitr, the charity of the fast. The prayers of the festival begin an hour after dawn, and afterwards visits are made to the graves of relatives. A big breakfast is eaten on the return home and this is followed by a round of visiting between neighbours by heads of families. The rest of the family remains at home to receive visitors. During Ramadan, Eid cards are sent to friends and relatives who live at a distance.

Eid-ul-Adha–on the tenth day of the pilgrimage a sacrifice of a lamb or goat is made, recalling Abraham's sacrifice of a ram in place of his son Isaac, and this is celebrated all over the Muslim world. There are special prayers at the mosque, and after returning home, a lamb is sacrificed. Some of the meat is eaten by the family and some is given to neighbours or to the poor. Visits are made to neighbours as at Eid-ul-Fitr, and greetings cards are sent.

33.4 Dietary Laws

The Qur'an decrees that pig-meat is forbidden and also, as in Judaism, the blood of any animal. There are strict rules governing the slaughter of animals and it must be carried out by a Muslim. Alcohol in any form is forbidden.

33.5 Rites of Passage

1 Birth

When a child is born, the 'call to prayer', the **Adhan**, is said softly by the father into the right ear, and the **Igamat**, or minor Adhan, the second call when the congregation is ready to pray, is whispered into the left ear. The child is given a name on the seventh day at a celebration called the **Ageegah**; the child's name is usually one derived from one of the names of God, or the Prophet. Sheep or goats are slaughtered and a feast is held at which relatives and friends are invited. The baby's head is shaved and gold or silver equal to the weight of the hair is given to the poor.

The Khitan, or circumcision, of a male child then takes place. Children are especially valued and it is the duty of parents to give them a good education and to bring them up to be devout Muslims.

2 Marriage

Islam lays great emphasis on the value of family life and the stability of marriage, which was instituted by God for the purpose of bringing children into the world. Marriage is not regarded so much a religious ceremony, as a legal contract between two parties, the families rather than the two individuals: thus when a man wishes to marry a girl, the negotiations are conducted through the two families. The man and woman must freely accept each other and if negotiations succeed and the dowry to be paid by the man is agreed, the engagement is announced and the wedding day is fixed.

Marriage can be contracted anywhere and it is not necessary for an Imam to officiate, but it is usual for him to do so in this country. Two adult male witnesses are necessary and a woman cannot marry without the consent of her father or male next of kin; she may not marry a non-Muslim, though a man may.

Polygamy is allowed under certain circumstances but, where practised, it is now mainly in country districts. Up to four wives are allowed but they must all be treated impartially.

Purdah (the wearing of the veil) is practised in some Muslim countries, but in others many women feel free not to submit to it.

Divorce is very much a last resort and there is a machinery for reconciliation by the families, but if this fails, divorce can take place. There are degrees of divorce, however:

(a) A divorce which can be revoked, allowing a husband to take his wife back without having to go through a new contract. This seems rather like our 'legal separation'.

(b) A divorce which makes a new marriage contract necessary if the husband wishes to take his wife back.

(c) A divorce which cannot be revoked unless the wife marries again and her second husband either divorces her or leaves her a widow. She may then remarry her first husband.

3 Death

A dying person should be encouraged to repeat the creed, 'There is no God but Allah, and Muhammad is the prophet of Allah', and also to pray for forgiveness and mercy. Sura 36 is read by someone present. At death relatives and friends visit the home to comfort the bereaved and pray for the deceased.

The body is washed three times with soap and water in the same sequence as for the ablution which precedes prayer. The body is wrapped in three pieces of white cotton and carried on a bier or in a coffin to the mosque or direct to the burial ground for the funeral prayer. The dead should always be buried in a Muslim cemetery, preferably at the place of death. Post-mortems are allowed only in very special circumstances and cremation is forbidden. The mourning period varies from seven days in some countries to three months in others.

33.6 The Mosque

Most are rectangular in ground plan with an attached courtyard and a covered way. Facilities for ritual ablutions are provided. Inside the main building is an arched recess in the wall which indicates the **Qibla, the direction of Mecca**. This recess is called the **mihrab** and the Imam stands there to lead the prayers. Immediately to the right of the mihrab is the minbar, the pulpit from which the Imam delivers the sermon. There is also a bookstand or lectern, **Kursi al-Qur'an**, from which the Qur'an is read. There are no pictures or statues but mosques are often lined with beautiful ceramic tiles carrying texts from the Qur'an, and the floors are deeply carpeted. **Minarets** are slender towers which rise high above the mosque, from which the **muezzin** makes the call to prayer, in most cases nowadays, with the help of a public address system. There are no chairs or any seating arrangements. The **Imam** is the prayer leader at the mosque.

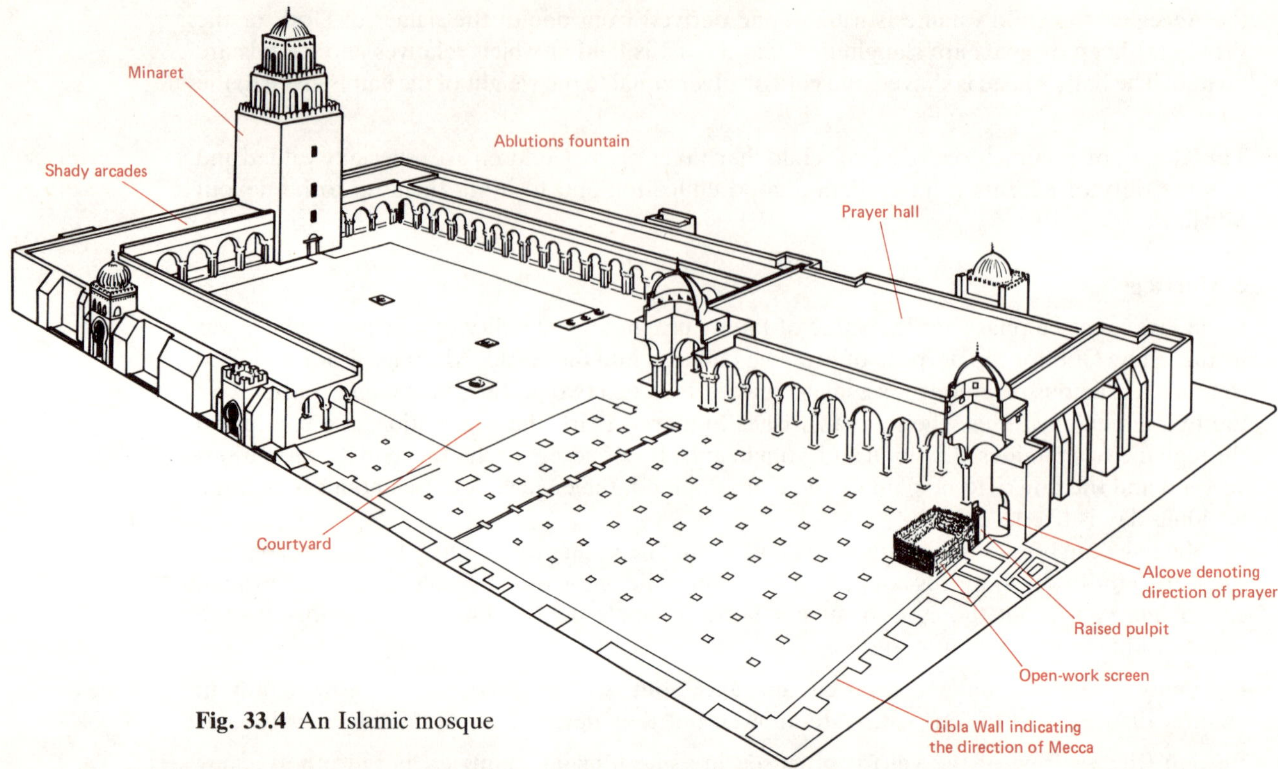

Fig. 33.4 An Islamic mosque

33.7 The Qur'an

This is the **holy book of Islam** and is the record of the words spoken to Muhammad by the angel Gabriel, the messenger of God. To the Muslim, the Qur'an is the very word of God. Some of the suras were written down during the lifetime of the Prophet and others were faithfully committed to memory and written down by **Zaid**, Muhammad's servant, after the death of the Prophet.

The Qur'an is the main source of belief and behaviour and its main object is to convey the truth that God is sovereign over all. Since it is the very word of God, it is held in great reverence: it is read five times daily at prayer and many holy men (**Hafiz**) know it all by heart. It has a place in every Muslim home and is treated with the greatest respect.

The second source of authority in Islam is the **Sunnah**, which is a record of what Muhammad himself said and did. After his death it was written down, and the stories about his life are a source of great inspiration.

34 Hinduism

34.1 Background

The word is derived from the Persian word 'Sindhu' meaning a river. The Persians referred to the people of India as 'those who lived beyond the Sindhu' (the Indus); since, in ancient Persian, the letter 's' was often pronounced 'h', we have 'hindu', hence 'Hindus'. It began as the name of a people which later became the name of a religion.

Unlike the other world religions **Hinduism has no founder, no creed, and it is not possible to date its origin**: it simply grew. It is a religion of bewildering variety, ranging in form from devotion to many gods worshipped through priests and complicated rituals, to a rejection of all outward expressions of faith and a concentration upon personal contact with God in meditation.

The Hindu view of the universe is that it is a constant state of change; even the things which look most permanent like the stars, the sea, the mountains, are all subject to change, yet they are not destroyed but become something else, as the mountain by erosion becomes the soil. All life becomes reborn at death and the soul lives only temporarily in man, but not only in human beings, also in plants and animals. The Hindu believes that there must be a permanent reality behind this changing material world, something that itself does not change. The world in which we live is not real, in the sense that it is transitory, it is an illusion. The 'real' is that which causes all things, pervades and permeates all things, it is that in which all things exist. Hindus call this reality **Brahman**, and in the Hindu Scriptures the word is neuter: Brahman is an 'it' and therefore not a personal God but some Hindus do think of God as the supreme, personal reality.

For the Hindu there is one reality, some believing that it is personal and some that it is impersonal, yet Hinduism in practice appears to indulge in the worship of countless gods. Some devotees attach themselves to a particular god or goddess, or to several, each of which is distinct, but each god expresses only one aspect of the whole, the reality, and the Hindu finds it practical to worship just one aspect of the reality at a time.

Brahman is the unchanging reality, that which is in all things yet is not subject to change. This is called pantheism or 'God-is-all-ism', and yet this term can give rise to misunderstanding: Hinduism does not say that God is identical with the universe but that all is in Brahman–a very different idea.

The best way of explaining it as far as it is capable of explanation is by the 'parable' of the salt in the Hindu sacred Scripture, the **Chhandogya Upanishad**. What it says in effect is, that if salt is placed in clear water it dissolves and can no longer be seen, yet every drop of the water tastes of salt. The atman, the soul of man, is like the water and the salt is like Brahman or Atman–with a capital A; they cannot be distinguished just as the salt cannot be distinguished from the water in which it is dissolved. Brahman pervades the atman, the soul of every man, and indeed everything that exists, exists in Brahman. That is not to say that when we look at a material object we are looking at a piece of Brahman, but it does mean that by looking at it we can realize the presence of Brahman, though that presence cannot be defined or precisely located.

The Upanishads teach that because Brahman pervades and permeates the soul, the self, the way to understand Brahman is to understand the real self, and that is why the form of meditation which involves looking inward upon the self, is so much emphasized in Hindu religion. This is obviously a solitary exercise because Brahman is a personal experience, and that is why Hindus have never attached much importance to congregational worship.

34.2 Basic Concepts

Samsara (Rebirth)

In the West we think of time as being linear–one event succeeds another in a single line, there was a beginning and there will be an end; but Hinduism thinks of time in terms of a revolving wheel, in which life repeats itself, and the only way of ending involvement in this cycle of time is to find union with Brahman. Hindu belief in rebirth or reincarnation or transmigration of souls is called Samsara. The body dies but the soul cannot die because it is Brahman and therefore eternal. The soul lives on and is reincarnated in another body or a succession of bodies; the status of the reborn soul, its caste, depends on the quality of life lived in the previous existence.

Maya (Illusion)

Being reborn after death has no attraction for the Hindu because he believes that this material world is maya, that is, inferior, it is less than real, an illusion: men and women become so involved with the things of this life that they lose touch with reality and are seduced by the attractions of maya, accepting it as the reality instead of Brahman. The goal is to be absorbed into Brahman, the one unchanging reality.

Karma (Action, deeds and their consequences, guilt)

In order to avoid rebirth the Hindu strives to detach himself from the things of this life which build up karma, the actions which attach him to this world of illusion and hinder him from discovering that atman and Brahman are one. Man's deeds have good or evil effects and his bad deeds build up his karma and have a culminative effect. He can control his karma by disciplining his mind and body and this will have an effect on his status at rebirth. The Hindu strives to live in such a way that he will have as little karma as possible, so that he will earn a better rebirth even if he does not achieve Moksha.

Moksha (Release, salvation)

This is the goal of Hinduism, release from the cycle of rebirth by the absorption of the self into Brahman. The certainty of this is experienced in this life, the sense of release being experienced when the devotee becomes conscious that the atman and the Atman, the Brahman, are identical. It is comparable in some ways to the kind of certainty some Christians experience at conversion.

Because death can be the occasion of either a rebirth or absorption into Brahman, it is an especially important event. Many aged Hindus go to Benares, to the sacred river, where, by ritual ablutions, they wash away karma, for it is karma which will decide the next incarnation.

Yoga (Discipline)

This is very much more than the discipline of relaxation and controlled exercise as commonly practised in the West. It is a means of disciplining the mind and body so as to shut out the world and concentrate on Brahman. Karma is built up by worldliness, so the means of controlling karma is to focus the mind on Brahman.

This is achieved by sitting on a low platform with crossed legs in a relaxed and comfortable position. The technique is to focus the eyes on the tip of the nose and in this way to put out of focus most of what might cause distraction. (To close the eyes would be to invite the mind to day-dream and so make concentration more difficult.) In this way the world can be shut out and the mind concentrated on Brahman, the only reality. Masters of the art can practise incredible physical control, they can even appear to stop breathing for long periods, but the object of the discipline is to discover the true self, the atman, and therefore to find God within the self.

34.3 The Stages of Life

There are four stages called ashramas.

1 **Brahmacharya**–the stage of the student, of education and discipline.

2 **Girahastan**–married life, work, in which obligations to family and the caste are fulfilled.

3 **Vanaprahastan**–retirement; detachment from the world is being increased. It is the time to guide and advise younger people.

4 **Sannyasin**–complete renunciation of the world, the life of the recluse, the abandonment of all possessions and concentration on achieving union with Brahman.

There are three legitimate objectives in Hindu life:

1 **Dharma**–right conduct within the context of the caste. (Each caste has different duties.)

2 **Artha**–the life of economic prosperity.

3 **Kama**–the enjoyment of the good things of life.

When these three basic objectives have been achieved the aim should be to achieve Moksha (release) and union with Brahman.

34.4 Caste

To be a Hindu is to belong to a caste into which one has been born, and caste decides to a large extent the place the individual occupies in society. The Brahmins, those of the highest caste, teach that Brahma created the first man, **Purusha**, who was sacrificed, and out of his body were made the four varnas, or orders of men.

1 The mouth of Purusha became the **Brahmins**, who were given the privilege of teaching the Scriptures.

2 The arms became the **Kshatriyas**, the warriors and rulers, whose duty it is to maintain law and order and to protect the country from enemies.

3 The thighs became the **Vaisyas**, the farmers, traders, money-lenders. Their duty is to maintain economic welfare.

Each of these are **dvija**, orders or classes, and all may study the Scriptures though only the Brahmins may teach them. In each, there is an initiation ceremony by which the member is introduced into the religious and social duties of the order and given the sacred thread which is worn ever afterwards next to the skin and over the left shoulder.

4 From the feet of the Purusha came the **Shudras** whose duty it is to serve the other orders. They are outside the dvija class and do not wear the sacred thread.

Outside the class system altogether are those who are derived from no part of Purusha, the **Pariahs** or Untouchables.

There are, then, four orders or classes in Hindu society and each is divided into castes. The main features of the caste system are:

1 it is not possible to change the caste one is born into;

2 marriage must be within the same caste;

3 there are strict regulations regarding the acceptance of food from any other caste.

Each caste protects itself from pollution by lower castes by strict adherence to these rules.

Many reformers from Buddha to Gandhi have protested about the caste system and it has now been made illegal under the Indian Constitution. The Untouchables, whom Gandhi called 'children of God' are now legally freed from discrimination, and some have gained high office in the state. Inter-marriage between the castes has now been legalized. However, the law cannot abolish deeply rooted social and religious conventions overnight, and caste is still a potent force, especially in the villages where eighty per cent of the people of India live.

From the religious point of view caste is still very important, for it is easier for a Brahmin than for the Shudra to achieve oneness with Brahman because he has the advantage of being able to study and teach the Scriptures. It is true that if you are at the bottom of the scale of castes you could, by Herculean effort, achieve Moksha, but it would be very much more difficult than for a high caste Hindu. The more practical aim would be by keeping the **dharmer** of the caste to which you belong (that is by fulfilling the social and religious duties expected of you), to be reborn at death into a higher caste where the opportunities of achieving Moksha would be greater.

This teaching serves as an instrument of social control making it a virtue humbly to accept the hardships of one's lot, and it is also a method of explaining away the fact of suffering. If you suffer, it is because of past sins, your karma.

34.5 Worship (Puja)

The most important place of worship is the home. There may be a room set aside or more usually an alcove or corner where pictures and images of the chosen god are set up. The mother of the family usually carries out the daily puja. She begins by bathing and putting on clean clothes, and the worship may take the form of washing or decorating the statue and offering incense, food and prayers. The statue both symbolizes and represents God, and while it is not worshipped it is offered the same reverence as God Himself.

The food offered (**prasad**) is carefully selected and prepared, and expresses the fact that since food is given to us by God this should be acknowledged by offering it first to Him before taking it ourselves. It is the devotion that matters not the gift itself, for God is not hungry: what He values is the love which prompts the gift.

The Western way of life makes it difficult to perform complicated rites every day, so incense sticks and a **diva**, which consists of cotton wool dipped in ghee (clarified butter), are lit, and a prayer is said. On the way to work, **Om**, which in Hindu tradition is the best name for God, may be chanted repeatedly, and the **Gayatri mantra** is said, the most holy prayer in Hinduism, 'Let us meditate on the most excellent light of the creator; may he guide our intellects.'

In the evening the diva is lit and families come together to pray. There are family rites for every occasion from conception to death and afterwards.

Worship at the temple is far less important than in the home. Temples in India are often large and splendid. In the middle is the central shrine which is the home of the chief God. Fruit and flowers are offered and there are diva and incense sticks for burning. The god is awakened at dawn in his bedroom and is carried to the shrine where he is washed and dressed. Offerings are made to him and he is entertained with music.

34.6 Sacred Scriptures

There are three groups of Scripture, the Vedas, the Upanishads and the Epics.

The Vedas

These consist of four sacred books of which the **Riq-veda** is by far the most important; it consists of over a thousand hymns to thirty-three gods. It was originally handed down orally in **Sanskrit** and dates from about 1200 BC. The gods of the Riq-veda can be classified in three groups;

1 Gods of the heaven: sky gods and sun gods

2 Gods of the atmosphere

3 Gods of the earth and water

They are all nature gods consistent with a very early stage of religious development, and make no moral demands on their devotees. In the later vedas, however, it is made clear that there are not many gods, but that the various deities are different forms of the one God, the ultimate reality.

The Upanishads

These are in the form of question and answer between pupil and teacher (guru) and contain the most important teaching of Hinduism, **belief in Brahman**.

The Epics

Unlike the Upanishads which express a complicated and abstract set of beliefs difficult for any but the intellectual fully to understand, the Epics emphasize devotion to Brahman not as an abstract principle, but as a person. This emphasis on personal devotion to God is called **bhakti**. The God of the Epics is **Vishnu,** and in the **Bhagavad Gita, Krishna**. The great epics which express bhakti are the **Ramayana** and the **Mahabharata**, and they express in story form the principles of the Upanishads in a way in which unsophisticated minds are able to understand.

Avatars Hindus believe that life involves a constant conflict between good and evil and that most of the time these forces are in balance; sometimes, however, evil gains the upper hand and this is unfair to man. Vishnu intervenes by coming to earth in an avatar or incarnation, and both Rama and Krishna are avatars of Vishnu. This very important teaching was introduced in the Epics, and led to considerable development in Hindu religious ideas.

The Ramayana The ten-headed demon-king **Ravana** who ruled in Sri Lanka had been promised by **Shiva** that he could not be killed by gods or demons. Armed with this promise he proceeded to tyrannize both gods and men. The gods asked Vishnu to destroy the demon so he came to earth as **Rama**, because Ravana had not asked that he might not be killed by a man.

Rama was King of the Northern Kingdom and was unjustly exiled. Sita, his beloved wife, an avatar of Lakshmi the wife of Vishnu, was captured by Ravana who carried her off to Sri Lanka. She was discovered by **Hunaman**, chief of the monkeys, who with his subjects built a bridge of rocks and planks across the sea over which Rama passed to destroy Ravana and rescue his wife. In the end Rama and Sita returned from exile, ascended the throne, and Rama's reign brought peace and happiness to all his people.

Rama is the ideal husband, a great warrior, a good king, a generous and forbearing person, a perfect man, and Sita is the perfect, loving wife. The story is a source of inspiration and instruction to all Hindus. Since Hunaman helped Rama, monkeys are now regarded as sacred in India.

The Mahabharata contains 100 000 verses and is the longest poem ever written; it emphasizes the importance of morality and of personal devotion to God (bhakti).

The most important section is the Bhagavad Gita which is considered to represent the flower of Hindu philosophy. It is the story of **Arjuna**, son of **Indra**, the Vedic nature god of the heavens. He was a princely member of the Pandava family and was at war with the Kaurava family who were related to him. The five Pandava brothers were noted for their faith in dharma (moral righteousness) while the hundred Kaurava brothers were evil. A battle was fought in the Punjab at Kurukshetra near Delhi and the Pandavas won. The story is about the resolution of a very serious moral problem. Arjuna's army is in battle order but he cannot make up his mind whether it is right to kill friends and relatives who are in the army of his enemies. He discusses the issue with his charioteer, Krishna, who turns out to be an avatar of the god Vishnu. Krishna tells him that it is his duty as a Kshatriya (a member of the soldier caste) to fight against evil, regardless of the consequences; it is the obligation of his caste. He tells him that since the soul cannot be killed there can be no real death for it will either be reborn or become absorbed into Brahman. Arjuna must therefore do his duty without thought of reward or honour but in accord with yoga, discipline.

There is **jnana-yoga**, the way of knowledge, **raja-yoga**, the way of mental discipline and there is **karma-yoga**, the way of good deeds, but the best and easiest form of worship that can be offered is **bhakti-yoga**, devotion to a personal loving God. The abstract, impersonal Brahman, the 'It', the all-pervading reality, becomes a loving person in Krishna, his avatar. Everyone who loves truth is dear to Krishna and he is kindly enough to accept any devotion even if it is offered to other gods.

At the end of the story Arjuna asks Krishna to reveal himself in his true form, and Krishna gives him a 'Divine Eye' so that he will not be blinded by the sight of the god's glory. Even with this aid, the sight is terrifying and Arjuna asks Krishna to return to human form, at which he is no longer afraid.

The Upanishads meet the needs of the thoughtful Hindu who is capable of appreciating the abstract, but the Gita, by introducing devotion to Krishna meets the needs of ordinary men. No wonder the Gita and the Ramayana are so popular and are loved throughout Hinduism.

34.7 Deities

The principal deities of Hinduism are **Vishnu** and **Shiva**, but there are three groups of devotees, one of which worships Shakti, which is the female aspect of Shiva. There is no friction between them, for the devotees of one god look upon the others as minor manifestations of their own god. The less educated consider all gods to be various aspects of the one reality.

Vishnu

Vishnu is widely worshipped. It is he who preserves the universe and protects the good from evil-doers. He has descended to earth in both human and animal avatars at times of crisis; there are ten avatars (incarnations) of Vishnu.

1 The Fish (Matsya) When the world was overwhelmed by a flood he took the form of a golden fish and saved mankind by towing a ship with a rope.

2 The Tortoise (Kurma) During the flood, amrit, the nectar of the gods, was lost under the sea; Vishnu came down as a tortoise to help in the search for it.

3 The Boar (Varaha) A demon cast the earth once more under the flood, and he had obtained the promise of Brahman that he could not be killed by any god or animal. He forgot, however, to mention the boar, so Vishnu became a boar, killed the demon and lifted the earth out of the sea on his tusks.

4 The Man-Lion (Nara-simha) A demon tortured his son to persuade him to give up the worship of Vishnu, but the son refused to yield. The demon had been promised by Brahman that he could not be killed by god, man or animal, by day or by night, inside or outside his house. Vishnu became a man-lion and killed the demon at twilight on the threshold of his house.

5 The Dwarf (Vamana) The demon king Bali had driven the gods out of their home. They asked Vishnu for help and he came down to earth as the dwarf Vamana. He asked Bali to give him as much land as he could cover in three strides, and as soon as Bali agreed Vamana grew to so great a height that he measured the whole of Bali's kingdom in two strides. Since there was no other place to put the third stride, it was put on Bali's head. Bali was forced to surrender and he was exiled while earth and heaven were restored to gods and men.

6 The Rama with Axe (Parasurama) Vishnu took human form as Parasurama to free the Brahmins from the tyranny of the Kshatriyas.

7 Rama This is the story of the Ramayana (See Unit 34.6).

8 Krishna This is the most highly regarded of all the avatars. It is believed that in Krishna, Vishnu manifests himself fully, and we find in him the Hindu ideal in every way. He gave the Bhagvad Gita which marks the summit of Hindu teaching on morals.

9 Buddha (See Unit 36)

10 Kalki This is yet to come. It is believed that the world will eventually become so evil that not even Vishnu will be able to save it. He will appear as Kalki, riding upon a white charger and waving a sword, to destroy the world so that a new one may be built.

Shiva

Whereas Vishnu is kindly and seeks to help man, Shiva has attractive and unattractive aspects. He is often represented as the **Lord of the Dance**. He has four arms; in one hand he holds a drum, symbol of sound as the first element in the Creation (rather than light as in Judaism); and in the opposite hand a flame, symbol of the final destruction of the universe. Of the other hands, one is raised palm forward, and the other lowered with the palm reversed. The gesture indicates the delicate balance between life and death. The right foot stands on the back of the demon of forgetfulness, while the left is raised in the attitude of dance. The dancing figure stands in an arch, representing the universe, which is subject to Shiva who creates and destroys. The dance represents the harmony and balance with which Shiva runs the world in the rhythm and flow of the seasons.

Fig. 34.1 Krishna **Fig. 34.2** Shiva

Shiva is also the god of those who spend their lives in **meditation**, and he is too, the god of **fertility**.

Shiva's wife, **Shakti**, is an important goddess, and is known by several names. While the majority of Hindus regard her as a minor deity, her devotees believe that after Shiva created the universe, his energy became expressed through his wife, for the creation came into being through the intercourse of Shiva and his wife. Her devotees feel that Shiva does not need to be worshipped, but his wife is the great **Mother Goddess** to whom all worship should be addressed; she is the supreme God.

Shakti

She has many names and many forms. In her fierce form she is called Durga or Kali. In her representations Durga is seated on a lion and has ten hands. On her right she holds a disc, a sword, a bow and a rosary. The disc represents the wheel of existence which is in her control, and those who worship her will not be subject to time; her sword is the weapon which destroys ignorance; the bow protects her devotees from demons; the rosary means that prayers should be said with a rosary in hand. In her left hands Durga holds a conch, a mace, a lotus and a trident: the conch represents the four kinds of sound described in the Vedas; the mace is the weapon of destruction and indicates that all who submit to her need have no fear of death. The lotus is a powerful symbol: the roots live in the mud which represents this life, while the stem which reaches upwards through the water represents desire, a reaching up for the spiritual existence, and the flower lying on the water represents that existence. The lotus flower is the emblem of purity, love and beauty. Some of the festivals of Durga are observed throughout India.

Ganesha

This is the elephant-headed god, the eldest son of Shiva, and the bringer of good fortune. At all important functions and enterprises he is first worshipped.

He has four arms and in his right hand holds a rosary and a goad and in the left hands an axe and some sweet-meats. The rosary is a symbol of prayer and also the circle of time, and reassures devotees that Ganesha controls death. As an elephant driver uses the goad to direct the elephant, so Ganesha directs and guides, and the fate of everyone is in his hands. The axe symbolizes the destruction of ignorance and the sweet-meats are the reward of those whose ignorance is dispersed.

The snake under his chin represents death and proclaims that he is the controller of death. His huge ears indicate that he has heard many Scriptures and his small eyes indicate that he meditates.

There are many thousands of village gods whose worship is confined to one place and there are many sacred objects. The sacred cows which can be seen wandering everywhere, even in the busiest towns, are symbols of Mother Earth. Monkeys are sacred to Rama and snakes are holy also. Certain trees are sacred as is also the River Ganges.

34.8 Rites of Passage

1 Birth

It is important to remember the date and exact time of birth so that a Brahmin priest can work out the horoscope. Astrology is considered important. A boy's first haircut is a joyful family occasion as it is believed that the cutting of the hair removes any karma remaining from the previous life.

2 Upanayana

This is the sacred thread ceremony. When a boy is old enough to receive religious instruction from a spiritual leader (guru) he is given a thread with three strands, white, red and yellow, which is worn over the left shoulder and across the body. They remind the wearer of the three principal Gods: Brahma, Vishnu and Shiva. It is a symbol of the second birth, the first being when he was born into the world and the second when he undertakes religious study. Only members of the first three classes or orders wear the sacred thread.

3 Marriage

Family life in India is very different from that in the West. Grandparents, parents, children, uncles, aunts, cousins live together, if possible in one house under the headship of the father or grandfather. All belongings are shared and each member looks after the needs of the others; orphans, widows, the aged, all have their place in the family caring. Success of one is shared by the others. The head of the family has authority even over his married sons.

It is easier to understand the practice of **arranged marriages** when we consider this background. A new bride is very much the concern of the whole family; she will have to take her part in caring for the whole family so account is taken of caste, relationship, education, social background and financial standing, not forgetting the horoscope. The relatives do the arranging, but the parties are kept informed of the progress of negotiations and have the right of refusal, though pressure is strong if the families approve of the match.

The bride is the one on whom the greatest demands are made because she has to fit in to an established family.

Hindu marriages are as spectacular as the families can make them; the ceremony takes place at the bride's home and the date and time are fixed by the priest. It is illegal to give a dowry but the bride's parents buy the bride many saris and also give saris to the bridegroom's relatives. The hands and soles of the feet of the bride are decorated with the dye henna and her wedding dress is usually a red sari embroidered with gold thread.

The marriage service begins with a prayer offered to Ganesha. The ancestries of both bride and groom are recited, then the bride is given away by her father and mother, the sacred fire is lit and offerings are made while prayers are said. A white cord is attached to the shoulders of the couple and they then take seven steps around the fire, each step representing a particular blessing on the marriage. The bride renounces her attachment to her parents and promises allegiance to the family of the groom. Prayers for good fortune bring the ceremony to an end. A joyful celebration follows.

4 Death

Hindus believe that the bodies of the dead should be cremated not buried. The eldest son lights the funeral pyre, but in the West he stands by the coffin and watches it pass into the furnace. It is believed that the soul leaves the body and is reborn into a new one, so there is nothing final about death. On the third day the ashes are cast into a river, preferably the sacred Ganges. The last rite is the offering of rice and milk on the tenth, eleventh and twelfth days after cremation.

34.9 Festivals

Hindus observe the lunar calendar as Muslims do, so it is possible to give only approximate dates for the festivals in the solar calendar. The lunar months are divided into two fortnights, one bright and the other dark. There are many festivals and they vary from area to area. It is interesting to look at three of the most colourful.

1 Holi (March–April)

This is a harvest festival at which people eat the new corn, and it is held in the bright fortnight of the month Falgun. Bonfires are lit in every village and there is much noise and merry-making.

On the second day of Holi people throw coloured powder and water over each other and people of all castes enjoy themselves together on this day. It is said that Krishna began this custom by spraying the cow-girls in fun on this day. At his birthplace in Mathura the festival goes on for several weeks and pilgrims from all over India attend.

2 Ganesha Chaturthi (September–October)

The elephant-headed god is the bringer of good fortune and is worshipped before any important enterprise, such as marriage, begins. His incarnation day is celebrated with great enthusiasm all over India. An image of Ganesha fashioned out of clay is brought into the home and special prayers are offered; he is worshipped with music, feasting and dancing for ten days and then is taken in procession to a river or the sea, and, with pleas to return again soon, is immersed in the water.

2 Diwali (October–November)

This is a festival of lights and is celebrated throughout India in honour of **Lakshmi**, the wife of Vishnu. She is often represented as a beautiful woman, either sitting or standing on a lotus. She is a devoted wife and is the goddess of happiness and good fortune, who loves light.

For months before the festival Hindu women make candles and lamps to decorate their homes for the three nights of festivity when lights burn in every window and doorway. The tradition is that Lakshmi circles the earth on the third night as an owl, inspecting the homes to see if they have been cleaned and if lights are burning in her honour. If she is pleased she will grant the home prosperity for the coming year which begins next day. Pictures and statues of Lakshmi are placed in the homes. In some parts of India cows are especially honoured at this time because they represent **Lakshmi, the Earth Mother** who provides milk and butter.

Diwali is a time to remember that as darkness is dispersed by light, evil is conquered by goodness. The story of Rama and Sita and the victory over the demon Ravana is told, sometimes in the dance. The next day is the first of the New Year and people make New Year resolutions; family vows are renewed also at this time. Old clothes are put away and new ones put on and greeting cards are exchanged. The ‘mountain of food’, **Annakoot**, is cooked and offered to Krishna to commemorate how he sheltered the people from heavy rain sent by Indra by holding a mountain over them. The grateful people cooked various foods for him and, to remember the event, people provide a ‘mountain of food’ which is later distributed to the people.

34.10 Dietary Laws

Strictly speaking Hindus may not eat meat, fish, eggs, garlic or onions. The cow is a sacred animal and beef is especially prohibited. The rules have been considerably relaxed, however.

34.11 Pilgrimage

The principal Hindu pilgrimage is to the holy city of Benares on the sacred river Ganges. There, pilgrims may wash away karma. Dead bodies are cremated on the steps leading down to the river and the ashes are thrown into the river.

34.12 Revered People

In modern times no Hindu has been as revered as **Mahatma Gandhi**, who ranks as one of the great men of world history.

His main teaching was **ahisma**, non-violence, and, paradoxical as it may seem, he used this as an effective weapon. He believed that non-violence in the face of injustice imposed by physical force would lead eventually to an appreciation of the truth. He went further by saying that the truth would be revealed by inflicting suffering upon the self rather than on an opponent. In his struggle to gain the independence of India from British rule, he exerted great pressure by undertaking to fast to death. His critics would say that this was only a subtle way of bringing physical force to bear, because if he had died as a result of any of his many fasts there would probably have been an uncontrollable outbreak of violence throughout India. However, there can be no doubt of the sincerity of Gandhi, and India owes its political independence more to him than to any other man.

He believed that untouchability could not be defended and that it was contrary to the true spirit of Hinduism. He called the untouchables **'Harijan'** 'the children of God', and he did much to improve their lot. Untouchability is now illegal in India and the Harijan may enter temples and stand for parliament.

Gandhi did not merely teach ahimsa: he was the living embodiment of it and men saw it in action in a life dedicated to God.

35 Sikhism

35.1 Background

Muslims had been invading India from the eleventh century, and by the sixteenth century the Mogul Empire was established over most of the country. There was a good deal of animosity between Hindu and Muslim but, since Hinduism is a very tolerant religion which does not find it difficult to absorb valuable beliefs from other faiths, there were several movements aimed at making a synthesis of the two religions composed of the best elements of each. Sikhism was a successful attempt to do this.

Kabir (1440–1518) was a poor Muslim weaver who went about teaching that there is truth in all religions, and that though God may be called by many names, there is only one God. Kabir called himself a child of Rama and Allah, for his mother was a Hindu. He became the disciple of an Indian guru who taught that God was not to be served by complicated rituals or abstract philosophy, but in personal devotion. Thus Kabir was introduced to the idea of **bhakti**, which later became an important element in Sikhism. The Muslims were persecuting the Hindus and Kabir could not support this: he taught a religion of love in which Hindu and Muslim could be brothers. Caste observance, ritual and ascetic discipline were not required, for all that was needed was bhakti, personal devotion to God.

Kabir was a very remarkable man, a great teacher and poet, and more than 500 of his songs are preserved in the Sikh Scripture. Though he probably never met Nanak, the founder of the Sikh religion, he greatly influenced him.

Nanak (1469–1539) was a Hindu of the Kshatriya, the warrior caste, who believed that God was to be found in the home, in working for a living, as well as in solitude. He observed the similarities between the bhakti-yoga way of devotion in Hinduism and the emphasis on the need to love God being preached by the Muslim mystics, the **Sufis**, who also, untypically of Islam, recognized that other religions could also be ways to God.

When he was about thirty and a married man with two children, he received a call from God when meditating after taking a bath in the river Bein. God held out a cup of nectar (amrita) and told him to go and repeat the divine name and make others do so. He obeyed the call at once and taught his disciples to repeat the word 'Nam' which is punjabi for 'name'.

Nanak rejected some of the Hindu ideas he had been taught as a youth and accepted some of the central truths of Islam. To demonstrate his aim of bringing Hinduism and Islam together he wore the yellow robes of the Hindu holy man and the turban and prayer beads of the Muslim.

Nanak believed that:

1 There is one God who is personal and loving; he rejected the abstract, philosophic God of the Upanishads.

2 He rejected the caste system, saying that all men and women are the children of God, and therefore equal.

3 He rejected the Hindu belief in avatars, incarnations of God.

4 He continued to believe in Samsara (the cycle of rebirth) and karma (actions and their consequences which build up a record of guilt).

5 He objected to idols, so his followers were not allowed to enter Hindu temples.

6 He did not believe in yoga as a way of salvation, for to believe in the Name was all that was needed.

7 Worship had no meaning for Nanak if it did not express itself in good conduct and compassion for the unfortunate; in connection with this, he rejected the life of the solitary, recluse or monk, because it failed to provide opportunities for love and service of mankind.

8 He discarded the institution of priesthood on the grounds that it encouraged formalism in worship.

9 He believed that family life was central to the approach of God, and the obligations of family life could be fulfilled by a three-fold discipline:

(a) the remembrance of God, the practice of realizing God's continual presence;

(b) honest daily work;

(c) sharing the proceeds of honest work with others, for wealth is the gift of God and thanks should be offered to Him for providing opportunities for serving our fellow men.

Guru Nanak originated **Sangat** (congregational worship) and **Pangat** (common dining), practices which were further developed by his successors. He called his followers 'Sikhs' (disciples); he was their first guru and the founder of their religion.

The name guru (teacher) is confined in Sikhism to the first ten leaders of the movement. It is believed that Guru Nanak was reborn in each of the nine gurus who succeeded him. They are all regarded as perfect men and Sikhs find their inspiration by being united with the spirit of the gurus, feeling that the gurus live within them, filling them with spiritual power.

The second Guru, Angad (1539–1552), is notable for beginning the compilation of the **Granth Sahib** (Holy Book), including in it poems by Kabir and the work of Nanak and himself. He also built gurdwaras (temples).

The third Guru, Amar Das (1552–1574), stressed the importance of doing away with caste distinctions by insisting that all his visitors ate together in the langar, the common kitchen. He divided up the country into missionary areas, some of which were in the charge of women. He denounced the practice of suttee, in which widows were burned alive on their husband's funeral pyre. Sikh women were not veiled as were Muslim women.

The fifth Guru, Arjan (1581–1606), completed the Granth Sahib and built the golden Temple at **Amritsar** in the centre of a huge pool dug out by his father, Guru Ram Das. This temple, known as **Harimandir** (the House of God), is the principal Sikh shrine and an important centre of pilgrimage. He was killed by the Emperor Jehangir and became the first Sikh martyr.

The tenth Guru, Gobind Rai (16775–1708) made a very important contribution to Sikhism. His father, Guru Teg Bahadur, was executed for resisting the Emperor's attempt to suppress the religious freedom of all non-Muslims, so Gobind decided to organize his followers into a military force capable of protecting themselves from persecution. He founded the order of **Khalsa**, a Sikh brotherhood bound by vow to keep the **'Five Ks'**:

1 Kesh–never to cut hair or beard, for this is an act of interference with nature and therefore a failure to submit to God.

2 Kangha–to wear a comb to fasten the hair under a turban.

3 Kachha–to wear shorts, a symbol of purity.

4 Kara–to wear a steel bracelet on the right wrist, the symbol of a Sikh's unbreakable attachment to his religion.

5 Kirpan–to wear a sword, a symbol of freedom from oppression.

There were five original members of the Khalsa, but they were soon joined by many thousands more and became a formidable fighting force quite capable of defending themselves; as a result, in 1764 the Sikhs gained independence.

Guru Gobind Rai changed his name to **Singh** (lion) and all Sikh men have this name. The day before his death in 1708 he declared that from henceforth the Granth, the Holy Scripture, would be the Guru of the Sikhs. The Guru Granth Sahib was to be the symbolic representative of all the ten gurus, and the guide and inspiration of the Sikh religion.

35.2 Distinctive Beliefs

1 There is one God

(a) Complete knowledge of Him is impossible but though He is invisible He can be found in every form of life. God is immanent.

(b) Since there is only one God, both incarnations and idol worship are ruled out.

(c) Only God knows how and when the universe came into being, but everything in it depends on God for its existence.

2 Man

(a) Man owes his existence to God.

(b) Man is the crown of creation because he alone has been given freedom to choose how to act.

(c) The body dies but the soul survives until the time when, after many rebirths, it merges with God.

(d) God and the soul of the individual are the same, as a drop of water taken from the ocean has the qualities of the whole of the ocean. The body is the temple of God and must therefore be kept pure by good deeds and holy thoughts, and should not be despised. Fasting and ascetic exercises harm the body, which should be cared for in order that God's praises may be sung and his people served.

(e) Every man reaps what he sows in accordance with the law of karma. A man's past deeds decide whether or not he will be born into a good position in life, but man is free to choose between right and wrong and he has the power to improve his lot in the next rebirth. If he chooses evil he may fall even to becoming an animal.

(f) Death is not the end but only a stage in man's journey to ultimate union with God. Only God can end the cycle of rebirth and man's present life, his good deeds and devotion to God determine his future.

Fig. 35.1 The Golden Temple of Amritsar *Reproduced by courtesy of the J. Alan Cash Photo Library*

(g) Heaven and Hell exist in this life only.

(h) To become like God, man must root out his self-centredness.

Sikhism is a religion of tolerance, believing in the Fatherhood of God and the Brotherhood of mankind. People of all creeds are welcomed to their Gurdwaras (places of worship) and to the free common meal in the langar (kitchen). Work is as important as worship since it is the means of helping others, and idleness meets with strong disapproval. Wealth may be acquired, but it must be used responsibly and to help the poor. Great emphasis is laid on family life rather than the solitary life of the holy man, and it is within the family that the way of God is best practised.

Sikhs feel the use of force to be justifiable in defence of freedom and when all other means have failed.

Leisure is thought to be just as necessary as work and worship. Gambling and drinking are frowned upon.

35.3 Worship

A Sikh place of worship is called a **gurdwara** and, whether it is an elaborate building or a room in a dwelling house, it acquires sanctity because the **Guru Granth Sahib**, the Sikh Holy Scripture, is there. The building is not orientated to any point as are mosques and most Christian churches. Outside flies a triangular yellow flag bearing a device made up of a **khanda** (two-edged sword), a **chakar** (quoit) and two **kirpans** (swords). The gurdwara is open to everyone, but a visitor should remove his shoes and cover his head as a matter of good manners. The Guru Granth Sahib rests on a raised platform under a canopy; a Sikh will kneel or prostrate himself when approaching it and make an offering. The floor is carpeted and everyone sits cross-legged, men and women sitting separately.

The attention of the service is centred upon the Guru Granth Sahib and it is handled with the greatest reverence. As the readings proceed, a **chauri**, a symbol of royalty, is waved over the book. It consists of a handle of wood or metal to which are attached nylon threads or yak hair, or even peacock feathers. The book is treated with the same reverence as if it were one of the ancient gurus, for it is itself, the eleventh and last, Guru.

Sikh public worship is congregational and a matter of obligation. Morning and evening services are held at most gurdwaras, but Sunday is the day when most Sikh families attend.

A large part of the Guru Granth Sahib is in poetry and is sung to the accompaniment of musical instruments. After the singing comes a lecture or discourse and the service ends with prayer and the eating of **karah prashad**, a mixture of semolina, flour, sugar, water and butter. The karah prashad symbolizes equality and brotherhood.

Gurdwaras have other uses besides worship. They are places of spiritual education, provide shelter for travellers, and are used for making social contacts.

Private prayer

A Sikh is expected to get up early in the morning and, after taking a bath, to meditate on the Name of God. The morning prayer (**Japji**) is known by heart and may be recited while preparing for the day's work. There are set prayers for the evening, but prayers may be said at any time.

35.4 Dietary Laws

In the Punjab the Sikh diet consists of cereals, vegetables and milk; meat is eaten only occasionally since animals are kept largely for work and producing milk. Meat may be eaten, but not if it is killed in the Muslim way, because during Muslim rule 'Halal' meat was associated with conversions to Islam.

Fasting is not regarded as a useful spiritual discipline. Alcohol, tobacco and intoxicants of any kind were forbidden by the gurus, though many Sikhs ignore the prohibition of alcohol.

35.5 Rites of Passage

1 Naming the child

The family assembles in front of the Guru Granth Sahib either at home or at the gurdwara, and after prayer the book is opened at random and a lesson is read. The first letter of the first word of the reading is taken as the first letter of the name of the child. The name chosen is then announced to the congregation and karah prashad is distributed after the service. A steel wrist band is given to the child at, or soon after, the ceremony. Parents have an obligation to bring up their children in the faith and to see that they are taught to read and write **Gurmukhi**, the script in which, after Guru Angad, the Punjabi language was written.

2 Amrit Pahul (Initiation Ceremony)

The ceremony takes place before the Guru Granth Sahib, five Sikhs called the **Panj Pyaras**, the **Granthi** (reader) and the congregation. All those immediately concerned in the ceremony take a bath and bring with them the 'Five Ks'. The candidates are reminded of the principles of their religion and when they consent to them, prayer is offered and the Scripture is read. Water is then poured into a steel bowl and sugar is added and stirred with a two-edged sword, while the prescribed Scriptures are read. The candidates drink from the bowl and the liquid is afterwards sprinkled on their hair and eyes. The ceremony concludes with the karah prashad.

3 Anand Karaj (Marriage)

Marriages are arranged between families. Sikh boys or girls who marry outside the fold are the subjects of the greatest disapproval from family and community. Chastity is highly prized and especially demanded of women, who are expected to be virgins when they marry. Faithfulness within the marriage bond is expected and divorce is very much discouraged.

Marriage takes place before the Guru Granth Sahib. The ceremony begins with a meeting between the man's father, grandfather and mother's brother and their counterparts in the woman's family. After light refreshments the bride and groom sit facing the Holy Granth. Hymns are sung and an address on the duties and responsibilities of married life is given; the father of the bride then ties the edge of her head-dress to the bridegroom's sash. The marriage-hymn is then read, and it is sung while the couple walk around the Holy Granth. This is repeated four times while four more hymns are sung.

After the final prayer karah prashad is distributed and the congregation leaves the gurdwara for the wedding celebrations.

4 Death

The dead are cremated as soon as conveniently possible. A prayer for the peace of the soul is said and the pyre is ignited by a close relative. The bed-time prayer is recited by the mourners while the cremation is taking place. Karah prashad is distributed. The ashes are collected later and thrown into the running water of a stream.

Sikhs are forbidden to raise monuments in memory of the dead, since the quality of their lives is their only true memorial.

36 Buddhism

36.1 Historical Background

Unlike Hinduism, Buddhism had a historical founder, **Siddhartha Gautama**, who lived from about 563–483 BCE. Our knowledge of his early life comes from legendary stories in the Buddhist Scriptures. He was a Hindu born into the Kshatriya (soldier) class and his father was ruler of a territory just inside what is now known as Nepal. When he was very young it was predicted that he would become an ascetic, a holy man, if ever he became confronted with sickness, old age and death, so his father did his utmost to shelter him from such unpleasant sights. He was brought up in a very protected environment, married a beautiful woman and had a son, whom he called Rahula, meaning 'the fetter', which suggests that he had intimations that all was not well, that his idyllic existence somehow lacked reality.

One day he managed to evade his father's restrictions and rode out on his own; he was staggered to be confronted in succession by a sick man, an old man and a dead man. Enquiry assured him that this was a process which happened inevitably to everyone. Believing as he did in reincarnation, the thought of enduring this process perhaps thousands of times in the future made him determined to find a way to avoid it, and the sight of a wandering holy man decided him. At the age of twenty-nine he stole away from his family by night and began the life of an ascetic.

For a time he sat at the feet of various religious teachers but found no answer to his problem, so he then practised the most severe austerities, inflicting terrible hardships on his body, to the extent of almost starving himself to death. Five ascetics were so impressed by his self-discipline that they followed him, but when, at the point of death, he broke his fast, they were disillusioned and left him. On recovering, he saw how pointless his austerities had been and tried another way.

Eventually he came to Uruvela and sat under a Bo-tree, vowing never to leave it until he had found the true meaning and goal of existence–enlightenment. He put himself into a trance and while in this state, saw all his previous rebirths in succession, and discerned that physical death and rebirth are governed by the moral quality of former lives; he also saw the causes of suffering. At last he passed through the stages of meditation to complete spiritual 'enlightenment': he now became the Buddha, **the Enlightened One**.

He was sorely tempted to accept release from earthly life to enter **Nirvana**, the final goal of existence, but nobly chose instead to remain on earth and to accept the arduous task of teaching all men the way of deliverance he had discovered.

Buddha first sought the five holy men who had deserted him; he found them in Benares where, in the Deer Park he preached to them his first sermon, and they became his first disciples.

36.2 The Deer Park Sermon

The Dhamma (teaching) of Buddha was summed up in **the Four Noble Truths**:

1 Life is full of suffering (**dukkha**). The term includes not only illness and physical pain but also consciousness of the meaninglessness of existence.

2 The cause of suffering is **tanha**, desire, that is, the desire to cling to the life of this world, to be attached to, rather than detached from, things which can give no lasting satisfaction.

3 When we cease to desire we cease to suffer, so to end suffering we must eradicate desire. Buddha saw suicide to be running away from the problem, for to desire to die is as much tanha as desire to live.

4 The way to remove desire, to end suffering and enter Nirvana, the goal of existence, is to follow the Noble Eight-Fold Path (see Unit 36.4).

The first three of the Noble Truths are based on Hinduism, but the fourth is original to Buddhism.

36.3 Nirvana

Man is chained to the cycle of **Samsara**: an endless round of existence, birth followed by death, followed by birth and so on. The assumption of Indian philosophy is that existence involves suffering and the desirable goal is to cease to be subject to this endless round of misery. What keeps this cycle going is **karma**, literally 'deed' or 'action', a force which is continually being built up by man's failings in word and deed. Hinduism and Buddhism share the belief that a man's actions in this life establish the station he will occupy in his next existence, but Buddhism introduced a quite different idea of the meaning of karma and of the goal of existence.

Buddha believed that karma arose from men's desires and that the way of release was to quench the flame of desire. He saw rebirth as the lighting of one lamp from another and so on. Thus, if at death a man has not quenched his desire, the flame will continue indefinitely to be passed on. Nirvana means literally 'going out', as of a flame, so this condition is reached when there is no longer fuel to feed the flame.

Another difference is that Hinduism believes that Brahman, the unchanging reality, permeates the soul of every man as salt dissolved in water permeates every single drop of the water, so that when the body dies the soul cannot die because it is Brahman, and therefore eternal. There are therefore large numbers of eternal souls or selves. Buddhism, on the other hand, holds that the persons and the things which exist in the world have three qualities:

1 experience of suffering (dukkha);

2 impermanence (anicca);

3 absence of self (anatta).

The human personality is not static, but subject to the process of change. If we look at a series of photographs taken at different times in our lives, we see that this is so. The changes are physical, as may plainly be seen, but the personality changes also–habits, attitudes, outlooks, ways of thinking–have all changed. To the Buddhist, all life is a process of becoming something new. Nothing in the world is permanent; everything is in a constant state of change. The only permanence is Nirvana. When applied to persons this means that there is no permanent element in man, no permanent eternal soul, so there is no soul to 'carry over' from one existence to the next, no reproduction of an individual who has existed before. There is no 'cosmic self', no supreme personal reality which lives in all things; **the only reality is Nirvana**.

This makes the idea of rebirth rather difficult to understand. It is not the soul which is reborn, for there is no soul, but rather the stage in the mastery of desire which a person has achieved by the time he dies, which causes a new sequence of states of desire in the next life. Each man is made up of five constituent parts called **skandhas**, namely

1 the body;

2 feelings;

3 awareness of the world;

4 thoughts, ideas, wishes;

5 the consciousness which brings together and combines the constituent parts to form a person.

The soul is not part of this and does not exist.

At death the skandhas survive, but no longer in combination–they 'fall apart', but they are then put together in a new combination. A useful analogy is the way in which the same basic parts in a construction kit like 'Lego' can be used to create a large number of different designs or objects. The new shape in which the skandhas are combined depends upon the quality of life achieved by the person at death: it could be a better or a worse existence. When Nirvana is reached, there is no further use for the skandhas.

An analogy Buddhists use is of a candle whose flame is passed on to another candle and so on in a long line of succession. The flame of desire is reborn and will continue to be reborn until the stage is eventually reached when there is no desire left to keep it alight and Nirvana is achieved. **It is the desire which is reborn, not the individual.**

It must be conceded that this is very difficult for anyone unused to Eastern ways of thinking to understand. Not least among the difficulties is the fact that many believe the Buddha's teaching on the existence or non-existence of the self is, itself, unclear.

Nirvana is difficult to define. It is a state in which all desire is extinct and in which the cycle of Samsara no longer operates: a man is no longer subject to rebirth. It would certainly be misleading to describe it as heaven.

36.4 The Noble Eight-Fold Path

The last of the four Noble Truths indicates the way in which release from desire and suffering may be achieved. Buddha declared this to be 'The Middle Way', that is, the course between the self-torture of asceticism, and over-indulgence in the things of this world, the Noble Eight-Fold Path. Each of the principles laid down begins with a word which is translated as 'right' in English: it means 'balanced' rather than 'correct', that is, rightly balanced, free from selfish desires. These principles can be conveniently divided into three groups:

1 The Way of Wisdom

2 The Way of Conduct

3 The Way of Concentration

1 The Way of Wisdom

(a) Right views, understanding. Belief in the Four Noble Truths and The Noble Eight-Fold Path.

(b) Right thought and intention. To know the truth is not enough: resolve is needed to follow it.

2 The Way of Conduct

(c) Right speech. Control of the tongue: to desist from saying what is hurtful or untrue.

(d) Right action. Exercise of self control, the practice of good towards others and abstinence from what is evil. Obedience to the Five Precepts.

(e) Right occupation. The work done for a living must be one which benefits others. Work involving violence to others is ruled out (soldiers) and also that involving violence to animals (butchers and fishermen).

3 The Way of Concentration

(f) Right effort. To abstain from misuse of energies and to direct them towards rooting out selfish desires.

(g) Right mindfulness. To control the mind so that it becomes sensitive to the needs of others and a means of contemplation.

(h) Right concentration. The training of the mind to become calm and peaceful, not subject to distraction, but subject to the will.

36.5 The Buddhist Community (The Sangha)

An essential element in Buddhism is the **sangha**. This means the community of all those who can conscientiously repeat the Buddhist prayer called **'The Three Jewels'** every day:

I take refuge in the Buddha
I take refuge in the dharma (the doctrines or teachings of Buddhism)
I take refuge in the sangha (community)

The term is, however, sometimes restricted to those who take the three special vows of poverty, chastity and non-violence.

1 Poverty

A Buddhist monk may possess only his yellow robe, the girdle which holds the three pieces of the robe together, a needle for repairs, a razor, a 'begging' bowl and a water strainer to ensure that he does not consume any living creature which may be in his drink. He does not actually beg, for the devout are only too pleased to provide him with food; he does not thank them when they do so, for by their charity they gain merit. Books, paper and pens etc, are the common property of the sangha.

2 Chastity

A monk is forbidden to have anything to do with women, especially to be alone with a woman. Buddha at first refused to allow women to become members of the sangha but later relaxed this rule. Both his wife and step-mother took vows.

3 Non-Violence

All living things are to be loved and never made the subject of violence. The killing of animals is forbidden.

36.6 The Ten Precepts

These are the ten rules to be observed by all bhikshus (monks and nuns). They are forbidden to:

1 take the life of any living creature;

2 steal;

3 indulge in sexual misconduct;

4 tell lies;

5 indulge in alcohol or drugs;

6 eat after mid-day;

7 watch dancing or any such spectacles;

8 use cosmetics or perfume, or wear any form of adornment;

9 sleep on a raised or soft bed;

10 accept money.

Those Buddhists who have not taken vows observe the first five of the precepts and on holy days some observe them all except the tenth.

36.7 The Spread of Buddhism

Buddha devoted himself to teaching to the end of his life. He never claimed to be anything other than a man, but his followers look on him, with justification, as a very extraordinary man. The only essential difference is that he received 'enlightenment', and this is available to all who follow his teaching. Because of this it could be argued that he never intended to found a religion in the accepted sense.

The foundation of Buddhism as an international religion owes a great deal to **King Asoka** who lived about 260 BCE in northern India. He was a lover of the hunt and also a great soldier, but one day, filled with remorse over the great suffering his activities had caused, he met some Buddhist monks who converted him. He devoted the rest of his life to his new religion, building monasteries, caring for the poor and sending out missionaries at home and abroad. Many of the Buddhist inscriptions which Asoka set up all over India survive to this day.

The new religion took root in Sri Lanka and one of the most successful missionaries there was Asoka's son, Mahinda. In the fifth century CE Buddhism was introduced to Burmah and from there, over the centuries, to Thailand, Cambodia (Kampuchea), Malaysia, Java and the Philippines. Another important centre of missionary activity was China. Buddhism reached there from India in the first century CE and though progress was slow, because China already had two religions of her own, Confucianism and Taoism, by the ninth century CE it had become powerful.

From China Buddhism spread to Korea in the fourth century, Japan in the sixth century and Tibet in the seventh. In each of these cases pure Buddhism became somewhat contaminated by the existing religion of the land. In Tibet and Nepal certain magical practices were introduced, and in Nepal many ideas drawn from Hindu mythology. In Japan there are a number of Buddhist sects, the best known of which is Zen. In each of these countries Buddhism has moved a long way from that which we find in Sri Lanka, which is probably closest to the original.

36.8 Theravada and Mahayana Buddhism

There are two main branches of Buddhism, Theravadins, who claim that they alone remain faithful to the teaching of the founder, and the Mahayanists, who claim to be more liberal and broader in their interpretation. The main differences in belief are:

1 Theravadins believe that if a man is to reach Nirvana he must do so by his own efforts. Man needs first to know The Middle Way as proclaimed by Buddha, and he must then resolve to follow it without help from anyone, for no one can help him to reach this intensely personal experience. Only monks can possibly reach Nirvana for they alone have the time to practise the necessary disciplines.

The Mahayanists believe that man is not alone in the universe, and help can be obtained by prayer. They believe that Buddha laid greatest emphasis on compassion rather than knowledge.

2 The Theravadins see the Buddha as a very great man, but still a man, while the Mahayanists see him as a super-human being; this accounts for the fundamental difference in outlook of the two. The former see Buddha as an example of a man who found enlightenment by his own efforts and passed on knowledge of the way to achieve it in order that others by their own efforts, might also attain it. There is no other way. The latter, by contrast, believe that since Buddha was compassionate enough to postpone his entry into Nirvana so that he could pass on the secret of enlightenment to his fellow men, compassion is the noblest of all the virtues; his compassion for man is still available to those who seek his help. For this reason, the Mahayanists have places where men may offer praise and pray for help.

The ideal man for the Theravadins is the **Arhat** (worthy one), he who with heroic perseverance and zeal strives to reach Nirvana by his own single-minded effort, while to the Mahayanists the ideal is the **Bodhisattva** (Buddha-to-be) who, while striving for Nirvana, does not neglect his duty of compassion for others.

36.9 Distinctive Theravada Beliefs and Traditions

1 Scriptures

These are read mainly by monks and are in three groups, the Three Baskets:

(a) The Basket of Discipline–rules of life

(b) The Basket of Discourses–the teachings of Buddha

(c) The Basket of Philosophical Writings

2 Religion

Theravada Buddhism has no concept of God either as a person or impersonal force and therefore there is no worship. Buddha is revered as the greatest of all teachers.

3 Meditation

Meditation occupies a significant part and is based on the belief that body and mind are closely related and that control of the body helps to control the mind.

36.10 Mahayana Beliefs and Traditions

1 A Bodhisattva is an important figure in Chinese, Tibetan and Japanese Buddhism. He or she is one who, like Buddha, has achieved enlightenment and won the right to enter Nirvana, but who, again like Buddha, chose to remain in the world to help others. When they have done all they can they will enter Nirvana and for the present are known as 'Buddhas-to-be'. They do not wait to be asked for help but take the initiative in offering it.

A famous Bodhisattva called Amitabha in China and Amida in Japan is especially loved. When, after a long life devoted to discipline and love of men, he achieved enlightenment, he declared that he did not wish to become a Buddha unless it could be for the sake of all who sought his help so that they could join him in Paradise. This is vastly different from the idea of Nirvana as a spiritual state which is impossible to describe.

2 Sukhavati (Heaven or Paradise)

This is described as an idyllic place, like a Garden of Eden, full of fruit and flowers. It is also called the 'Pure Land' and the Buddhist sect which reveres Amitabha is called the Pure Land Sect.

3 Universal salvation

The Pure Land Sect in China, Tibet and Japan believe that salvation is not achieved by man's own effort in self discipline and good works, but rather that it is the gift of a loving God. It is also believed that Amitabha, by his perfect life, built up a store of merit far greater than was necessary for his own need: he is able to pass on this treasury to those who have faith in him.

4 A personal God

Mahayana belief in a personal and loving God indicates the distance which separates this form of Buddhism from the original Buddhist teaching that there is no cosmic self, no supreme personal reality, and that the only reality is Nirvana, a state of non-existence.

The Theravada tradition is followed in Sri Lanka, Burmah and Thailand; the Mahayana in China and Tibet.

36.11 Principal Feasts

The birth of the Buddha

In the Mahayana tradition the birth of the Buddha is celebrated in the month of Vesakh which falls at the end of April, his death in February and the enlightenment in December. In the Theravada tradition all three events are celebrated on the same day in the month Wesak. It is the most important Buddhist festival, observed with public processions through decorated streets and food and money are given to the poor.

The first preaching

This commemorates not only the sermon at the Deer Park, but also the moment when Buddha was conceived and the day he left home to seek the meaning of life. In Sri Lanka the first preaching of Mahinda, the missionary who brought Buddhism to the island, is commemorated. (He was the son of King Asoka.) Pilgrimage is made to a holy mountain near the ancient ruined city of Anuradhapura.

The meditation period

The Buddha arranged that during the monsoon period members of the sangha should spend the three months living together, occupied in study and meditation. Lay people try to spend some part of this season living with the monks. At the end of the period there is a festival called **Kathina** when the monks are given new robes by wealthy patrons.

Stupas

When the Buddha died, his cremated remains were distributed among many important cities and were placed in special buildings called stupas, meaning relic containers. These often take the form of a stone mound on which is erected a spire. These buildings are often very large and elegant. Another type of Buddhist building is the pagoda, one of the finest of which is the Shwe Dagon in Rangoon, the spires of which are higher than St Paul's Cathedral and are covered in gold leaf. These also often contain relics of the Buddha or of holy men.

Many local religious festivals are observed in connection with these holy places, the most notable of which is the Festival of the **Sacred Tooth** at Kandy in Sri Lanka. It takes place in August and lasts for ten days. The Sacred Tooth of the Buddha is carried in procession through the streets on the back of a huge elephant to the banging of drums and the clanging of cymbals.

Urabon

This is a very spectacular and moving Japanese festival held during the Meditation Period, and is devoted to prayer for loved ones who have died. Lanterns lit by the stones which have been set up to commemorate the departed are taken home for ten days; the spirit of the dead person goes home with the lantern. In the evening of the last day of the festival, each lantern, carrying the name of the dead person, is floated on a river or a lake, and as it moves away the spirit of the departed goes with it. The sight of thousands of lanterns floating on the water in the darkness is very moving.

36.12 Rites of Passage

Many boys spend a period in a monastery when they are very young. They are first dressed in elaborate costly robes to remind them both of the riches of this world and the fact that Gautama was once a prince. The robes are then removed and the boys' heads are shaved, the father cutting off the first locks, and after a ritual bath they put on the yellow robe of the monk. In the monastery they are taught the art of meditation and the rule of life laid down by the Buddha in the Eight-Fold Path.

36.13 Images of Buddha

These show him either sitting, standing or lying down.

1 Sitting

(a) The Lotus or meditation posture He sits with each foot placed on the opposite thigh in an

Fig. 36.1 Seated Buddha
Reproduced by courtesy of the Buddhist Society

attitude of meditation. The roots of the lotus flower (water lily) are in the mud while the long stem reaches upward to the surface where the beautiful flower grows. The lotus posture is taken to resemble the lotus flower and is powerfully symbolic.

(b) Calling the earth to witness The posture is as for the lotus except that the fingers of the right hand are touching the ground. The evil spirit, Mara, tried to persuade Buddha that his experience of enlightenment was an illusion and that even if it were true it would be better for him at once to enter Nirvana. Buddha called the earth to witness that there had always been people who devoted their lives to the welfare of mankind.

(c) The Wheel of Law A wheel with eight spokes representing the Noble Eight-fold Path is one of the chief symbols of Buddhism and a common image of Buddha represents him setting the wheel in motion. The position is as for the Lotus except that the fore finger and thumb of the right hand are joined to form a circle representing the Wheel of the Law, while a finger of the left hand points to it as if setting the wheel in motion.

2 Standing (Blessing)

The hands are raised in front of the breast, palms touching in the usual Indian form of greeting. Another version is with the right hand raised in blessing, palm outward, while the left hand points forward in a gesture of giving.

3 Lying down

This represents Buddha entering Nirvana.

Section III Self-test Units

Part I The Synoptic Gospels

O-Level Questions

1 Give an account of the political situation in Palestine during the ministry of Jesus, including in your answer some reference to Pilate, Herod Antipas and Caiaphas.

(Cambridge 1982 O Level)

See pages 1–3

2 **(a)** Write a short account of **three** of the main religious and political parties of the Jews at the time of the ministry of Jesus. (**15** marks)

(b) Describe **one** incident or parable in the Gospels in which any one of these parties is mentioned. (**5** marks)

(JMB 1981 O Level)

See pages 3–6, 33–40

3 Who were the Sadducees? In what important respects did they differ from the Pharisees? Why did the Sadducees oppose Jesus?

(Cambridge 1981 O Level)

See page 4

4 **(a)** Who were the Samaritans and why were they hated by the Jews?

(b) Describe **two** incidents, concerning Samaritans, in which Jesus was involved.

(c) Suggest a situation in the modern world to which the teaching contained in these stories might have relevance.

(JMB 1982 O Level)

See page 6

5 Write notes on:

(a) tax collectors

(b) Herodians and

(c) Samaritans.

In each case mention one occasion when they are referred to in the Gospels.

(Cambridge 1981 O Level)

See pages 1, 5, 6

6 **(a)** What is meant by the 'synoptic problem'? State briefly the sources used by the synoptic writers and give a possible solution to this problem.

(b) *(i)* Identify a story found only in Matthew and a story found only in Luke. *(ii)* State **one** characteristic or interest of the writer which each story illustrates.

(JMB 1982 O Level)

See pages 12–15

7 Discuss the theory that St Mark's was the first Gospel to be witten. What evidence is there in early writings about the authorship of this Gospel? **(16** marks)

(O & C 1981 O Level)

See page 13

8 What are the main differences between the Gospels according to St Matthew and St Luke? What may be gathered from these about the methods and purposes of the writers? **(16** marks)

(O & C 1980 O Level)

See pages 14–15

9 Relate Matthew's account of the wise men, and Luke's account of the shepherds. Mention two other ways in which the birth stories recorded in these Gospels differ. State briefly why you think these differences occur.

(Cambridge 1981 O Level)

See pages 20–23

10 (a) Which **five** incidents in the birth story of Jesus are linked by Matthew with references to the Old Testament? **(15** marks)

(b) Comment on Matthew's use of the Old Testament. **(5** marks)

(JMB 1981 O Level)

See page 20

11 (a) Outline the account in Luke of the events which took place in Bethlehem at the time of the birth of Jesus. **(12** marks)

(b) What does Matthew say about *(i)* the visitors to the infant Jesus, and *(ii)* the reason for which the family left Bethlehem? **(4** marks)

(c) What significance did the writers see *(i)* in the fact that Jesus was born in Bethlehem and *(ii)* in the different visitors to the infant Jesus? **(4** Marks)

(JMB 1982 O Level)

See pages 21–23

12 Relate how the birth of John the Baptist was foretold. What can be learned from this incident about the purpose of his work?

(Cambridge 1981 O Level)

See page 23

13 Give an account of Jesus' visit to the Temple when he was twelve years old. What can be learned from this incident, (a) about Jesus and (b) about his parents?

(Cambridge 1982 O Level)

See pages 23–24

14 Relate **three** occasions when Jesus spoke about John the Baptist. How did Jesus regard John?

(Cambridge 1981 O Level)

See page 25 and Luke 7: 24–28

15 Give an account of Jesus' visit to the synagogue in his home town, Nazareth. Compare this with the events and reception accorded to Jesus in a synagogue on any **one** other occasion, recorded in the set passages **(16** marks)

(O & C 1981 O Level)

See pages 23–24, 50

16 Give a careful account of what Jesus said in the Sermon on the Mount on the subject of making a show of religion ('practising piety before men' (RSV)) in acts of charity, praying and fasting ('giving alms, praying and fasting' (RSV)). Discuss the importance of this teaching **(12, 4** marks)

(O & C 1982 O Level)

See pages 27, 30–31

17 Give a full account of Jesus' teaching in the Sermon on the Mount about prayer. What further teaching about prayer did Jesus give in his parables in Luke's Gospel? (Do **not** relate the parables.)

(Cambridge 1981 O Level)

See pages 27, 31–32, 38

18 In the Sermon on the Mount, Jesus contrasts the law of Moses ('It was said to the men of old . . .') with his own teaching ('But I say to you . . .'). Relate **three** examples he used, and comment on the difference between his interpretation of the law and that of the scribes.

(Cambridge 1982 O Level)

See page 29

19 (a) Relate the teaching of Jesus in the Sermon on the Mount about anxious thoughts. **(12** marks)

(b) Narrate a parable told by Jesus in which the central character thought more of his material comfort than of God. **(4** marks)

(c) Relate a parable told by Jesus in which the central character showed a willingness to sacrifice everything for God. **(4** marks)

(JMB 1982 O Level)

See pages 32, 63, 130–131

20 'And when the devil had ended every temptation, he departed from him until an opportune time (until a fit opportunity arrived).' What was the *meaning* of these temptations? State **one** other occasion in the life of Jesus when he was tempted.

See pages 25–26, 55 (*London 1979 O Level*)

21 Give an account of Jesus' three temptations in the wilderness. What light is shed on his ministry by this story?

See pages 25–26 (*Cambridge 1982 O Level*)

22 Describe one parable from St Luke's Gospel about persistence in prayer and one about humility. How do these illustrate the value of parable as a method of teaching?

See pages 33, 36, 38 (*London 1979 O Level*)

23 Give a careful account of the parable of the rich man and Lazarus. What do we learn from the parable about **(a)** Jewish ideas at the time about life after death, and **(b)** St Luke's attitude to riches in general?

See page 39 (*London 1979 O Level*)

24 Relate the parable about the labourers hired to work in the vineyard at harvest time. What meaning was Jesus trying to convey in this story?

See page 35 (*Cambridge 1981 O Level*)

25 Narrate and explain the Parable of the Dishonest Steward. Describe briefly **two** other examples of the teaching of Jesus on wealth. (**12, 4, 4** marks)

See pages 37–39, 130–131 (*Oxford 1982 O Level*)

26 What was Jesus' reply to the question, 'And who is my neighbour?' Give a short account of **one** other narrative involving a Samaritan. What does Jesus say, in the Sermon on the Mount, about loving your neighbour? (**10, 5, 5** marks)

See pages 32–33, 38 (*Oxford 1980 O Level*)

27 Outline the parable of the Prodigal Son. Indicate why Jesus told the story and what it means. (**16** marks)

See pages 36–37 (*O & C 1981 O Level*)

28 Describe fully the raising of Jairus' daughter and the healing of the woman with haemorrhages (the woman with a flow of blood (RSV)). Comment on special features of interest. (**16** marks)

See pages 44, 45 (*O & C 1980 O Level*)

29 Describe the incidents when Jesus healed **(a)** the blind man at Bethsaida and **(b)** Bartimaeus. Comment on **two** important features in each incident.

See page 45 (*Cambridge 1981 O Level*)

30 **(a)** Describe the incident when Jesus cured the daughter of a Syrophoenician (NEB a Phoenician of Syria). (**8** marks)
(b) What guidance does this story offer to the first Christians about the attitude they should take to non-Jewish people? (**4** marks)
(c) Relate **one** parable in the Gospels which offers similar advice to the first Christians about the attitude they should take to non-Jewish people, and show clearly what attitude is being encouraged. (**8** marks)

See pages 42–43 (*JMB 1981 O Level*)

31 Describe **two** occasions on which Jesus showed himself to have authority over nature in different ways. (**12** marks)
What interpretations might be given of *each* of these incidents? (**8** marks)

See pages 46–47 (*AEB 1982 O Level*)

32 Give an account of the conversation at Caesarea Philippi between Jesus and his disciples up to the point where Jesus rebuked Peter. Comment on the importance of this occasion in the ministry of Jesus and on his request for secrecy.

See page 47 (*Cambridge 1981 O Level*)

33 Describe Jesus' transfiguration, and explain its importance. (**16** marks)

See page 48 (*O & C 1981 O Level*)

34 Give a full account of what took place on the two occasions when a voice from heaven proclaimed Jesus to be the Son of God, and explain the significance of each in his life. (**16** marks)

See pages 25, 48 (*O & C 1980 O Level*)

35 Give an account of the sending forth of the seventy disciples and of their return. Comment on **two** points of interest in these accounts.
See page 49 (*London 1979 O Level*)

36 Describe the preparations which Jesus made for his entry into Jerusalem, and how the people received him as he rode into the city. State why he went to the temple, and his criticism of those who were misusing it. (**24** marks)
See pages 49–50 (*WJEC 1980 O Level*)

37 How was Jesus questioned in Jerusalem about **(a)** his authority and **(b)** paying taxes? Comment on the aptness of Jesus' replies. (**16** marks)
See pages 50–51 (*O & C 1981 O Level*)

38 What personal request did James and John make to Jesus? In what circumstances was this request made? Give Jesus' answer and his teaching on greatness which followed.
See page 49 (*Cambridge 1981 O Level*)

39 Describe what happened in the garden of Gethsemane on the night of Jesus' arrest. Give some possible explanations of Judas' betrayal of Jesus.
See pages 54, 55 (*Cambridge 1981 O Level*)

40 Describe the Last Supper according to the Synoptic Gospels. Summarize what happened subsequently in the Garden of Gethsemane according to the Synoptic Gospels. (**14, 6** marks)
See pages 54–55 (*Oxford 1982 O Level*)

41 Give an account of the examination and trial of Jesus before **(a)** the Jewish council, **(b)** Pilate, *and* **(c)** Herod. What were the reasons for these three 'trials'?
See pages 56–58 (*London 1981 O Level*)

42 What have the Synoptic Gospels to say about the trials of Jesus? What part was played by Herod in these events? (**14, 6** marks)
See pages 56–58 (*Oxford 1981 O Level*)

43 Describe **(a)** the happenings immediately after Jesus died on the cross, and **(b)** the arrangements that were made for his burial and the guarding of the tomb. Account for the emphasis placed on these stories in the Gospels.
See pages 59–60 (*Cambridge 1982 O Level*)

44 Describe what happened on the way to and at Emmaus. What new insights came to Jesus' companions?
See page 60 (*Cambridge 1981 O Level*)

CSE Questions

45 **(a)** What were the disciples discussing on the road to Emmaus when they were joined by Jesus? What was the explanation that Jesus gave them?
(b) Describe what happened when they reached the village.
See page 60 (*LREB 1982 CSE*)

46 Matthew 3:1 1–17
(a) Why did people come to John to be baptized
(b) Why was John unwilling to baptize Jesus?
(c) What did the voice from heaven say?
(d) Suggest a reason why Jesus allowed himself to be baptized.
See pages 24–25 (*EMREB 1978 CSE*)

47 The Temptations
When Jesus was in the wilderness he was tempted by the Devil.
(a) What were the **three** temptations with which the Devil tempted Jesus?
(b) How did Jesus answer the devil?
(c) What is the significance of the three temptations?
See pages 25–26 (*LREB 1982 CSE*)

48 **(a)** On which day of the week did Jesus visit the synagogue? Why did he do so?
(b) 'When you pray, you must not be like the hypocrites; for they love to stand and pray in the synagogues and at the street corners . . .'
Why is Jesus criticizing these people? What did he tell his followers about the way they should pray, in contrast to the hypocrites?
See pages 26, 31 (*NWREB 1981 CSE*)

49 Relate the parable of the Prodigal Son. Explain what Jesus meant by this parable and why he chose to teach by way of parables. What value has this parable today?
See pages 33, 36 *(WJEC 1982 CSE)*

50 **(a)** Tell the parables which begin *(i)* 'The Kindgom of Heaven is like a grain of mustard seed.' *(ii)* 'The Kingdom of Heaven is like leaven.' *(iii)* 'The Kingdom of Heaven is like treasure hid in a field.'
(b) What did Jesus mean by 'the Kingdom of Heaven'?
See page 34 *(NWREB 1981 CSE)*

51 **(a)** Give a brief account of the parable of the bags of gold.
(b) *(i)* Explain the point of this parable when Jesus first told it. *(ii)* Explain how the parable is often used today to illustrate the use of talents.
(c) Suggest ways in which Christians can use their talents in service to *(i)* God, *(ii)* others.
See page 39 *(SEREB 1981 CSE)*

52 Relate the parable of the unjust judge and the parable of the Pharisee and the tax collector. Explain Jesus' teaching in **each** of these parables.
See pages 36, 38 *(WJEC 1981 CSE)*

53 **(a)** Give a brief account of the parable of the dishonest steward.
(b) *(i)* What was Jesus teaching in this parable? *(ii)* Suggest why Jesus used parables for most of his teaching.
(c) What methods of raising money for the Church do you consider to be *(i)* acceptable, *(ii)* unacceptable from a Christian point of view?
See pages 33, 37, 130–133 *(SEREB 1982 CSE)*

54 Select **one** of the following parables:
(a) The Rich Man and Lazarus,
(b) The Pounds (Gold Coins),
(c) The Great Supper (Great Feast).
Relate the story, **explain** its meaning and **state** why Jesus used parables as a method of teaching
See pages 33, 35, 39 *(EMREB 1982 CSE)*

55 In the grainfields
(a) Why were Jesus' disciples criticized for plucking ears of corn?
(b) What did Jesus say about David in his reply to them?
See pages 9–10 *(LREB 1982 CSE)*

56 Jesus asked the disciples, 'Who do people say I am?'
(a) What were the replies given by the disciples, including Peter?
(b) What was the significance of Peter's reply and Jesus' instructions to the disciples?
See page 47 *(LREB 1982 CSE)*

57 Write an account of the transfiguration of Jesus and give an explanation of the events that occurred.
See page 48 *(WJEC 1980 CSE)*

58 The disciples failed to heal the boy with a dumb spirit. Relate the conversation that took place between Jesus and the boy's father and how Jesus healed the boy. What important lesson did the disciples learn from Jesus that day?
See page 43 *(WJEC 1982 CSE)*

59 Describe the occasion when Jesus calmed a storm. Why do some people find this account difficult to believe?
See page 46 *(NWREB 1982 CSE)*

60 James and John asked Jesus, 'Teacher, we want you to do for us whatever we ask of you.' Relate what they asked, explain Jesus' answer and comment on the teaching that followed.
See page 49 *(WJEC 1980 CSE)*

61 *'The sabbath was made for man, not man for the sabbath.'* Relate the incident that led up to this saying. Explain Jesus' teaching on this occasion and his teaching when he healed the man with a withered hand on the sabbath.
See pages 10, 43–44 *(WJEC 1981 CSE)*

Part II The Old Testament

O-Level Questions

62 **(a)** What does the book of Genesis say about *(i)* the origins of man, *(ii)* the beginnings of sin, *(iii)* the covenant with Noah?
(b) Refer briefly to a passage you have studied to show that God's covenant with Israel made moral demands upon man.
See pages 65–67 *(JMB 1981 O Level)*

63 Why did Abram leave Haran? Who was with him at the time? Where did they go? What did Abram do to mark the places they visited?
See page 67 and Genesis 12:1–9 *(Cambridge 1981 O Level)*

64 **(a)** Tell the story of Abraham's attempted sacrifice of Isaac.
(b) What did Abraham learn from this incident about his relationship with God?
(c) What did **one** of the eighth century prophets say about animal sacrifice?
See pages 68, 91–104 and Amos 5:21–27 *(JMB 1981 O Level)*

65 **(a)** Describe the covenant which God later made with Abram.
(b) By referring to the covenant at Sinai and the new covenant described by Jeremiah, show how the idea of a covenant changed.
(JMB 1982 O Level)
See pages 75, 78, 109 and Genesis 15:1–21; Exodus 19:1–25; Jeremiah 31:31–34

66 Give a careful account of the way Jacob took the blessing from Esau. Comment on the importance of this story for the religion of Israel.
See page 69 and Genesis 19:27–34 *(London 1980 O Level)*

67 'Misunderstanding, jealousy and fear'. In telling the story of Joseph's early life up to and including his being sold into captivity, show how these three elements are to be found.
See page 70 and Genesis 37:1–36 *(London 1980 O Level)*

68 'It was not you who sent me here but God.' Show how the plan of God worked out in the life of Joseph.
See page 70 *(London 1981 O Level)*

69 Give a careful account of the life of Moses before he saw the burning bush. Show how this time prepared him for his later experiences.
See pages 72–78 and Exodus 2:1–25 *(London 1981 O Level)*

70 Examine the response of Moses to the Divine call and comment on God's reaction, as it is explained in the Biblical narrative.
See pages 72–73 and Exodus 3:1–15 *(SUJB 1982 O Level)*

71 **(a)** Write an account of the experiences at Mount Horeb of *(i)* Moses, *(ii)* Elijah.
(b) Explain why each was at the mountain.
(c) What mission was each given?
See pages 75, 90 and Exodus 19 and 20; 1 Kings 19 *(JMB 1982 O Level)*

72 **(a)** Show how the Ten Commandments indicate *(i)* man's duties to God, *(ii)* man's duties to his fellow men.
(b) Illustrate from passages you have studied, how **three** of these commandments were either kept or broken.
See pages 75–77 *(JMB 1981 O Level)*

73 How does the story of Gideon illustrate the problems which faced the Hebrew peoples after the settlement in Canaan?
See pages 78–79 and Judges 6 and 7 *(SUJB 1982 O Level)*

74 Outline the birth, boyhood and call of Samuel commenting on any **two** points of interest for the religion of Israel.
See page 83 and 1 Samuel 1; 2:12–36; 3:1–19 *(London 1981 O Level)*

75 **(a)** Describe briefly the events which led to the anointing of Saul as King of Israel.
(b) What signs did Samuel say would indicate that Saul was the Lord's anointed?
(c) What do we learn from these events about the nature of prophecy in early Israel?
See page 84 *(JMB 1981 O Level)*

76 Give a brief account of the relationship of David with Saul as revealed in the prescribed passages. What can be learned about the characters of both men?
See pages 85–86 *(London 1981 O Level)*

77 How was the news of Saul's death brought to David? What was his reaction? How did he treat the messenger?
See page 86 and 2 Samuel 1:1–16 *(Cambridge 1981 O Level)*

78 **(a)** What was the role of king in Israel?
(b) Relate **one** incident about David which shows that the writer regarded him highly.
(c) Relate **one** story from the reign of Solomon which reflects his wisdom.
(d) How successful do you think each of these was as king?
See pages 81–82, 86–88 and 1 Samuel 17:1–58; 24 *(JMB 1982 O Level)*

79 **(a)** Describe how David was anointed as King of Israel.
(b) Explain how Jeroboam became King of the Northern Kingdom.
(c) What does each incident tell us about the role of the prophet?
See pages 85, 88 and 1 Samuel 16:1–13; 1 Kings 12:1–17 *(JMB 1981 O Level)*

80 Give a careful account of Nathan's parable. Explain the circumstances which led to its being told and the consequences for its hearer.
See page 86 and 2 Samuel 12:1–14 *(London 1980 O Level)*

81 Give an account of the dedication of the Temple and the prayer Solomon made. Comment on the prayer.
See pages 87–88 and 1 Kings 5, 6 and 7 *(Cambridge 1981 O Level)*

82 Tell the story of Naboth's vineyard and of Elijah's meeting with Ahab afterwards. What principles were at issue in this incident? (**16** marks)
See page 90 and 1 Kings 21:1–26 *(O & C 1981 O Level)*

83 **(a)** Describe the contest on Mount Carmel between Elijah and the prophets of Baal.
(b) What can be learned from this incident about the differences between the religious customs and beliefs of the Canaanite prophets and those of Elijah?
See pages 89–90 and 1 Kings 18:22–46 *(JMB 1982 O Level)*

84 **(a)** *(i)* Relate the story of Ahab's consultation with Micaiah and the other prophets before the battle of Ramoth-Gilead. *(ii)* How was the true prophet of God distinguished from the false prophets in this story?
(b) Briefly describe the point of disagreement between Amos and Amaziah about the nature of the prophet's role.
See pages 91–92, 95 and 1 Kings 22:1–38 *(JMB 1981 O Level)*

85 What do you understand by the word 'prophet'? Illustrate your answer from the passages you have read and show the role of the prophet in society.
See pages 80–81, 89–118 *(O & C 1982 O Level)*

86 Describe the characteristics of the 'ecstatic prophets' illustrating your answer with reference to biblical material you have studied. (**15** marks)
What evidence is there to suggest that Elisha was an 'ecstatic prophet'? (**5** marks)
See pages 80–81, 89–91 and 2 Kings 6 *(AEB 1982 O Level)*

87 Describe Naaman's visit to Elisha. What lesson did **(a)** Naaman, and **(b)** Elisha's servant learn from it? (**16** marks)
See page 91 and 2 Kings 5 *(O & C 1981 O Level)*

88 By using the prescribed passages, show the qualities in the character of Elijah which make him 'one of the greatest prophets'.
See pages 89–90 *(London 1980 O Level)*

89 **(a)** Describe the call to be a prophet experienced by *(i)* Elisha, *(ii)* Amos, *(iii)* Isaiah of Jerusalem (Isaiah Chapters 1-39).
(b) Say in what ways these incidents are similar and in what ways they differ.
(JMB 1982 O Level)
See pages 90–94, 100–101 and 1 Kings 19:19–21; Amos 7:10–17; Isaiah 6:1–9

90 Describe life in the Northern Kingdom in the days of Amos, and state why the leaders were opposed to his teaching. (**24** marks)

(*WJEC 1981 O Level*)

See pages 92–95

91 By using the prophecies show how Amos saw the worship of God revealing itself in terms of social justice.

(*London 1980 O Level*)

See pages 92–95

92 Give an account of the teaching of Amos on the theme that God is just and demands justice from his people. (**20** marks)

(*AEB 1982 O Level*)

See pages 93–95

93 **(a)** *(i)* Tell the story of Hosea's marriage. *(ii)* What did his experience teach him about the nature of God and the future of the nation?

(*JMB 1981 O Level*)

See pages 96–97

94 'The ultimate truth about God is that He is a God of love.' Do you think this is a fair description of the message of Hosea? Use his prophecies in support of your answer.

(*London 1980 O Level*)

See pages 95–98

95 What did
either **(a)** Isaiah of Jerusalem (Isaiah Chapters 1-39),
or **(b)** Jeremiah
say about the survival of a remnant of the nation?

(*JMB 1981 O Level*)

See pages 104, 110

96 What use was made of parables by the prophets **(a)** Nathan *and* **(b)** Isaiah? Illustrate your answer with specific references to **one** parable for **each**, including the teaching presented through the parable.

(*AEB 1982 O Level*)

See pages 86, 103 and 2 Samuel 12:1–14; Isaiah 5:1–7

97 Give a careful account of the call of Jeremiah and show how its influences and meaning are seen in his prophecy.

(*London 1981 O Level*)

See pages 105–111

98 What lesson did Jeremiah teach through 'the potter and the clay'? Why, in your opinion, did he need to teach this lesson, and how do you think the people responded? (**24** marks)

(*WJEC 1981 O Level*)

See page 110 and Jeremiah 18:6

99 **(a)** Describe and explain *(i)* Amos' vision of a basket of summer fruit, *(ii)* Jeremiah's vision of an almond branch.
(b) Illustrate **three** other ways in which prophets you have studied received or communicated messages from God.

(*JMB 1982 O Level*)

See pages 83–118 and 2nd Amos 8:1–3; Jeremiah 1:11

100 Relate fully the vision which Ezekiel saw at 'the entrance of the gateway of the inner court'. Add explanatory notes on those aspects of the vision that refer to the condition of the people of Jerusalem. (**24** marks)

(*WJEC 1981 O Level*)

See pages 112–115

101 What was the allegory that Ezekiel spoke to 'the rebellious house'? Give the main content of the allegory, and explain carefully its message of judgment.

(*WJEC 1981 O Level*)

See pages 112–115

102 What did Deutero-Isaiah teach about
(a) the worship of idols, (**8** marks)
and **(b)** the importance of Cyrus? (**8** marks)
What does this teaching reveal about the prophet's view of God?

(*AEB 1982 O Level*)

See pages 115–118

103 Relate the teaching of Deutero-Isaiah on idolatry. Why did he teach that it was wrong for the people of Judah to depart from their God? (**24** marks)

(*WJEC 1981 O Level*)

See pages 115–118

Part III Christian Social Responsibility

O-Level Questions

104 What are the Ten Commandments? How far do you think they are appropriate as a basis for life today?

See pages 75–77, 119–154 (*London 1979 O Level*)

105 **(a)** What teaching on divorce is found *(i)* in the Gospels, *(ii)* in Paul's writings?
(b) What reasons concerning the remarriage of divorced persons in church are given today *(i)* by those who would permit it, *(ii)* by those who are opposed to it?

See pages 122–123 (*JMB 1982 O Level*)

106 'In the sweat of your face you shall eat bread' (Genesis). What is the place of work in a society where religion is important?

See pages 126–130 (*London 1980 O Level*)

107 **(a)** How far do you think that *(i)* strike action, *(ii)* working to rule, and *(iii)* the practice of a 'closed shop' can be regarded as right actions from a Christian point of view?
(b) Outline in each case the arguments which might be used by those who hold a different view from yours.

See pages 126–130 (*JMB 1980 O Level*)

108 'Sex and money are two of the strongest driving factors in human make-up.' Discuss the religious understanding of **one** of these factors in human life.

See pages 119–125, 130–133 (*London 1979 O Level*)

109 **(a)** Give a full description of the work of any **two** organizations concerned with overseas aid.
(b) For what reasons ought a Christian to support such aid?
(c) Suggest what reservations a Christian might have as to the use to which such aid might be put.

See pages 133–135 (*JMB 1980 O Level*)

110 An eighteenth-century judge, in passing sentence, said 'You are not being hanged for stealing a sheep. You are being hanged so that sheep will not be stolen.'
(a) What is the purpose of punishment illustrated by this statement?
(b) State and explain three other possible aims of punishment, illustrating each by one example.
(c) Comment upon all four aims in the light of Christian principles.

See pages 139–145 (*JMB 1980 O Level*)

111 Violence and juvenile crimes appear to be on the increase. What factors may account for this? (**10** marks)
Discuss ways in which Christians may try to improve this situation. (**10** marks)

See pages 139–145 (*AEB 1982 O Level*)

112 'It is lawful for Christian men, at the command of the Magistrate, to wear weapons and serve in the wars.' (Articles of Religion)
(a) What arguments from scripture might be brought *(i)* to support, and *(ii)* to oppose this statement?
(b) Consider how far it would be right for an individual Christian to follow the dictates of his own conscience rather than an official church declaration in this matter. Give reasons for your answer.

See pages 143–145 (*JMB 1980 O Level*)

113 Outline and discuss the arguments (religious and non-religious) for and against the 'just war'.

See pages 143–145 (*London 1979 O Level*)

114 It has been said that in recent years there has been a decrease in the number of adult smokers and that smoking has become less socially acceptable. At the same time there has been an increase in the number of children, some of them young, who smoke.
(a) Suggest **three** reasons for the prevalence of this habit amongst children.
(b) What do you think should be done by *(i)* parents, *(ii)* schools?
(c) For what reasons ought parents and schools to be concerned about this issue?

See pages 145–147 (*JMB 1982 O Level*)

115 What is Euthanasia? (**2** marks)
Discuss the arguments for and against euthanasia in the following cases:
(a) a baby born with severe mental and physical handicaps, (**9** marks)
and **(b)** a person suffering from an incurable illness. (**9** marks)
(*AEB 1982 O Level*)
See pages 150–152

116 'Population control is the major need in solving the hunger question'. Show some of the problems such a method can cause for religious people.
(*London 1980 O Level*)
See pages 152–153

117 One-third of the world's population consumes more than two-thirds of the world's food supply. The difficulties of rectifying this situation are such that we might end up with the 'hungry' nations even worse off. What policies are being adopted, and what else do you consider might be done, to ensure a juster distribution of resources?
(*SUJB 1981 O Level*)
See pages 152–153

CSE Questions

118 'Honour your father and mother.'
(a) What guidance do you find in the Bible about the relationship between parents and children?
(b) Discuss some of the special needs of elderly people, and suggest ways in which society can help them.
(*EAEB 1980 CSE*)
See pages 119–122

119 **(a)** Giving **two** examples for **each**, show how a young Christian might have responsibilities for *(i)* a widowed grandmother living alone, *(ii)* younger members of the family.
(b) Writing a separate paragraph on each, outline **three** of the main areas of conflict in the family.
(c) 'The family that prays together, stays together,' Giving your reasons, state whether or not you agree with this statement.
(d) Mark Twain said:
'When I was sixteen, my father didn't know anything. But when I reached the age of twenty-two I was amazed at what he had learnt in six years.'
What does this quotation teach us about the views some young people hold about their parents?
(*NIEB 1982 CSE*)
See pages 119–122

120 **(a)** What reason is given in one of the Ten Commandments for honouring parents? Explain what it meant.
(b) Does this reason still apply today? What other reasons can you give?
(c) In what ways could you honour your parents when you are old enough not to have to obey them any more?
(*ALSEB 1981 CSE*)
See pages 77, 122

121 **(a)** What effects might divorce have on the children of a broken marriage?
(b) What views are held by different Christian denominations or other religions about divorce and remarriage?
(*EMREB 1980 CSE*)
See pages 119–123

122 **(a)** Discuss **two** ways in which going to work fulfils a purpose in a Christian's life.
(b) Show that you understand the meaning of 'unemployment' and 'redundancy'.
(c) What problems do these cause?
(d) What are the advantages of belonging to a Trade Union?
(*EAEB 1982 CSE*)
See pages 126–130

123 **(a)** How did Jesus answer the question, 'Are we or are we not permitted to pay taxes to Caesar?'
(b) Suggest what Jesus meant when he said *(i)* 'Where your wealth is there will your heart be also'; *(ii)* 'You cannot serve God and Money.'
(c) What attitudes might Christians take towards *(i)* gambling; *(ii)* hire purchase?
(*SEREB 1981 CSE*)
See pages 51, 130–133

124 What led to the setting up of **(a)** the Samaritans **(b)** Dr Barnardo's Homes? Give an account of the present work of one of them.
(*EMREB 1980 CSE*)
See pages 133–135

125 **(a)** Describe the trial of Jesus after he had been sent back to Pilate by Herod.
(b) What is meant by *(i)* corporal punishment; *(ii)* capital punishment?
(c) Suggest what purpose corporal and capital punishment are supposed to serve, and say whether either can be supported from a Christian point of view, giving your reasons.
(SEREB 1981 CSE)

See pages 57, 139–143

126 **(a)** Describe the occasion before his arrest when Jesus prayed, 'Father, if it be thy will, take this cup away from me. Yet not my will but thine be done.'
(b) What is meant by *(i)* abortion; *(ii)* euthanasia?
(c) What problems might a Christian face in interpreting the will of God when making decisions about abortion and euthanasia?
(SEREB 1981 CSE)

See pages 55, 149–152

127 **(a)** *(i)* Write **four** sentences on *either* Alcoholics Anonymous *or* Gamblers Anonymous. *(ii)* State **five** main reasons why a young person today might become addicted to *either* alcohol *or* gambling.
(b) Write **two** sentences on *each* of the following terms: *(i)* Tolerance; *(ii)* 'Hooked'; *(iii)* LSD; *(iv)* Heroin; *(v)* Pusher.
(NIEB 1980 CSE)

See pages 145–148

128 **(a)** Write a paragraph on **each** of the following: *(i)* How a drug addict wrongs himself, *(ii)* How a drug addict wrongs his family, *(iii)* How a drug addict wrongs God.
(b) *(i)* In **four** sentences explain what is meant by solvent abuse or 'glue sniffing'. *(ii)* Write an extended paragraph explaining what a Christian might do if he knew that one of his class mates was 'glue sniffing.'
(NIEB 1982 CSE)

See pages 145–148

Part IV World Religions

Judaism

O-Level Questions

129 Describe the observance of Yom Kippur and explain its importance.
(London 1980 O Level)

See page 162

130 Describe the Jewish marriage ceremony and explain the significance of the various elements.
(London 1980 O Level)

See page 159

CSE Questions

131 **(a)** What is understood by 'Bar-Mitzvah'?
(b) For what purposes was the synagogue used?
(c) What items in a house would indicate that a Jewish family lived there? Why would these items be kept?
(YREB 1982 CSE)

See pages 156, 159, 160

132 Describe the home and family life of a practising Jew.
(EAEB 1981 CSE)

See pages 154–165

133 Describe **two** of the following Jewish festivals:
(a) Feast of Weeks,
(b) New year,
(c) Festival of Lights.
(EAEB 1981 CSE)

See pages 161–164

134 The family plays a very important part in Jewish life.
(a) Name **two** ceremonies in which this emphasis on family life can be seen in the Jewish community today.
(b) Choose any Jewish ceremony and write about it in detail. Show clearly how the whole family is involved.

(c) Explain why the Jews value family life so much.

See pages 154–165 (*EAEB 1981 CSE*)

135 Describe the interior of a synagogue, paying particular attention to its central feature.

See pages 156–157 (*EAEB 1982 CSE*)

136 'Remember the Sabbath Day, to keep it holy.' Discuss ways in which a Jew keeps this commandment in the synagogue and at home.

See pages 157–159 (*EAEB 1981 CSE*)

SCE Questions

137 'Passover, the great Festival of Freedom'.

(a) What events in history are commemorated at Passover?
(b) Why is unleavened bread eaten at this festival?
(c) Which members of the family play prominent parts in the Seder and what do they do?
(d) Why do Jews still attach so much importance to Passover?

See pages 71–78, 163–164 (*Scottish specimen 1983 O Level*)

138 **(a)** What words are given to foodstuffs fit and unfit for an orthodox Jew to eat?
(b) What requirements does the orthodox Jewish housewife have to observe in her kitchen?
(You might mention the range of foodstuffs, the buying of meat, utensils, etc.)
(c) What other responsibilities of a ritual nature does the orthodox Jewish mother have to undertake in the home?
(d) Describe any difficulties an orthodox Jew might have about keeping Jewish laws about food.

See page 160 (*Scottish specimen 1983 O Level*)

Islam

O-Level Questions

139 Explain the religious significance and use of the architectural features and furnishings of the mosque.

See page 171 (*London 1981 O Level*)

140 Write a short essay on Muslim worship, with particular reference to the weekly service in the mosque.

See pages 167–168 (*NIEC 1981 O Level*)

141 Show the importance of the following in Islam: **(a)** the Ka'aba **(b)** Arafat *and* **(c)** Mina.

See pages 165–170 (*London 1980 O Level*)

142 Explain the teaching and practice in Islam on charity and almsgiving.

See page 168 (*London 1981 O Level*)

143 Give a careful account of the Muslim rules about **(a)** food *and* **(b)** fasting.

See pages 168–169, 170 (*London 1981 O Level*)

144 What happens at **(a)** Friday prayers in a mosque *and* **(b)** Eid-ul-Adha? Show the importance of these observances.

See pages 168, 170 (*London 1981 O Level*)

145 Explain the significance of Jihad in the history of Islam and indicate any problem which may arise concerning this religious duty in the present day.

See page 167 (*London 1981 O Level*)

CSE Questions

146 Show how the life of Muhammad led to the founding of Islam.

See pages 165–167 (*EAEB 1981 CSE*)

147 **(a)** What does a Muslim believe about the nature of God?
(b) Describe the main features of the interior of a moslem mosque and the general atmosphere there.
(c) Briefly state the attitude of Muslims towards the people of other religions.

See pages 165–172 (*YREB 1982 CSE*)

148 What has your study of Islam taught you about
(a) home and family life,
(b) dietary laws?

See pages 168–171 (*EAEB 1981 CSE*)

149 (a) Why do Muslims make the pilgrimage to Mecca?
(b) What are the main ceremonies performed during the pilgrimage?

See pages 169–170 (*EAEB 1980 CSE*)

150 (a) What was the Hijra?
(b) Describe the events of the Hijra, including the story of how Muhammad evaded capture by the Meccans.
(c) *(i)* Explain why the Hijra was so important for Muhammad. *(ii)* Why is this event important for Muslims today?

See pages 165–167 (*EAEB 1981 CSE*)

157 (a) What are the five pillars of Islam?
(b) What does the giving of Zakat teach a Muslim about himself and about others?
(c) What does a Muslim do during Ramadan?
What does this teach him about himself and about others?

See pages 167–170 (*EAEB 1981 CSE*)

152 (a) What is the rite of initiation for a Muslim boy?
(b) Describe what you would see and hear in a mosque on a Friday at midday.
(c) Choose **one** Muslim festival and describe what takes place. What is the significance of these events to a Muslim?

See pages 168, 170–171 (*EAEB 1981 CSE*)

153 Explain why the following are all holy places to the Muslim:
(a) Mecca,
(b) Medina,
(c) Jerusalem.

See pages 165–167, 170 (*LREB 1982 CSE*)

SCE Questions

154 (a) What is the Arabic word for 'pilgrimage' and what does it literally mean?
(b) Describe what happens at the Ka'aba. Why is this ritual performed?
(c) Describe what happens at Mount Arafat and the significance of this for Muslims.
(d) In what ways does the pilgrimage strengthen the feeling of brotherhood in Islam?
(e) State two factors in the modern world which could hurt the Islamic brotherhood and show how they might do so.

See pages 169–170 (*Scottish Specimen 1983 O Level*)

155 Read the following extract and answer the questions below.

THE UNITY

Revealed at Mecca

In the name of Allah, the Beneficent, the Merciful.

1 Say: He is Allah, the One!
2 Allah, the eternally Besought of all!
3 He begetteth not nor was begotten.
4 And there is none comparable unto Him.

(a) Describe Muslim prayer, mentioning such details as preparation for prayer, times, words and actions.
(b) What circumstances in a non-Islamic country would make it more difficult for a Muslim to pray than in an Islamic country?
(c) What value does a Muslim place upon regular prayer?

See pages 167–168 (*Scottish Specimen 1983 O Level*)

Hinduism

CSE Questions

156 What is the caste system, and how has it affected the lives of Hindus?

See pages 174–175 *(EAEB 1981 CSE)*

157 **(a)** Describe the festival of Diwali.
(b) Outline the main features of one joyous festival observed in another religion.

See page 180 *(EMREB 1980 CSE)*

158 Write a paragraph about each of the following:
(a) Brahma,
(b) Shiva,
(c) Vishnu.

See pages 173, 177–178 *(EAEB 1981 CSE)*

SCE Questions

159 **(a)** What is meant by the term avatar?
(b) Name two avatars of Vishnu.
(c) Say who the four main characters are in the Ramayana and what part they play in the story.
(d) Describe how the Ramayana is acted out in India.
(e) Why do you think the story of the Ramayana is so popular in India?

See pages 176–177 *(Scottish Specimen 1983 O Level)*

160 Read the passage below and then answer the questions which follow it.

Never in modern history has any man been mourned more deeply and more widely.
L Fischer 'The Life of Mahatma Gandhi'

(a) How and when did Gandhi die, and what evidence supports Fischer's claim that no man in modern history has been mourned more deeply and more widely?
(b) Gandhi's funeral followed Hindu ritual. Describe a typical Hindu funeral.
(c) Explain why you think Gandhi was so popular.

See pages 179–181 *(Styled on Scottish O Level)*

161 The Aryans probably brought the caste system into India.
(a) What are the four castes and what were the duties set out for each caste?
(b) Give two reasons why you think the caste system has survived for so long.
(c) Imagine that two Hindus, one from the top caste and one from the bottom caste, came to Britain to live and work. What advantages and what difficulties might each find as he settles in Britain?

See pages 174–175 *(Scottish Specimen 1983 O Level)*

162 **(a)** State four items which could identify the picture [on p.178] as Krishna, and explain the significance of each.
(b) Krishna is shown in many other ways. Describe one of these.
(c) Hindu gods and goddesses are often shown with many arms. Why is this so?
(d) What is the mark that is worn by Krishna's followers?
(e) A picture such as the [one on p.178] could be used in puja. State four other items that would be used in puja and explain their use.

See pages 175, 177, 178 *(Scottish Specimen 1983 O Level)*

Sikhism

CSE Questions

163 **(a)** *(i)* Into what religion was Nanak born? *(ii)* Write down **one** thing that Nanak disapproved of in this religion.
(b) Write an account of the life of Nanak, including at least **two** stories told about him.
(c) 'There is truth in all religions but there is only one God. This one God is known by various names . . .' Write about two aspects of Sikhism which show that Sikhs today agree with this statement.

See pages 181–183 *(EAEB 1981 CSE)*

164 **(a)** What ceremony marks a young person's joining the Sikh community?
(b) What are the rules introduced by Guru Gobind Singh that all Sikhs are expected to follow?

See pages 182–183, 185 *(LREB 1982 CSE)*

165 **(a)** Why do Sikhs call their holy book the 'Guru Granth Sahib'?
(b) Write about **four** things that Sikhs do to show the great respect they have for the Guru Granth Sahib.
(c) Choose **either** reincarnation **or** the duty of a Sikh. Write about the subject you have chosen, showing *(i)* what a Sikh believes; *(ii)* how this belief affects his everyday life.

See pages 182–184 *(EAEB 1981 CSE)*

166 Write a paragraph about each of the following:
(a) Guru Granth Sahib,
(b) Gurdwara,
(c) Karah Prasad.

See pages 183–184 *(EAEB 1981 CSE)*

167 **(a)** Describe the naming ceremony that takes place a few weeks after the birth of a child.
(b) How is a Sikh wedding celebrated?

See pages 184–185 *(EAEB 1982 CSE)*

168 Marriage
(a) What must be present at a Sikh marriage?
(b) Describe the ceremony.

See page 185 *(LREB 1982 CSE)*

169 What are the Five Ks?
How did they originate, and what was their purpose?

See page 182 *(EAEB 1981 CSE)*

170 **(a)** What is an arranged marriage?
(b) Describe what you would see and hear at a Sikh wedding.
(c) Write about the advantages and the disadvantages of the Sikh way of marriage.

See page 185 *(EAEB 1981 CSE)*

171 Choose **two** festivals of the Sikh religion. Describe how they are observed and say why they are important.

See page 180 *(EAEB 1982 CSE)*

Buddhism

O-Level Questions

172 Analyse the Eight-fold Path.

See pages 187–188 *(NIEC 1980 O Level)*

CSE Questions

173 Write about the life of Gautama, and show how he founded the Buddhist faith.

See pages 185–186 *(EAEB 1982 CSE)*

174 **(a)** What do these words mean? *(i)* Kamma (Karma), *(ii)* Anatta, *(iii)* Khandhas, *(iv)* Nibbana (nirvana)
(b) Use these words as guides to describe the Buddhist belief in rebirth.

See pages 186–192 *(EAEB 1982 CSE)*

175 Amplify the statement: 'We can divide the life of the Buddha into three stages: early family life; the Buddha seeking Enlightenment; the preaching of the Buddha to his fellow men'.

See pages 185–186 *(NIEB 1980 CSE)*

176 Set out carefully the Buddhist doctrines of Rebirth and Nirvana.

See pages 186–187 *(NIEB 1981 CSE)*

177 Analyse carefully the Buddhist doctrine of man, stating:
(a) man's five components or parts;
(b) what is meant by anatta (literally 'no soul').

See pages 186–187 (*NIEB 1981 CSE*)

178 Describe the rules which govern one's behaviour to others as set out in the relevant part of the Noble Eight-fold Path.

See pages 187–188 (*NIEB 1981 CSE*)

179 What do the Four Noble Truths tell us about the Buddhist attitude to life?

See page 186 (*EAEB 1980 CSE*)

180 Outline the distinctive beliefs of Mahayana Buddhism.

See pages 189–190 (*NIEB 1980 CSE*)

181 Write notes on **two** of the following: the Buddhist scriptures; the sangha; meditation; Bodhisattva.

See pages 188, 190 (*NIEB 1981 CSE*)

Multi-Faith

O-Level Questions

182 In *each* of **two** faiths
(a) explain where and how the sacred scriptures are housed in the place of worship; (**6** marks)
(b) explain how the scriptures are used on the occasion of a festival; (**6** marks)
and **(c)** discuss the importance of the scriptures to the individual believer. (**8** marks)

See pages 154–192 (*AEB 1982 O Level*)

183 Show the importance of pilgrimage in **two** of the religions you have studied.

See pages 154–192 (*London 1980 O Level*)

184 In *each* of **two** faiths
(a) describe the geographical location of **two** places of pilgrimage, (**6** marks)
(b) explain why pilgrims might visit each of these centres, (**8** marks)
and **(c)** describe some of the activities in which they might share during their visits. (**6** marks)

See pages 154–192 (*AEB 1982 O Level*)

185 Explain the place of family worship in *either* **(a)** Hinduism *or* **(b)** Judaism.

See pages 157–159, 175 (*London 1980 O Level*)

186 Describe and show the importance for the religions concerned of **(a)** Diwali *and* **(b)** Yom Kippur.

See pages 162, 180 (*London 1981 O Level*)

187 Ritual objects play an important part in religious practice. Discuss the importance of such objects in **(a)** Hinduism, and **(b)** Judaism.

See pages 154–165, 173–181 (*London 1981 O Level*)

188 Explain the religious importance of fasting in **(a)** Judaism, and **(b)** Islam.

See pages 162, 168–169 (*London 1980 O Level*)

189 Describe the funeral customs of *either* **(a)** Hindus *or* **(b)** Muslims.

See pages 171, 179 (*London 1981 O Level*)

190 Describe and show the importance for the religions concerned of **(a)** Durga Puja, and **(b)** Eid-ul-Adha.

See pages 170, 175 (*London 1981 O Level*)

CSE Questions

191 How did the following books come to be written **(a)** Adi Granth **(b)** Torah? What do they contain? How are they used in worship?

(*EMREB 1980 CSE*)

See pages 155, 157–159, 163

Index

Bold type indicates whole units